Sedimentation and Mineral Deposits in the Southwestern Pacific Ocean

OCEAN SCIENCE, RESOURCES AND TECHNOLOGY
An International Series

Sedimentation and Mineral Deposits in the Southwestern Pacific Ocean

Edited by

D. S. Cronan

*Applied Geochemistry Research Group,
Department of Geology, Imperial College
of Science and Technology, London, UK*

1986

ACADEMIC PRESS

Harcourt Brace Jovanovich, Publishers

London   Orlando   San Diego   New York   Austin
Montreal   Sydney   Tokyo   Toronto

ACADEMIC PRESS INC. (LONDON) LTD
24–28 Oval Road,
London NW1

U.S. Edition published by
ACADEMIC PRESS INC.
Orlando, Florida 32887

British Library Cataloguing in Publication Data

Sedimentation and mineral deposits in the
Southwestern Pacific Ocean.—(Ocean science,
resources and technology)
1. Mineral industries—Oceania   2. Marine
mineral resources—South Pacific Ocean
I. Cronan, D. S.   II. Series
333.8'5'091647   HD9506.02/

Library of Congress Cataloging in Publication Data
Main entry under title:

Sedimentation and mineral deposits in the Southwestern
Pacific Ocean.

Bibliography: p.
Includes index.
1. Sediments (Geology)—South Pacific Ocean.
2. Sedimentation and deposition.   3. Geology—South
Pacific Ocean.   4. Ore-deposits—South Pacific Ocean.
I. Cronan, D. S. (David Spencer)
QE471.2.S42   1985   553'.091647   85-9221
ISBN 0-12-195870-1

*Printed by W. & G. Baird Ltd,
at the Greystone Press,
Caulside Drive, Antrim, Northern Ireland*

Contributors

D. S. Cronan   Applied Geochemistry Research Group, Department of Geology, Imperial College of Science and Technology, London SW7, UK

D. J. Cullen   New Zealand Oceanographic Institute, PO Box 12-346, Wellington North, New Zealand

N. F. Exon   Bureau of Mineral Resources, Geology and Geophysics, PO Box 378, Canberra City, Australia

G. P. Glasby   New Zealand Oceanographic Institute, PO Box 12-346, Wellington North, New Zealand

H. R. Katz   Pacific Geo Consultants, 6 Wairere Road, Belmont, Lower Hutt, New Zealand

M. A. Meylan   Department of Geology, University of Southern Mississippi, Box 9247, Hattiesburg, Mississippi 39401, USA

M. Ann Morrison   Department of Geological Sciences, University of Birmingham, PO Box 363, Birmingham, UK

R. N. Thompson   Department of Geology, Imperial College of Science and Technology, London SW7, UK

None of the world's oceans cast such a glamour of adventure over us as the Pacific, none has stirred man's imagination, and none affords more enticing problems to the student. (G. L. Wood and P. McBride, "The Pacific Basin." Oxford University Press, 1930)

Contents

4
Regional Geochemistry of Sediments from the SW Pacific   117
D. S. Cronan

5
Near-shore Mineral Deposits in the SW Pacific   149
G. P. Glasby

6
Submarine Phosphatic Sediments of the SW Pacific   183
D. J. Cullen

1

Introduction

D. S. CRONAN

Applied Geochemistry Research Group, Department of Geology, Imperial College of Science and Technology, London, UK

The subjects of sedimentation and mineral resources in the Southwest Pacific have been the focus of an increasing amount of work over the past decade and a half. Before 1970, they were hardly ever mentioned in the literature. However, with the increasing interest in marine mineral resources shown by the New Zealand Oceanographic Institute, coincident with their acquisition of the vessel *Tangaroa*, in the early 1970s, and the establishment of CCOP/SOPAC (Committee for Co-ordination of Joint Prospecting for Mineral Resources in South Pacific Offshore Areas) at about the same time, sedimentation and mineral resource studies have moved ahead rapidly in the South Pacific. Work organized and carried out by these two organizations has been supplemented by activities carried out by nations from outside the region, principally the United States and France. As a result of these endeavours, we now have a good general knowledge of the major features of sedimentation and marine mineral resource geology within the region, which serves as a framework for the more detailed resource-orientated studies currently underway there.

The Pacific Ocean is the largest feature on the surface of the Earth, comprising about one-third of the area of the globe. It contains numerous islands (about 25 000), mostly small, but a few larger ones, also. Although the total land area of the South Pacific countries is only 513 603 km^2, the 200-nautical-mile exclusive economic zone surrounding them comprises 12 700 000 km^2 (Schwass, 1981), more than 20 times as much as the land area. This gives some measure of the potential importance of sea-floor mineral deposits to the economies of these nations.

Sedimentation and Mineral Deposits
in the Southwestern Pacific Ocean
ISBN 0-12-195870-1

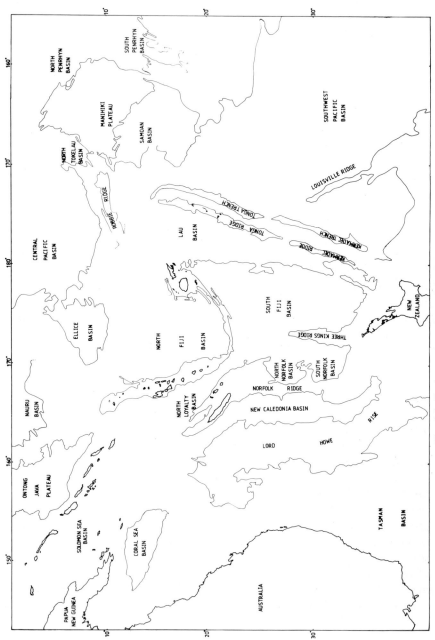

Fig. 1.1. Bathymetric regions of the SW Pacific.

The area dealt with in this book extends from the longitude of Australia and Papua New Guinea in the west to that of the Cook Islands Group in the east and from about 5°N to the latitude of New Zealand in the south. Within this region the bathymetry is extremely varied, comprising trenches, island arcs, inactive ridges, abyssal plains, deep basins and other features. The major bathymetric features of the region are shown in Fig. 1.1, which is based on the new bathymetric chart of the region (Kroenke *et al.*, 1983) and distributed by the Ministry of Lands, Energy and Natural Resources, Honiara, Solomon Islands, on behalf of CCOP/SOPAC.

The continent of Australia is bounded to the east by the Tasman and Coral Sea basins, between which are a number of plateaux. To the north of this area lies a complex of small basins and plateaux bounded to the north by the NW extremity of the volcanic arcs which cut across the region from NW to SE. To the east is the broad, inactive, Lord Howe Rise, which is separated from the volcanically active New Hebrides Arc by a number of plateaux and basins and, immediately next to the arc, by a complex of trenches. East of this arc/trench system are two large basins, the North and South Fiji Basins, bounded on the east by another arc/trench system and associated marginal basins, culminating in the Tonga–Kermadec Ridge and Trench. To both the north and east of this area lies the Pacific Plate, with a much more subdued bathymetry than the areas to the west. This is largely an area of uniformly deep water, but within it are a number of ridges and basins.

Sedimentation within the region is governed by a number of factors. Much of the area to the west and SW of the island arc systems is above the calcium carbonate compensation depth (CCD), and thus carbonate sedimentation predominates. However, this is of reduced importance in some of the deeper basins where the CCD is approached and sometimes exceeded, and in the vicinity of volcanically active island arcs where volcaniclastic sedimentation often comprises a significant proportion of the sediments. Hydrothermal sedimentary components are also found on a local scale, both on the arcs and in their marginal basins in some of which sea-floor spreading is thought to be taking place. Sedimentation on the Pacific Plate contains much less carbonate debris and a much greater proportion of clay, commensurate with its greater depth. However, in the equatorial region where biological productivity is high, siliceous ooze makes an appearance, and siliceous organisms sometimes comprise a significant fraction of the sediments as far south as 15°S. Phosphorites occur predominantly on seamounts and manganese nodules occur over large areas of the deep basins on the Pacific Plate.

As mentioned above, until recently the sediments and mineral resources of the SW Pacific have been relatively neglected. However, as a result of work done since 1975, the SW Pacific is now beginning to take on the appearance of a model area for sedimentation and marine mineral resource studies, the results of which may possibly be applicable in other areas in the future. In the present book the major features of this work in regard to sedimentation during the geological past in the area are considered, together with the composition of the sediments and the nature, distribution and origin of the potentially economic minerals and petroleum in them. The final chapter, while not dealing specifically with the SW Pacific, addresses a problem hitherto not looked at in that area but contains material undoubtedly of relevance there.

2

Stratigraphy of the SW Pacific

H. R. KATZ

Pacific Geo Consultants Ltd., Lower Hutt, New Zealand

2.1 Introduction

With the great complexity of geomorphic and tectonic units and provinces which make up the SW Pacific region—across one of the world's major plate boundaries (Indo–Australian to Pacific Plates)—it is to be expected that stratigraphic provinces are equally varied. Indeed, the age of oceanic basins ranges from Cretaceous (Tasman Sea and West Pacific Basin) to Recent (North Fiji Basin, Woodlark Basin), while continental rises and island arcs between these basins consist of rocks whose ages are as far apart as Late Paleozoic and Recent. The widely different tectonic settings and evolution of the various units, in particular, and the distribution of volcanism and its contribution to the sedimentary column have led to very different lithologic and facies characteristics from one place to another. Correspondingly, sediment thicknesses vary enormously, too, while average sedimentation rates may be different by as much as a factor of 100.

Apart from Papua New Guinea, Australia and New Zealand, which are not considered here, fragments of older continental crust—and therefore containing the oldest sediments in the region—are only found in the Lord Howe Rise and New Caledonia–Norfolk Ridge, both of which join up with New Zealand. These two rises or ridges—together with New Zealand—have been detached and drifted away from Protoaustralia in the course of the Gondwana breakup. All other areas in the SW Pacific are either younger oceanic basins underlain by oceanic crust—so-called marginal basins

Sedimentation and Mineral Deposits
in the Southwestern Pacific Ocean
ISBN 0-12-195870-1

(except for the Pacific Basin itself)—or young island arcs which are thought to have formed above subduction zones, both ancient and present day. In consequence, stratigraphic sequences to an overwhelming extent are of Tertiary age, and particularly Neogene.

The subduction-related, young island arcs are certainly the most characteristic, indeed outstanding, features of the SW Pacific. From the New Ireland and Manus islands in the northwest, they extend in a zigzagging, more or less diagonal line far to the southeast, i.e., across the Solomons, the New Hebrides, the Fiji Islands and the Lau and Tonga–Kermadec ridges: this is across 35° latitude and 40° longitude (0–35°S, 150°E–170°W)—a young orogenic belt of large dimensions. The history of these island arcs, as seen from their stratigraphic record, may go back to the Late Cretaceous, but mostly dates from the Eocene–Oligocene to the present; the greatest part of the stratigraphic column invariably is found in the Miocene. Total sediment thickness generally is many thousands of meters, which in itself indicates considerable tectonic instability. In this, the island arcs contrast sharply with intervening oceanic basins, which largely remained relatively stable and contain very thin sediments, or none at all.

In the sedimentary environments and types of rocks of the island arcs, however, little variation is found across the stratigraphic column. Basically there are only two components, which together constitute the entire lithologic frame: a volcanic, basaltic–andesitic component and an organic, carbonate component. Whether primary or secondary (i.e., reworked and derived from pre-existing volcanic or carbonate terrains), the rock sequence invariably is composed of one or the other, or any mixture of these two components. Only rarely is there a regional, older basement affected by erosion, thus providing a sediment source. In such cases, however (as for instance in the Solomon Islands), basement is formed mainly by primary volcanics, i.e., basaltic lavas; the materials supplied, therefore, are not much different. In most cases an island arc is built up in an open oceanic setting, thereby creating its own environment: it is the environment of active andesite volcanoes in an open-marine archipelago setting, which operates a self-sufficient sedimentary system where the active volcanic edifice (both submarine and subaerial) provides the source and environmental conditions to generate and move clastic debris by gravitational collapse and slumping, rubble avalanches, mass flows and turbidity currents from areas at or near sea level to bathyal depths (Jones, 1967). Resulting rock types are mainly primary, basaltic to andesitic volcanics (lavas and pyroclastics and their auto- and epiclastic derivatives), and finer volcaniclastic sediments such as greywacke arenites and siltstones often developed as turbidites. In the unstable conditions of active island arcs, such finer material can be moved about repeatedly and form first and second cycle deposits etc., and finally collect to very great thicknesses in the rapidly deepening sinks which often develop between volcanic buildups (typical examples are found in Fiji, e.g., Koro Sea or Bligh Water).

This environment of andesite volcanoes, both active and extinct, also creates the conditions for reef growth and the formation of organic limestones, which are mainly algal and coralline, or algal–foraminiferal. Calcirudites and calcarenites, as well as pelagic calcilutites, are associated products. Back-reef lagoonal and fore-reef slope environments, which may extend into bathyal depths, control a variety of local conditions. Calcirudites are very common amongst island arc deposits: probably because of the unstable, rapidly changing conditions around growing volcanic islands, the collapse of a reef mass and slumping of reef talus are processes which are widely observed. The frequent occurrence of mixed rudites may result from joint collapse of lava-derived and reef material, perhaps as a consequence of sudden loading of a reef by lava flows. In this connection it should be emphasized that the deposition of reef limestones *per se* does not indicate a period of volcanic quiescence—this can be seen quite clearly in many modern volcanic island arc settings. Even where reefs are destroyed by, and prevented from growing during, volcanic eruptions, such events quite generally are very local, temporary and short-term and are of no wider, geologic significance: life quickly restores to its former presence, and reef-building organisms usually renew their activity with little delay. While in some cases grand concepts of tectonic evolution of island arcs have been construed, postulating alternating periods of volcanic activity (and thus subduction!) based mainly, if not solely, on the presence or not of reefal limestone formations in the stratigraphic column, such arguments are based on unrealistic, simplistic views of geological processes which do not conform with actual facts.

In summary, the source and depositional environments in an active island arc setting are shown to be contemporaneous and coextensive, and to create a sedimentary system that is restricted to indigenous, mixed volcanic and organic components. And while the latter are predominantly represented by carbonate rocks of mainly coralgal origin and derivation, some organic material may also be land-derived and included in lagoonal and shallow marine environments, and perhaps even to greater depths. Since volcanic islands—at least in tropical latitudes such as in the SW Pacific—will rapidly be covered by thick vegetation, this may become a source for carbonaceous shales and siltstones, and rarely even coaly deposits. Such sediments, however, are always subordinate and of local occurrence. Where carried into deep-water environments, land-derived organic material is most likely strongly diluted within the bulk of mainly volcanic-derived material. Considerable dilution is also to be expected of any pelagic–oceanic oozes that may settle in the basins of a wider, offshore environment between volcanic islands. Pelagic limestone, marl and/or shale-siltstone—although they do exist here and there in island arc sedimentary columns, particularly in areas and time intervals of limited or no volcanic activity—are rarely widespread or of appreciable thickness. This is probably a result also of their relatively low sedimentation rate, as compared with volcaniclastic sedimentation.

2.2   Northwestern Sector: East and North of New Guinea, Papuan Peninsula, Louisiade Archipelago, Pocklington Rise

The tectonic evolution and relationships of this maze of small oceanic basins and active and extinct islands arcs (Fig. 2.1) is still not well understood, but for stratigraphic purposes one can distinguish three main provinces: (1) the young oceanic basins (Woodlark Basin, Solomon Sea and Bismarck Sea), (2) the island arcs (Schouten–New Britain or South Bismarck arc, Manus–New Ireland or North Bismarck arc, and the Solomon Islands arc), and (3) the older oceanic basins of the Caroline and Pacific plates to the north and east of the West Melanesian Trench (East Caroline Basin, Lyra Basin and Ontong Java Plateau).

2.2.1   The young oceanic basins

The *Woodlark Basin* between Pocklington Rise and Woodlark Rise exposes rough sea-bottom morphology characteristic of actively spreading basins. Magnetic lineations indicate that it has been opening for the past 3·5 m.y. (Weissel *et al.*, 1982a). In consequence, sediments are very thin; they are lacking near the central spreading rift but thicken away from it. Also, they are thicker in the western part (about 200 m) than in the east, probably because of sediment transport from nearby Papua New Guinea (Luyendyk *et al.*, 1973).

Of somewhat greater age is the *Solomon Sea Basin* to the northwest of the Woodlark Rise. Its crust, which is being consumed to the north and northeast along the New Britain and Bougainville Trenches, and to the south in the Trobriand Trough (Honza *et al.*, 1984; Lock, 1984), is assumed to be at least as old as early Neogene. Seismic sections [e.g., Hamilton (1979), Figs. 151 and 152] show a series of high and sharp basement ridges separated by small basins, which are filled with sediments whose thickness ranges from 0·5 to 1·0 sec reflection time (about 400–900 m).

The *Bismarck Sea* (Fig. 2.1) to the north of New Britain includes two NW

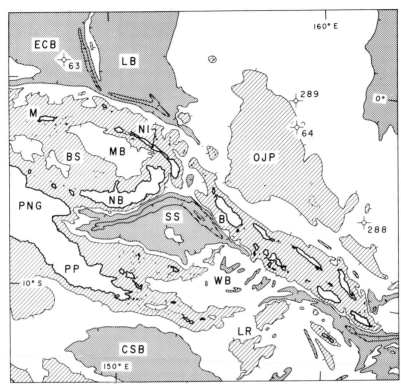

Fig. 2.1. Bathymetric map of the NW sector (Caroline Basin to Coral Sea), with location of DSDP holes 63, 64, 288 and 289. Contour interval 2000 fathoms. B, Bougainville Island; BS, Bismarck Sea; CSB, Coral Sea Basin; ECB, East Caroline Basin; LB, Lyra Basin; LR, Louisiade Rise; M, Manus Island; MB, Manus Basin; NB, New Britain; NI, New Ireland; OJP Ontong Java Plateau; PNG, Papua New Guinea; PP, Papuan Peninsula; SS, Solomon Sea; WB, Woodlark Basin.

trending basins (the New Guinea Basin in the west, 1800 to 2200 m deep, and the Manus Basin in the east, 2000 to 2700 m deep), which are separated by the Willaumez–Manus Rise. A complex earthquake lineament traverses the Bismarck Sea in a roughly east–west direction; centred around it there is a substantial sediment-free zone up to 60 km wide (Connelly, 1974). This seismic lineation is composed of several segments of both actively spreading oceanic basement and NW-trending transform faults; based on magnetic anomalies, the Manus Basin has been opening for the past 3·5 m.y. (Taylor, 1979).

In both basins of the Bismarck Sea, sediments thicken from the sediment-free area southwards. In the western part of the New Guinea Basin the seismic lineation coincides with a chain of seamounts that probably has acted as a barrier to sediment transport from the south: immediately south of this chain of seamounts, sediments are 1 km thick, but they increase to 2 km near the coast of mainland New Guinea (Connelly, 1974). Most of these sediments are Neogene or Recent continental rise deposits; high sedimentation rates are indicated because of the proximity of the Sepik, Ramu and other rivers. Also, the very active volcanic chain, nearly 1000 km long, which girdles the south side of the Bismarck Sea from the Schouten Islands to Gazelle Peninsula of New Britain, would have supplied considerable amounts of ash and other volcanic products to these sediments. Widespread, dark-olive mud and silt and volcanic sand have been reported by Krause (1965) from bottom samples in the SW Bismarck Sea, between 145° and 148°E. Underneath these younger, continental rise deposits, there appear to exist thick and more strongly deformed Paleogene rocks, which are supposed to be part of an ancient, complex volcanic arc–subduction–collision zone in this area [Hamilton (1979), Figs. 128 and 129]. Further east in the southern New Guinea and Manus basins, relatively thick sediments are also found along New Britain. But in some parts of the Manus Basin off the eastern New Britain volcanic arc, basement outcrops are numerous and recent intrusions have locally deformed the sediments (Connelly, 1974).

2.2.2   The island arcs

2.2.2.1   *North and South Bismarck arcs*

These two arcs have probably been part of one and the same island arc system as it existed prior to the opening of the Bismarck Sea (Taylor, 1979). In fact, the geologic histories and sedimentary sequences in both the North and South Bismarck arcs exhibit important similarities. In Manus Island as well as New Britain, the sequence begins with Eocene, calc–alkaline island arc volcanics with intermediate to basic pillow lava, agglomerate, breccia and tuffaceous sediments, including numerous lenses of corralline limestone (Dow, 1977; Jacques, 1980). These rocks indicate the presence of active volcanic islands with fringing coral reefs. Although foraminifera are common, a conspicuous lack of lower and middle Oligocene fossils (T_c and T_d stages) suggests a widespread hiatus through much of the region. Also, the Eocene Baining Volcanics in New Britain are more highly deformed and locally metamorphosed to the greenschist facies; they appear to be unconformably overlain by upper Oligocene rocks (Dow, 1977). This situation probably reflects the late Eocene to early Oligocene orogenic activity along the adjacent marginal trough of eastern New Guinea. In New Ireland, however, such a hiatus does not seem to exist: the basal Jaulu Volcanics, about 2000 m thick, contain limestone lenses whose foraminiferal fauna spans the lower to middle Oligocene T_c to T_d stages (Hohnen, 1978). During the rest of the Oligocene, the environment in the entire region was similar to that which had prevailed in the Eocene: active island volcanoes supplied subaerial and marine volcanic products, lavas and breccias in a complex mixture with volcanolithic sediments and reef limestones.

During the Oligocene and into the lower to middle Miocene, there also occurred numerous intrusions of mafic to intermediate plutons in New Britain (Page and Ryburn, 1977), New Ireland (Hohnen, 1978), and Manus Island (Jacques and Webb, 1975), which include "porphyry copper" mineralization. These intrusives probably originate from the same magma—as its more slowly crystallized portion—which gave rise to the widespread Oligocene volcanics (Hohnen, 1978).

After a short break with erosion around the Oligocene–Miocene boundary, extensive marine shelf limestone was deposited on the island arcs around the Bismarck Sea. At this time, too, volcanic activity had ceased quite abruptly throughout the region. Thus the limestone encroached around and over the extinct and eroded volcanic edifices, its base being markedly diachronous across a considerable topographic relief of the pre-Miocene, volcanic basement. At the base of the limestone, there are locally arenaceous to conglomeratic beds developed, which represent the erosive products from the underlying volcanics (e.g., Lossuk River Beds, 150 m thick, in New Ireland; Hohnen, 1978).

No Miocene limestone exists in New Hannover Island, while the Mundrau Limestone on Manus Island, of lower to middle Miocene age, is only about 200 m thick. Over most other parts of the exposed island arcs around the Bismarck Sea, however, limestone deposition continued during practically the whole Miocene, and partly into the Pliocene. The total thickness varies enormously, but is generally many hundreds of meters, and in some cases much over 1000 m. Sedimentary conditions were surprisingly uniform: the predominant rock type is a well bedded to massive, coralgal biomicrite, very pure and whitish to cream-coloured, often recrystallized. Calcarenite, calcilutite and other fine-grained calcareous sediments also occur. The depositional environment probably was that of a large fringing reef over a gradually subsiding area, including intra- and back-reef lagoons into which foraminifera were washed from the open sea. In the final stages (Pliocene), a reefal limestone platform had completely capped the remaining, older volcanic islands as atolls.

Block faulting with graben formation and renewed volcanic activity occurred in the latest Miocene and Pliocene (Page and Ryburn, 1977). Over 4000 m of Pliocene strata were deposited on Manus Island; they are mainly volcanics and volcaniclastic sediments but include some calcareous foraminiferal sediments. Volcaniclastics are the only Pliocene rocks in New

Hannover. In general, the Pliocene is thinning to the east, while at the same time the facies changes to more prominent limestone deposition. The Rataman Formation in New Ireland is composed of 500 m of "andesitic and dacitic crystallithic tuff, volcanolithic labile arenite and lutite, foraminiferal marl and limestone" (Hohnen, 1978). On the basis of abundant foraminifera and locally also molluscs, the age is N.18 or younger. Marls and arenites–lutites are generally well-bedded, including turbiditic graded bedding deposits in some outcrops.

Deep-water foraminifera in some of the rocks suggest that part of the Rataman Formation is a deeper-water equivalent of the late Miocene Lelet Limestone. With further subsidence during the Pliocene, more limestone was deposited, such as the 200–1000 m thick Punam Limestone in New Ireland which consists of only moderately recrystallized, chalky limestone and foraminiferal calcarenite, together with coralline calcirudite (Hohnen, 1978).

Faulting and uplift occurred in the Late Pliocene, obviously in connection with the opening of the Bismarck Sea–Manus Basin with its transform faults. Some of these transform faults cut across southern New Ireland and the Gazelle Peninsula of New Britain (D'Addario *et al.*, 1975). From Pliocene into Holocene, strong volcanic activity has developed all along the South Bismarck arc; in the North Bismarck arc, Quaternary volcanoes only occur in the Admiralty Islands and northeast of New Ireland. In New Ireland, further uplift and tilting to the NNE caused a series of raised reef terraces to develop. This tilting, however, has preserved the extensive sedimentary basin lying offshore the present New Ireland–Manus arc, i.e., between it and the West Melanesian Trench. About 160 km wide and 1200 km long, this New Ireland Basin has a sedimentary fill of over 2·5 sec reflection time, i.e., of about 3 km (de Broin *et al.*, 1977). The sediments probably are of the same age as those now exposed on New Ireland, i.e., mainly Neogene. Forming a gentle syncline, they clearly exhibit two unconformities within the sequence which onlaps, or abuts against, older rocks of the North East Ridge (de Broin *et al.*, 1977) immediately above the trench. A chain of Tertiary–Quaternary volcanic islands to the NE of New Ireland (Tabar, Lihir, Tanga and Feni Islands), which are mainly composed of alkalic rocks (Johnson *et al.*, 1976; Johnson, 1979), are lying within this basin. It has recently been covered by a reconnaissance seismic survey by CCOP/SOPAC (Exon and Tiffin, 1982) between 145° and 154°E, and 0° to 4°S. Exon and Tiffin conclude, from their study of previous Gulf seismic lines in conjunction with the CCOP/SOPAC survey, that the total thickness of sediments in the offshore New Ireland Basin may be more than 5 km. According to their interpretation, a Miocene limestone sequence could be up to 2000 m thick, representing mainly a carbonate platform environment but including reefal buildups in many areas.

2.2.2.2 Solomons arc

Physiographically, the Solomons island arc is the direct continuation of New Ireland and thus of the North Bismarck arc (Fig. 2.1). However, there are profound differences in its tectonic evolution, so that the stratigraphic sequence is very different, too; but there is no well defined geological boundary between the two arcs. This is to be expected since their relative positions must have shifted considerably since Oligocene times (Hamilton, 1979). Although there is still no clear picture regarding the correct reconstruction of the Bismarck–Solomons region—which is one of the most active and complex plate-interaction complexes in the world—it is quite obvious that fragmentation of island arcs, displacements along transform faults and the changing patterns of subduction, perhaps including polarity reversals, have very seriously affected the structure and history of the various segments.

   In their stratigraphic record, Buka and Bougainville Islands (Fig. 2.2) represent an area of transition between New Ireland and the Solomons. The extensive, calc–alkaline, andesitic and basaltic volcanics of lower Tertiary age (Buka Formation and Kieta Volcanics; Blake and Miezitis, 1967) can readily be correlated with the Eocene and Oligocene volcanics of both New Britain and New Ireland. The lower Miocene Keriaka Limestone in Bougainville, over 1200 m thick and representing complex barrier reef environments with reef, back-reef and fore-reef as well as lagoonal facies, can be closely compared and correlated with the Surker Limestone in southeast New Ireland, which is 500 m thick in its type section but thickening to 1300 m further southeast (Hohnen, 1978). On the western Bougainville shelf, however, the Mio–Pliocene section drilled in L'Etoile-1 well (Fig. 2.2) found no limestone, although all seismic reflectors which might represent the upper boundary of a carbonate reef zone had been penetrated. Instead, 1225 m of mainly tuffaceous sandstone and silty mudstone were encountered. The well, with a total depth of 1682 m, bottomed in thick volcanic agglomerate of probably lower Miocene age [total rock K–Ar analysis of rock fragments, excluding matrix, gave an age of $18 \cdot 2 \pm 7$ m.y., i.e., early to middle Miocene (Oceanic Exploration Co., 1975)]. These volcanics may correlate with the probably Miocene to Pliocene, younger volcanic suite of the Shortland Islands immediately southeast of Bougainville (Turner and Ridgway, 1982), which is composed of typically calc–alkaline island arc assemblages. Together with the Pleistocene to Recent andesitic volcanism as occurs on Bougainville itself, they apparently are associated with subduction from the west along the Bougainville or North Solomon Trench.

   Further south in the Solomon Islands, evidence is lacking of extensive arc volcanism of early Tertiary age, such as took place in the Bismarck island

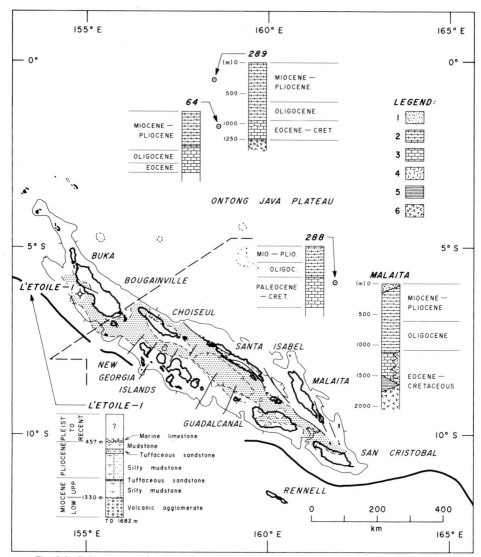

Fig. 2.2. Structural map of the Solomon Islands arc [after Katz (1980)]. Dotted area: Mio–Pliocene sedimentary basin. Bathymetric contour 1000 fathoms; dashed contour in closed basin south of Santa Isabel is 800 fathoms [from Mammerickx *et al.* (1974)]. Heavy line: axis of New Britain and South Solomon trenches. Dashed line: International boundary between Papua New Guinea and Solomon Islands. Locations and well sections of oil exploration well L'Etoile-1 (Oceanic Exploration Co., 1975) and of DSDP holes 64, 288 and 289. Legend: (1) shallow marine Hauhui Conglomerate in Malaita (Pleistocene), (2) nannoforam ooze and chalk, (3) limestone, chalk, nodular chert, siliceous limestone with radiolaria, (4) alkali basalts ("younger basalts"), (5) siliceous mudstone with radiolaria, (6) oceanic tholeiites ("older basalts"). Stratigraphic column of Malaita after Hughes and Turner (1977).

arcs further north. Instead, the early development of this area is characterized by an oceanic environment with the extrusion of extensive, submarine tholeiitic "flood basalts" associated with intrusions of gabbroic and ultramafic rocks at depth, during the late Mesozoic to early Tertiary. These igneous rocks are associated with deep-marine, pelagic limestone deposits. Thus the entire area was clearly an extension of the Ontong Java Plateau, perhaps representing its western margin (Ramsay, 1982). However, the subsequent development, in the Eocene–Oligocene, was very different here. South of a line from Santa Isabel to San Cristobal Islands— i.e., excluding the NE side of Santa Isabel and the whole of Malaita Islands (Fig. 2.2)—intense tectonism resulted in local metamorphism of the basement volcanics, which pass into dynamometamorphically altered metabasalts of greenschist to amphibolite facies (Hackman, 1973, 1980; Ramsay, 1978, 1982). They exhibit intense shearing, thrusting and faulting. In Guadalcanal, three or four different s-surfaces of schistosity have been recorded, the tectonic climax probably occurring in Eocene to lower Oligocene time (Hackman, 1980); a metamorphic age of schist from Choiseul Island is 44 ± 18 m.y. (Richards et al., 1966). Of about Oligocene age is probably also the emplacement of ultrabasics in Guadalcanal and in the southeast of Choiseul.

Thus while in the Eocene and Oligocene extensive, calc–alkaline volcanism took place in the Bismarck Sea area to the NW, an intense orogenic phase affected a belt of Cretaceous to Paleocene oceanic crust in the Solomons. This resulted in uplift and deep erosion, and the initiation of sporadic island arc volcanism from the late Oligocene onwards, together with basin formation and arc-related sedimentation (Katz, 1980). The earliest volcanics are the 2500 m thick Suta Volcanics in Guadalcanal— tholeiitic basalt and basaltic andesite, probably mainly of late Oligocene age—which in their high portions were extruded into and between fringing or offshore reef environments of lower Miocene limestones (Hackman, 1980). Also of late Oligocene age is the Poha Diorite in Guadalcanal (Chivas and McDougall, 1978), which is composed predominantly of hornblende and andesine, although quartz and biotite occur in a tonalitic variety (Hackman, 1980). The Poha Diorite thus represents a more acid phase of late Oligocene magmatic activity. It is overlain unconformably by lower Miocene limestones. The general basin history—although marine transgression had encroached locally over the deeply eroded basement already in late Oligocene time—is mainly of Mio–Pliocene to Pleistocene age. A synthesis of this basin history was presented by Katz (1980, 1982, based both on land and offshore (marine seismic) data (Fig. 2.2).

Particularly from the lower Miocene onwards, regional subsidence affected more and more of the entire region, except in some marginal areas which continued to supply sediments into the basin. Thus the late Oligocene to Mio–Pliocene sequence, which unconformably overlies basement, is to a large extent derived from the latter's erosional products. Local depressions in the old surface are filled with thick, monomict orthobreccias composed of large, angular basement clasts enclosed within a matrix of basaltic grit (Hughes, 1979). In more rapidly subsiding, deeper, marine areas, non-calcareous grit, sand and shale were deposited, often as turbidites. Around still emergent islands, on the other hand, coralgal fringing reefs and extensive carbonate platforms were established. Such reefs and their associated calcarenites were deposited locally throughout the basin's history, but are particularly prominent in the early Miocene. Individual calcareous bioherms may be several hundred meters thick (Hackman, 1980).

Although the non-calcareous, detrital sediments are mainly of volcanic composition and generally classed as volcaniclastics, they are derived principally—as mentioned above—from eroded basement material. There is scant evidence of contemporaneous volcanism which—although sporadically and locally occurring through most of Mio–Pliocene times—was far from significant and contributed little to the volume of the sedimentary succession (Katz, 1980). Caution should thus be applied when using volcaniclastic sediments as a proof of contemporaneous volcanism, and of the existence of an active volcanic island arc as postulated by Turner and Hughes (1982).

Basin subsidence continued at an accelerated rate into the Pliocene. Block faulting and local uplift in marginal areas created a continuing source for large volumes of clastic sedimentary material. Total thickness is many thousands of meters, but is difficult to establish precisely because of local variations and very rapid facies changes. Also, no basement reflector has been obtained in marine seismic records (Katz, 1980). In the north of Guadalcanal, transport by mass flow and turbidity currents and deposition in deep-sea fans was extensive (Fig. 2.3) (van Deventer, 1971). Thus the basin was gradually filling up and becoming shallower in the Pleistocene (Fig. 2.4). Extensive Pleistocene reef complexes formed, and were later uplifted stepwise to several hundred meters on the islands which more and more have been emerging to their present form. In the same process, other Pleistocene and younger reefs have been downfaulted in the Central Solomons Trough which strongly subsided between the two island chains (Katz, 1980) (Fig. 2.4).

Plio–Pleistocene to Recent time saw increasing volcanic activity. It is centred in Bougainville and along the southern side of the Solomon Islands as far as Guadalcanal, including occurrences within the Central Solomons Trough [e.g., Savo Island (Katz, 1980)]. Although this volcanism is generally referred to as calc–alkaline andesitic, it comprises a highly complex

Fig. 2.3. Well-bedded, volcaniclastic silt–sandstone: Pliocene deep-sea fan deposit in the Lungga River 10 km southeast of Honiara, Guadalcanal [Mbetivatu Sandstone member of the Toni Formation of Hackman (1980)].

group of volcanic rocks which range from high-alumina, low-titania andesites in Bougainville (Blake and Miezitis, 1967) to picritic and olivine basalts and hornblende andesites in Guadalcanal (Hackman, 1980). As was pointed out by Coleman (1976), the geochemically puzzling and very unusual assemblage of these volcanics has affinities as much with midoceanic igneous rocks as with those of island arcs. Related to this young volcanism are plutonic intrusions of diorite–tonalite, which in Bougainville (Panguna; Page and McDougall, 1972) and Guadalcanal (Koloula; Chivas and McDougall, 1978) are associated with porphyry copper mineralization.

2.2.3   The older oceanic basins to the north and east of the West Melanesian Trench

2.2.3.1   *East Caroline Basin*

From seismic sections [Hamilton (1979), Figs. 143, 147 and 148] and DSDP hole 63 (Winterer *et al.*, 1971), the East Caroline Basin is underlain by a very uniform, pelagic sequence of lower–middle Oligocene to Quaternary age, 561 m thick at the drill site, which with normal sedimentary contact overlies extrusive basalt basement (Fig. 2.5).

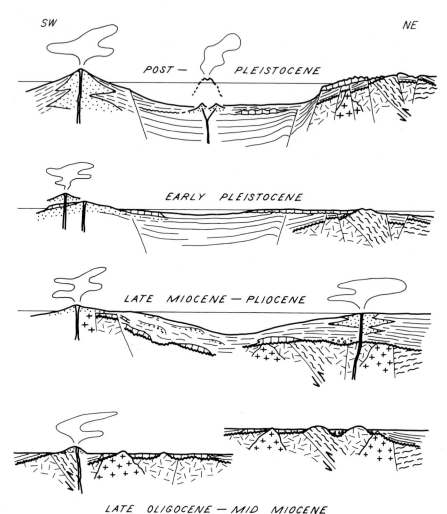

LATE   OLIGOCENE — MID   MIOCENE

Fig. 2.4. Diagrammatic cross-sections showing tectonic evolution of the Mio–Pliocene sedimentary basin in the Solomon Islands, and of the Pleistocene to Recent, Central Solomons Trough. [After Katz (1980), modified.]

The sediments are mainly nannofossil chalk ooze, with chert stringers and a marl unit near the base; this carbonate sequence changes into pelagic clay in the uppermost 35 m. Thus depositional conditions have little changed during the past 30 m.y. Sediment accumulation rates—neglecting compaction—gradually slowed from 30 B* in the Oligocene to about 7 B in the latest Miocene to Quaternary times. With increasing depth of burial, the calcar-

*B stands for Bubnoff units, 1 mm/1000 years, 1 m/million years (Fischer, 1969).

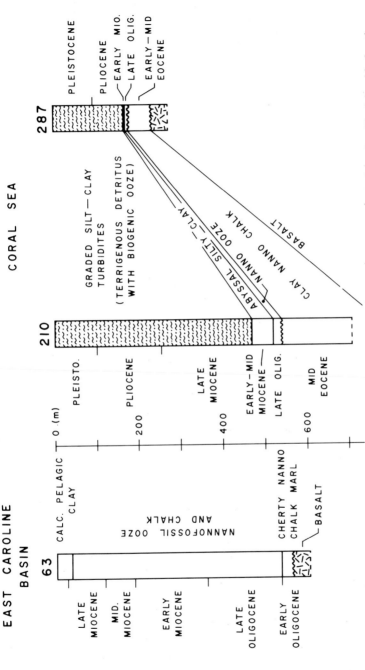

Fig. 2.5. Stratigraphic columns of DSDP holes 63, 210 and 287 in the East Caroline Basin and Coral Sea. For legend see Fig. 2.9 and for locations see Figs. 2.1 and 2.6.

eous oozes show progressive increases in induration and in the degree of silica mobilization and recrystallization. Some of the persistent reflecting layers seen on seismic reflection profiles can be correlated with changes in induration or the presence of flinty chert in the chalks.

2.2.3.2 Lyra Basin and Ontong Java Plateau

The early Tertiary crust of the Caroline Basin is bordered to the east by the complex Mussau Terrain (Mussau Trench to Lyra Trough; Hamilton, 1979), beyond which the Lyra Basin and Ontong Java Plateau are underlain by the Early Cretaceous crust of the western Pacific. Basin and plateau alike are covered by a continuous, pelagic sequence 1000–1500 m thick (Kroenke *et al.*, 1971; Kroenke, 1972). This was completely penetrated in DSDP hole 289 (Fig. 2.2; Andrews *et al.*, 1975). In this hole, tholeiitic plateau basalt lava flows of Aptian age or older are overlain, at 1262 m below sea floor, by 293 m of middle to late Cretaceous to Paleocene–Eocene radiolarian-bearing limestone, nanno–foram chalk, chert and tuff. The sequence is interrupted by at least six substantial, short periods of non-deposition or erosion; in particular, there is a long hiatus from late Aptian into Campanian. Upwards the sequence continues with 969 m of lower Oligocene to Quaternary nanno–foram ooze and chalk. Based on the excellent biostratigraphic age determinations, it has been possible to calculate sedimentation rates very accurately. Assuming an initial porosity of 70%, the corrected figure for the late Cretaceous rate of deposition is 23 B, and for the middle to late Eocene 30 B. From the early Oligocene to the Pleistocene most of the rates are between 30 and 50 B, but there are more extreme fluctuations between a high of 67 B for the late Oligocene and 15 B for the Pleistocene.

DSDP hole 288 (Fig. 2.2; Andrews *et al.*, 1975) penetrated 989 m of a similar, Aptian to Quaternary sequence in which, however, the Cretaceous to Paleocene section is more complete and thicker (489 m), while the Eocene is represented by a major hiatus. The early Oligocene to Quaternary section is only 500 m thick here. In consequence, sedimentation rates in the late Cretaceous are about double of what they are at site 289, i.e., 25 B increasing upwards to 40 B. On the other hand, the rates from the early Oligocene upwards generally fluctuate between 19 and 42 B, and thus are lower than at site 289. Extreme rates of 17 B and 77 B are both found near the top of the sedimentary column.

The same pelagic sequence continues south into the upraised Malaita fold belt, which appears to be an obducted part of the Ontong Java Plateau (Kroenke, 1972; van Deventer and Postuma, 1973; Hughes and Turner, 1977). It is exposed on Malaita Island and on the north side of Santa Isabel (Fig. 2.2) (Coleman *et al.*, 1978; Katz, 1980). Here, too, an oceanic, tholeiitic basalt basement is overlain by a carbonate mud sequence including siliceous marl with chert, white aphanitic limestone with brown chert nodules, nannofossil chalk and, higher in the sequence, blue-grey and brown calcisiltites. In central and northern Malaita, the basal part of the sediment sequence consists of highly siliceous mudstone with radiolaria (Hughes and Turner, 1976), suggesting a depositional environment below the carbonate compensation depth. The oldest limestones found in NW Malaita are of Albian age (van Deventer and Postuma, 1973), whereas in southern Malaita the base of the sedimentary succession is late Cretaceous (Hughes and Turner, 1976). The succession, in which younger basalt flows are intercalated locally in the Eocene (Hughes and Turner, 1976, 1977), is continuous to the Pliocene (McTavish, 1966).

Throughout the late Cretaceous to Pliocene interval, sedimentation was pelagic and mostly deep-marine, but shallowing conditions are indicated in the upper part of the sequence. Sedimentary structures are rare in the older deposits, which suggests quiet depositional environments; bioturbation is very common. From late Oligocene to Miocene time, however, an influx of volcanogenic silt–sand detritus becomes noticeable (Coleman *et al.*, 1978), and distal turbidites and slump deposits in the otherwise still pelagic environment are not uncommon in the upper part of the sequence (Hughes and Turner, 1976, 1977).

These features may indicate proximity to the tectonically unstable belt of a proto-Solomons island arc. Final collision with this arc took place in the late Pliocene, when the leading edge of the oceanic Malaita–Ontong Java province (Katz, 1980) was tightly deformed into an impressive fold belt, and amalgamated with the modern Solomons island arc. Pleistocene conglomerates, of deltaic to shallow-marine environments and derived from subaerial erosion and river transport in an early Malaita island setting (Hughes and Turner, 1976), rest unconformably on older rocks (Fig. 2.2).

2.3 Coral Sea to New Hebrides Basin: Oceanic Basins South of Papuan Peninsula–Pocklington Trough, and South and West of South Solomons and New Hebrides Trenches

This highly complex, oceanic area (roughly between 10° and 20°S) is made up of very different, physiographic–structural units. Their tectonic significance and relationships, however, are not fully understood, while the nature and thickness of their sediment cover is only known from rather sparse seismic sections, besides DSDP holes 210 and 287 in the abyssal plain of the Coral Sea Basin, and 286 in the New Hebrides Basin (Fig. 2.6).

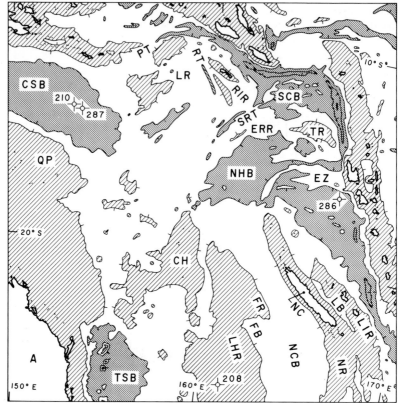

Fig. 2.6. Bathymetric map from Coral Sea Basin to New Hebrides Island arc, with location of DSDP holes 208, 210, 286 and 287. Contour interval 2000 m. A, Australia (Queensland); CH, Chesterfield Plateau; CSB, Coral Sea Basin; ERR, East Rennell Ridge; EZ, d'Entrecasteaux Zone; FB, Fairway Basin; FR, Fairway Ridge; LB, Loyalty Basin; LHR, Lord Howe Rise; LIR, Loyalty (Islands) Ridge; LR, Louisiade Rise; NC, New Caledonia; NCB, New Caledonia Basin; NHB, New Hebrides Basin; NR, Norfolk Rise; PT, Pocklington Trough; QP, Queensland Plateau; RIR, Rennell Island Ridge; RT, Rennell Trough; SCB, Santa Cruz Basin; SRT, South Rennell Trough; TR, Torres Rise; TSB, Tasman Sea Basin.

From west to east, the major physiographic features from the Coral Sea to the New Hebrides Basins are: the Louisiade Rise, which is cut by a linear, NE-trending trough system (the "Louisiade Fracture Zone", Landmesser *et al.*, 1975), the Rennell Trough, the Rennell Island Ridge (including Bellona and Rennell Islands and Indispensable Reefs), the South Rennel Trough, the East Rennell Ridge to Torres Rise, and the d'Entrecasteaux Zone.

The oldest known sediments throughout this area are Eocene. In the eastern *Coral Sea Basin*, which is over 4600 m deep (Coral Sea or Papua Abyssal Plain, DSDP hole 287; Andrews *et al.*, 1975), oceanic basalt is overlain by early to middle Eocene clay nanno chalk, 57·4 m thick, which was deposited above foram solution depth (Fig. 2.5). However, recent investigations around the margins of the western Coral Sea Basin (Symonds *et al.*, 1982) have indicated that the oldest part of the Coral Sea oceanic basement may be Cretaceous, being overlain by seismic formations of late Cretaceous to Paleocene age.

An erosional hiatus separates the middle Eocene from the late Oligocene. This is a regional feature across much of the SW Pacific, and is probably produced by a major bottom water current that swept north past eastern Australia in the early and middle Oligocene (Kennett *et al.*, 1972; Edwards, 1975).

During the late Oligocene, pelagic sedimentation resumed with the deposition of 7·7 m of yellowish-brown to grey-orange nanno ooze, followed by a second period of nondeposition and/or erosion, which probably extended into the early Miocene. The basin deepened during this time to below the nanno solution depth, and brown abyssal clay was deposited later in the early Miocene, to a thickness of 0·6 m—but 51·6 m of similar clay were deposited at the nearby DSDP site 210, of early to middle Miocene age (Fig. 2.5; Burns *et al.*, 1973). In the late Miocene the sedimentation regime changed drastically, and turbidity currents began transporting terrigenous debris from the Papuan Plateau into the Coral Sea Basin (DSDP site 210) and gradually spread further across its deeper parts. This detrital input, which in the Pliocene arrived at an accelerated rate, is derived from the Papua New Guinea mainland to the north, first from the Owen Stanley Range and later probably from the folded and uplifted Aure Trough (Rickwood, 1968; Dow, 1977). At site 287, 171·6 m of turbiditic, green silty clay and graded cycles of silt and clay with interbeds of nanno clay were deposited during the Plio–Pleistocene; at site 210 some distance further west, terrigenous turbidite deposition of graded silt and clay was from late Miocene to Pleistocene, amounting to 470 m.

Overall, the stratigraphy in this area indicates a very quiet, tectonically stable environment—apart from gradual subsidence of the basin. Lithological changes and those in sedimentation rates are mainly due to changes which occurred in the source area of sediments: slow pelagic sedimentation in the Eocene to early Miocene, at a rate of 25 B in the Eocene but probably much less in the late Oligocene and early Miocene when the basin was considerably deeper, gave way to terrigenous turbidite deposition which in the Quaternary reached a rate of 120 B (Andrews *et al.*, 1975).

The *Louisiade Rise* to the east and northeast of the deep Coral Sea Basin (Papua Abyssal Plain) is a dome-shaped structural feature whose flanks are characterized by normal faulting away from the rise crest (Landmesser *et al.*, 1975). It rises to 1829 m below sea level, and is cut into two by a fault-

B

bounded trough which trends NE. It is not known whether the Louisiade Rise is a remnant of continental crust like the Queensland or Papua Plateaux, or whether it is underlain by oceanic crust; Landmesser *et al.* (1975), in their discussion of this question, have stressed the fact that the characters of seismic profiles across the rise resemble those across the Queensland Plateau. From correlation with DSDP hole 287, the sediment sequence probably consists of the same Eocene–Oligocene, pelagic biogenic strata, but reflectors within the sedimentary column dip away from the rise crest where they appear truncated. Since erosion has left only a relatively thin veneer of sediment covering the basement near the rise crest, Landmesser *et al.* (1975) have noted that this would be a favourable area for future sampling of the older sediments.

The Pocklington Ridge in the northernmost apex of the Louisiade Rise, i.e., near the junction of Pocklington Trough with South Solomons Trench, is a feature of rather more complex bathymetry. It possibly represents a faulted basement surface which is overlain by a much thinner sediment section, or possibly a complex volcanic edifice. The basement itself crops out in a number of places.

Towards the Rennell Island Ridge in the east, the Louisiade Rise gradually bends down into the *Rennell Trough* which immediately west of Rennell Island is 4500 m deep. The acoustically transparent sediment cover of the Louisiade Rise closely follows the basement configuration down into the Rennell Trough, where it is unconformably overlain by a wedge of flat-lying, stratified sediments which are up to 1·3 sec thick (two-way time; Gulfrex profiles SI-42 and SI-43). These sediments probably represent turbidites of Miocene to Recent age. The up-faulted horst of the *Rennell Island Ridge*, which is capped by Bellona and Rennell Islands (Taylor, 1973) as well as Indispensable Reefs—which latter may be a similar, en echelon structure controlled by strike–slip displacements—mainly consists of pre-Miocene basement which locally is covered by more or less flat-lying, but fault-displaced sediments up to 1·1 sec thick. Atoll–reef formations on the islands show a reef complex exceeding 500 m in thickness, which obviously was formed during slow subsidence of a basement platform. Recent block faulting has uplifted part of the coral atoll including areas of former lagoon environments, which contain important bauxite deposits (Taylor and Hughes, 1975). More detailed marine seismic surveys by CCOP/SOPAC (Saphore and Exon, 1981), covering part of the Rennell Island Ridge, may define more closely its structure and conditions of underlying sedimentary formations, as well as possible economic potential.

While present data are still rather limited, the Rennell Trough (which is less than 300 km long) and Rennell Island Ridge have been interpreted by Récy *et al.* (1977a) as a fossil subduction zone–island arc couple.

The *South Rennell Trough*, which from the north of the Chesterfield Islands curves NE and opens up into the Santa Cruz Basin near the junction of South Solomons and North New Hebrides Trenches, is about 700 km long and up to 5000 m deep. A thin sediment fill is only found in its deepest part, but in general the trough as well as its upraised shoulders have basement exposed with no sediments. From its morphology, magnetism, gravity and dredge samples, Larue *et al.* (1977) have concluded that the South Rennell Trough is a fossil, oceanic spreading zone.

Eastwards across the *New Hebrides Basin*, which includes several deep basins below 5000 m, there are two large structures of high relief with elevations to much less than 1000 m below sea level: the *Torres Rise* (West Torres Massif; Collot *et al.*, 1985) and the *d'Entrecasteaux Zone*. From magnetic anomalies in the New Hebrides Basin, both to the NW and SE of the d'Entrecasteaux Zone, the age of its crustal formation is identified to range from 80 to 42 m.y., decreasing southwards (Lapouille, 1982). The d'Entrecasteaux Zone would be a younger feature that intervened only later in this configuration. Thus the New Hebrides Basin as a whole would be comparable in age (late Cretaceous to Eocene) to the Coral Sea Basin. DSDP hole 286 immediately south of the d'Entrecasteaux Zone confirmed a minimum age of middle Eocene for the oceanic basement, while magnetic lineations in this southern part of the New Hebrides Basin, which Weissel *et al.* (1982) have tentatively correlated with anomalies 18 through 23 of the geomagnetic reversal scale, suggest an age of the seafloor at this site which is greater than 55 m.y., i.e. late Paleocene. Since the 452 m of rapidly deposited, volcanic conglomerates and sand–siltstones of middle to late Eocene age in hole 286 suggest the presence of active, andesitic volcanism nearby, Maillet *et al.* (1982) conclude from this and other evidence that an active volcanic island arc and subduction zone existed during the Eocene along the present d'Entrecasteaux Zone, continuing into Loyalty Ridge–New Caledonia. If such a plate boundary did exist during the Eocene, the oceanic basins to the north and south of the d'Entrecasteaux Zone (which latter also is oceanic and uplifted at a later date; Maillet *et al.*, 1982) could not have originally been one and the same basin, as conceived by Lapouille (1982).

The stratigraphic section drilled in DSDP hole 286 (water depth 4465 m) is interesting in a number of aspects. While the deposition of coarse andesitic volcanic debris occurred in middle Eocene time directly on basaltic flows of the oceanic basement, this deposition was in relatively deep water (but above, and later near the foram solution depth), in a submarine fan that was supplied through debris flows and turbidity currents from a source area in shallow water (sediments include abundant shallow-water fossils). The rate of deposition was high: initially 230 B, it declined to 80 B in the late Eocene when the supply of volcanic material declined. Mainly biogenic sediments were deposited from late Eocene through Oligocene time, but small amounts (generally less than 10%) of glass shard ash are found throughout this interval. The sediments, 114 m thick, are yellow-brown nanno ooze and chalk, and exhibit moderate to strong bioturbation. The depositional surface was between the foram and nanno solution depths; sedimentation rates had decreased to 16 B in the late Eocene, and to 3–6 B in the Oligocene. Contrary to other areas in the SW Pacific, there is no Eocene–Oligocene erosional unconformity at this site. In the latest Oligocene or possibly Miocene, 19 m of abyssal red clay were deposited, which contains glass shard ash. It is barren of forams and contains only residual nannofossil elements, and thus was probably deposited at or below the nanno solution depth. Reworked older and shallow-water-derived fossils (bryozoan and algae) indicate intermittent transportation of material from outside the area, and sedimentation within this abyssal clay.

Sedimentation in the Miocene was either abnormally slow and represented by an extremely condensed section, or—more likely—interrupted for some time before the overlying, 64 m of Plio–Pleistocene ash with radiolaria- and nanno-rich ooze and clay were deposited. Seismic sections in this area (Gulfrex NH-10 D, NH-11, Mobil Oil 72-250/251) do in fact suggest an unconformable relationship between a younger series, 0·1–0·2 sec thick, and an older and slightly deformed series with a thickness of 0·5–1·0 sec (this matches well with a sonobuoy profile taken at site 286, where the 0·11-sec reflector marks the base of the 64 m thick, Plio–Pleistocene series). The new influx of volcanic material, and also of fossils derived from shallow water, indicate a source area of the sediments from the now active, volcanic island arc of the New Hebrides to the east.

The Eocene–Oligocene (to possibly Miocene) sedimentary series, 585 m thick at DSDP site 286, can be mapped regionally from seismic records, where it is found unconformably overlying an undulating basement surface at a variable thickness of about 0·5–1·0 sec (Katz and Daniel, 1981; Daniel and Katz, 1981). Locally basement is exposed, with sediments wedging out against the flanks of basement highs. The same series is also preserved in a shallow basin in the middle of the d'Entrecasteaux Zone immediately north of Sabine Bank (profiles 9 to 10 of "Chain"; Luyendyk *et al.*, 1974). Towards its eastern end, the basin is about 30 km wide from south to north and filled with 1·0 sec of well-layered sediments. A transparent top interval 0·15–0·2 sec thick, of slightly unconformable behaviour, probably represents the Plio–Pleistocene ash layer as found in DSDP hole 286 to the south of Sabine Bank. Sabine Bank itself, which protrudes up to nearly sea level, seems to be intrusive into the Eocene–Oligocene series, and thus could not be the source of it as suggested by Maillet *et al.* (1982). On the other hand, the boundary ridge at the northern margin of d'Entrecasteaux Zone consists of older basement against which the Eocene–Oligocene sediments thin and wedge out in a normal, onlapping situation.

In the northern part of the New Hebrides Basin, the most enigmatic feature undoubtedly is the *Torres Rise*. Its crest is only 809 m below sea level, i.e., virtually as high as the New Hebrides island arc structure to the east, from which it is separated by the North New Hebrides Trench which in this area is 7500–8000 m deep. The Torres Rise is a smooth flat feature, dome-shaped in its highest part but elongate in an overall east–west direction, descending westward towards the South Rennell Trough. Across this whole area, i.e., from a shoulder above the North New Hebrides Trench at 3500 m, over the rise crest at 800 m and west to the edge of the South Rennell Trough at 4000 m, a continuous sediment cover is found which is uniformly about 0·4 sec thick (Gulfrex seismic profiles SI-43 and NH-1 to NH-4, on open file at the Vanuatu Geological Survey, Port Vila). Apart from minor normal faulting with displacements at the most of a few hundred metres, which affects basement and sediment layer alike, there are no other structural disturbances discernible. From magnetic anomalies, however, the basement could include small, shallow intrusions. In general, and from the very limited data available, this feature of the Torres Rise is not very dissimilar to the Louisiade Rise, and could well be an old, continental relic. The age of the sediment cover is of course crucial for a closer interpretation; the nearest that can be speculated here is an Eocene–Oligocene age, correlating with the sediments of the d'Entrecasteaux Zone and further south, and most probably also of the Louisiade Rise.

North of the Torres Rise, the sea floor descends to the 4000–5000 m deep *Santa Cruz Basin*, which is devoid of sediments and characterized by a very rough basement surface, reminiscent of young, oceanic crust. To the north it is bordered by a somewhat higher ridge. A steep drop both north and east leads to the South Solomon and North New Hebrides Trenches, respectively, which both are devoid of sediments, too.

2.4   The Loyalty Ridge and Loyalty Basin, New Caledonia–Norfolk Rise, New Caledonia Basin and Lord Howe Rise

These regional elements, generally parallel and aligned due SE and S comprise two narrow and elongate, oceanic basins and two rises which consist of continental crust (Fig. 2.6).

The *Loyalty Basin*, which stretches from the d'Entrecasteaux Zone in the north for about 1300 km to the southeast and south, is separated from the main New Hebrides Basin by the *Loyalty Ridge*. Thanks to the very detailed and comprehensive, geophysical-geological studies by O.R.S.T.O.M. in Noumea (Bitoun and Récy, 1982), our knowledge of the geologic evolution of this area is well advanced. The following presentation is mainly based on these studies.

The Loyalty Ridge appears as an old rise of the sea floor that already existed in pre–late Eocene time. Later magmatism accentuated the ridge morphology, building up a chain of volcanic edifices. The final products of this magmatism are alkali basalts rich in titanium, as they crop out in the island of Maré where they have been dated as 11·2 and 9·3 m.y. old by Baubron et al. (1976), and between 13·5 m.y. and 9·6 m.y. old by Bonhomme (1979). The oldest known sediments in the Loyalty Islands are lower Miocene coral limestones which interdigitate with tuffs and lavas (Paris, 1981). After cessation of volcanism, the volcanic edifices were finally capped, from the Pliocene onwards, by reef formations of coral atolls, which more recently have been uplifted to as much as 140 m above sea level.

Since early time, therefore, the Loyalty Ridge has formed a barrier between the Loyalty Basin in the west and the New Hebrides Basin in the east and north. In its present form, however, the Loyalty Basin came into existence through the obduction and overthrust, in the late Eocene (Paris et al., 1979), of the large ophiolite sheet in New Caledonia; its erosive products immediately began filling the Loyalty Basin in the east. Because of the barrier and sediment trap formed by the Loyalty Ridge, the sequence in the Loyalty Basin is much thicker and very different from that in the New Hebrides Basin (e.g., DSDP hole 286, see above), which is contemporaneous but relatively thin, consisting mainly of Eocene to Oligo–Miocene pelagites.

The Loyalty Basin is underlain by oceanic crust (peridotites, gabbro and basalt with seismic velocities of 8·4, 6·8 and 5·8 km/sec, respectively), which can be followed westwards into the root zone of the overthrust sheet close to the east coast of New Caledonia. Except for a relatively thin sediment layer immediately overlying basement (and not always clearly visible on seismic sections) which is regularly inclined parallel to the basement and thus may be older than late Eocene, i.e., older than the nappe overthrust, the principal sediment fill of the basin is obviously later than that overthrust and a direct consequence and product of it—i.e., of its erosion. On land it has been shown that such erosion had its effect immediately after the nappe emplacement, with chromite and serpentinite of the overthrust ultrabasic complex being included in the late Eocene sediments (Paris et al., 1979). Thus the sedimentary filling of the Loyalty Basin, which clearly is derived from the ophiolites in the west, would essentially have begun in the Eocene, too.

Basically, the sequence is divided into two formations of which the older

(Formation II) consists of well indurated sediments with a seismic velocity of around 4·9 km/sec; its thickness is up to 4 km. The younger formation (Formation I) is further divided by two clearly marked reflectors into three members of variable thickness and more or less unconformable disposition. The two top members are of low velocity (2·1 km/sec) and hardly consolidated, while the lower one is somewhat more consolidated (2·9–3·0 km/sec); the total thickness of Formation I is about 1·3–4·0 km. Thus the cumulative basin fill may attain 8 km of sediments, but it is to be noted that the depocenters of each formation and member are not at the same place. Correlation with tectonic events in the New Caledonia area allows a close match of sedimentary distribution in the basin, as well as respective age assignments of the various formations. Thus Formation II would range from late Eocene to early Miocene, and is thickest where the New Caledonia block has been uplifted most; in particular, its great thickness in a southward direction beyond the present extent of New Caledonia correlates with other evidence of a southward extension of the overthrust sheet to at least 23°30′S, and of deep erosion and later submergence of this block. On the other hand, the limited thickness of Formation II in the north reflects the lesser intensity of erosion there, due to a much weaker relief of emergent lands in the area of the present-day northern lagoon.

Differential easterly downfaulting of the New Caledonia block from the early Miocene onwards, caused by a regional basin subsidence of 1500 m, is responsible for great variations in thickness, coupled with local unconformities, in the lower part of Formation I. The age of this lower part therefore is early to middle Miocene. The thickness distribution in the upper part of Formation I, on the other hand, does not seem to be related to a potential terrigenous source for its sediments. The general decrease of mechanical erosion on land in the late Miocene (as opposed to chemical erosion) and the establishment of an eastern barrier reef would both effectively reduce the availability of sediments from a land-derived source. As a consequence, the sediments in the upper part of Formation I become more and more pelagic. However, the overall thickness of this youngest part of Formation I, which is of late Miocene to Recent age, is probably still too great for an entirely pelagic derivation: during the time of volcanic activity in the Loyalty Ridge (late Miocene and possibly before), there was probably a considerable input of volcanic debris into the basin, from this easterly source. At present, on the other hand, sedimentation is entirely pelagic in the central Loyalty Basin.

Southwards, the Loyalty Basin passes into the *North Norfolk Basin* (Fig. 2.7), which is considered a marginal oceanic basin that would have opened, according to magnetic anomalies, in the late Cretaceous from about

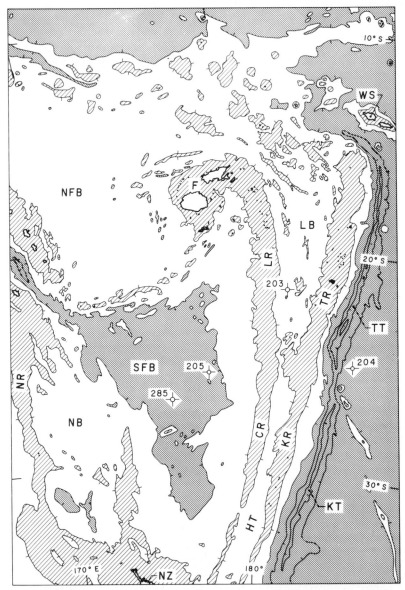

Fig. 2.7. Bathymetric map of area between North Fiji Basin, Tonga and New Zealand, with location of DSDP holes 203, 204, 205 and 285. Contour interval 2000 m. CR, Colville Ridge; F, Fiji Islands; HT, Havre Trough; KR, Kermadec Ridge; KT, Kermadec Trench; LB, Lau Basin; LR, Lau Ridge; NB, Norfolk Basin; NFB, North Fiji Basin; NR, Norfolk Rise; NZ, New Zealand; SFB, South Fiji Basin; TR, Tonga Ridge; TT, Tonga Trench; WS, Western Samoa.

85 to 76 m.y. (Launay *et al.*, 1982). Sediment thickness is much less in the Norfolk Basin than in the Loyalty Basin, since it did not receive—as did the latter—abundant debris from the large, overthrust periodotite mass in New Caledonia which underwent strong degradation and deep erosion. A seismic refraction survey in the northwest of the Norfolk Basin (Shor *et al.*, 1971) has shown 800 m of sediments with a velocity of 3·97 km/sec; below this are two basement layers of 6·05 and 7·04 km/sec, respectively.

West of the Loyalty–Norfolk Basins is the *New Caledonia–Norfolk Rise*, which southwards merges into the New Zealand continental block (Figs. 2.6 and 2.7). Like New Zealand it has an older continental crust, which in the axial part of *New Caledonia* is 35 ± 4 km thick (Dubois, 1969). A detailed geological synthesis of New Caledonia has recently been presented by Paris (1981). Thus a core of polymetamorphic schist has been assigned a pre-Permian age (Fig. 2.8). Thick Permo–Triassic and Jurassic sediments have close faunal and facies links with those in New Zealand (Lillie and Brothers, 1970; Fleming, 1968; Stevens, 1977), and have traditionally been interpreted as forming part of a continuous geosynclinal belt which extended from New Caledonia to New Zealand. In detail, however, the paleogeographic settings of New Caledonia and New Zealand Permian to Jurassic rocks have been shown to be considerably different (Paris and Bradshaw, 1977).

Fig. 2.8. Pre-Permian, metamorphic schist north of Col d'Amieu, New Caledonia—oldest rock in the Southwest Pacific Island region. Scale: case for sunglasses (top-right).

Following the regionally extensive neo-cimmerian orogeny (Paris and Lille, 1977), Senonian deposits are transgressive and unconformable and mainly represented by terrigenous, detrital sediments such as conglomerates, sandstones and carbonaceous shales. Coal seams are abundant in certain areas ("formation à charbons"). However, while paleoenvironments were often lagoonal–estuarine and brackish to littoral, marine sediments also exist and have yielded a rich ammonite and bivalve (*Inoceramus*) fauna which is mainly of Campanian to Maastrichtian age. Tuffaceous volcaniclastics and basaltic and rhyodacitic intercalations give rise to an additional, locally significant facies. Mainly along the western side of New Caledonia, the Senonian deposits are associated with submarine lava flows of an extensive basalt formation, which is supposed to range through Late Cretaceous to mid Eocene, and which includes tholeiitic and calc–alkaline elements (Paris, 1981).

During Paleocene to lower Eocene times, tectonic quiescence and deepening of the sea resulted in the formation of very homogeneous, fine-grained siliceous sediments or phtanites, which are associated with lenticular, micritic Globigerina limestones. Uplift and erosion separates these sediments, however, from unconformable, middle to late Eocene deposits, which are profoundly different and indicate strong changes in the paleogeographic setting: these reflect the first and important, tectonic movements of the Alpine orogeny (Paris, 1981).

The marine transgression of middle to late Eocene age extensively immersed the New Caledonia block. A strong relief remained, however, in particular with two high zones staying emergent, i.e., mainly along the axial part and on a western paleo-structure around the Bay of Saint Vincent. Correspondingly, two main facies types resulted, represented by (a) thick flysch deposits, locally including large olistoliths, which developed in the deeply subsiding parts of the basin (e.g., Noumea and Bourail), and by (b) epicontinental limestone deposition along the borders of emergent land. These limestone formations suggest attempts at reefal buildups, thus indicating a warm climate and shallow, agitated waters.

Along the west coast of New Caledonia, volcaniclastic flysch of the very latest Eocene is discordant over the folded, middle to late Eocene flysch and carbonate rocks; but at Népoui these deposits also overlap outliers of the great peridotite nappe, the emplacement of which, therefore, is precisely dated at this place (Paris *et al.*, 1979). Further to the SE, however, the final advance of this huge overthrust sheet did not occur until after the terminal Eocene.

The peridotite sheet, which in the "Great Southern Massif" (Grand Massif du Sud) consists of two main lithological units 1000–1500 m thick each (Guillon, 1975), covers 8000 km^2 of New Caledonia but in its submerged southern prolongation probably extends for another 100 km or so (Bitoun and Récy, 1982). It consists mainly of little differentiated ultramafic rocks, with harzburgite as the predominant type; dunite and pyroxenite, though ubiquitous, are minor constituents. Locally this main peridotite unit is overlain by voluminous cumulate units passing from dunite to noritic gabbro. In general, the New Caledonia peridotite nappe represents part of an ophiolite series, i.e., obducted oceanic crust; its missing effusive members higher in the sequence may have been eroded away, or tectonically truncated (Paris, 1981). Nickel, cobalt, and chrome are the most economically important minerals associated with these peridotites. In particular, nickel and cobalt concentration in the deeply weathered surface layers of the peridotite, often below thick, iron-rich laterites, are the type of deposits which have become basic to the New Caledonia economy. They are the result of long, subaerial exposure due to uplift and deep dissection of the axial part of New Caledonia—ongoing since the end of Eocene times—and may also occur offshore to the SE and possibly NW, provided that submergence in these areas has taken place only relatively recently, so that enough time was available prior to submergence, for alteration of the peridotite and the formation of these thick, weathered residuals. This, in fact, seems to be the case in the southern prolongation of New Caledonia, where subaerial exposure and intense weathering has lasted until at least the lower Miocene (Bitoun and Récy, 1982), and where relatively rapid subsidence of at least 400 m has occurred mainly in post-Miocene time (Daniel et al., 1976).

The various stages of transgression and regression, of flysch deposition followed by tight folding, uplife and erosion which began in the middle Eocene, and which culminated in the latest Eocene to Oligocene with the regional obduction and overthrust of oceanic crust represented by the ophiolite sheet, have been grouped together in a tectonic sense as representing the Alpine orogeny (Paris, 1981). This major orogenic event is also represented by compressive thrust tectonics associated with high P/T metamorphism in a thick pile of Cretaceous to Eocene sediments in the extreme NW of New Caledonia (Lillie, 1970; Brothers and Blake, 1973; Black and Brothers, 1977; Briggs et al., 1978). As a consequence of this orogeny, the entire region was high and emergent during the Oligocene—and most of it has stayed high ever since. Only very marginally, along the present west coast, did the sea encroach again later.

Thus the folded, late Eocene flysch is covered here with an angular unconformity by a lower Miocene series of sandstone–conglomerate, indicating an intense phase of erosion. Seawards and upwards, these coarse detrital sediments are followed by calcarenite and micritic limestone including reef-associated deposits (Paris, 1981).

The *Norfolk Rise* south of 24° seems to have had a somewhat different

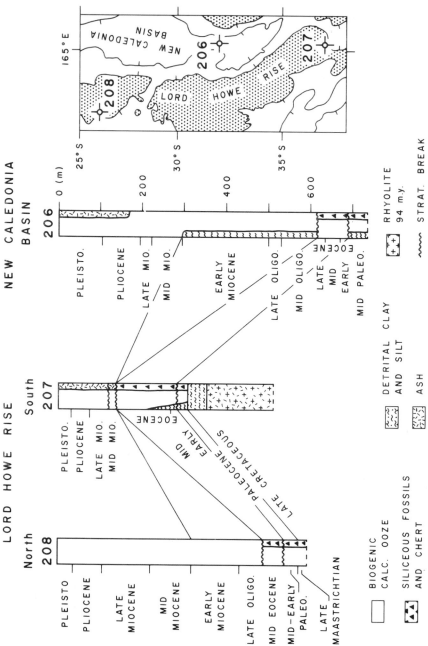

Fig. 2.9. Stratigraphic columns of DSDP holes 206, 207 and 208, Lord Howe Rise and New Caledonia Basin.

history. Seismic profiles (Dupont *et al.*, 1975) show relatively thick sediments draped more or less comformably over the rise and extending into and across the adjacent basins, where they can tentatively be correlated with those in DSDP hole 208 (Figs. 2.6 and 2.9), thus spanning an age range from late Cretaceous to Recent. It is probable that acoustic basement is not the real basement, but rather represents the erosion surface above older formations which in the neo-cimmerian (New Caledonia) or Rangitata orogeny (New Zealand) have become highly distorted and indurated. As interpreted by Dupont *et al.* (1975), the middle of three marker reflectors A, B and C would represent the early Oligocene gap as widely recognized in the SW Pacific. Reflector C could approximate the Paleocene–Cretaceous boundary, but more probably corresponds to a facies discontinuity within the Cretaceous. The sedimentary series above reflector C correlates with the first layer of Shor *et al.* (1971), which from refraction measurements at 27°S was found to be 1630 m thick, with an average velocity of 2·73 km/sec. The greatest sediment thickness, however, was found by Dupont *et al.* (1975) some distance further north (at 25°30′S), with over 3000 m of sediments.

Younger intrusions and numerous faults characterize the recent structure of the Norolk Rise, which in its overall aspect is an uplifted fault block of continental affinities, separated from the adjacent, oceanic Norfolk and New Caledonia Basins. The intrusions are probably related to the late Tertiary volcanic activity as manifested on Norfolk and Philip Islands (Aziz-Ur-Rahman and McDougall, 1973).

The *New Caledonia Basin* (Fig. 2.6), west of and parallel to the New Caledonia–Norfolk Rise, is one of the major physiographic features in the SW Pacific. It is about 3000 m deep and stretches over 2500 km from the NW of New Caledonia SE and south towards New Zealand. However, it has been subdivided structurally into various portions, notably the Fairway Basin, Fairway Ridge and New Caledonia Basin *sensu stricto* in the north (Fig. 2.6), and Wanganella Bank, West Norfolk Ridge and South New Caledonia Basin in the south (Dubois *et al.*, 1974; Ravenne *et al.*, 1977b). In spite of its limited width (150–300 km), the basin is generally considered to be underlain by oceanic crust, with a crustal thickness from seismic refraction studies (Shor *et al.*, 1971) of 10–17 km, and from gravity data (Woodward and Hunt, 1970) of 9 km.

Sediments are 2000–3000 m thick (Ravenne *et al.*, 1977b). Correlation of seismic reflectors with DSDP holes 206 and 208 (Fig. 2.9; Burns *et al.*, 1973) suggests an age range from the late Cretaceous to Recent. A regional unconformity that corresponds to the late Eocene to early–middle Oligocene gap is well marked, especially on the ridges and edges of the basin, while in more central parts the series are often conformable, and perhaps more complete (Ravenne *et al.*, 1977b). Thus in DSDP hole 206, which is located in the southern New Caledonia Basin on the flank of a local high structure, this gap encompasses the late Eocene and early Oligocene; also there is another hiatus here around the middle–late Paleocene including the Paleocene–Eocene boundary. In addition, the overall thickness is much reduced here (only 734 m from the bottom of the hole in the early Paleocene), which probably is due to this location being in a regional narrow pass of the basin, or saddle, which according to Dupont *et al.* (1975) and Ravenne *et al.* (1977b) marks the southern end of the Fairway Basin. Further south, on the other hand, the total sediment thickness in the South New Caledonia Basin is nearly 4000 m according to refraction data (Bentz, 1974); the section measures about 1200 m down to the Oligocene unconformity, as against 614 m for this interval in DSDP hole 206. However, considerable slumping down the eastern flank of the Lord Howe Rise has occurred here, resulting in stacking of sediments on the floor of the New Caledonia Basin (Bentz, 1974). Lithologically, the sediments in the New Caledonia Basin are nearly exclusively pelagic oozes. They represent rather uniform conditions of deposition throughout this time, where very little subsidence has occurred in the basin, and only so from the Paleocene to the Oligocene; throughout the Neogene, depths have been similar to those of the present day, i.e., lower bathyal. In fact, all deposition occurred in the bathyal zone above carbonate compensation depth. And while there are some sedimentary disturbances in the Paleocene–Eocene section of hole 206, including an age reversal probably due to slumping, the section from the Middle Oligocene to Recent contains one of the most complete biostratigraphic records of planktonic foraminifera and calcareous nannofossils in transitional latitudes of the Southern Hemisphere. The figures of calculated sedimentation rates are very relevant, therefore, and show a general upward increase which becomes particularly pronounced in the late Pliocene and Pleistocene. These rates, corrected for compaction, are 14 B in the middle–late Miocene, 21 B in the early Pliocene, 37 B in the late Pliocene and 55 B in the Pleistocene. Since the sequence is almost exclusively biogenic, this increase in the rate of sedimentation probably reflects higher production due to changes in oceanic turnover, particularly in the Pleistocene (Burns *et al.*, 1973). More precise data on paleooceanography, and paleoenvironmental changes in general, are extremely important therefore. DSDP leg 90 (December 1982 to January 1983) was designed specifically to study these problems of Cenozoic paleooceanographic, paleoclimatic and biotic evolution in the SW Pacific (Anonymous, 1983c).

West of the New Caledonia Basin is the *Lord Howe Rise*, which is a large segment of continental crust about 600–1200 km off the east coast of Australia, from which it is separated by the 4000–5000 m deep, oceanic basin of the Tasman Sea. As outlined by the 1500–2000 m isobaths, the Lord Howe Rise extends for some 2000 km from the Coral Sea region to New Zealand, with an average width of 300 km. It is surmounted by ridges and banks of only 750–1200 m water depth, or even less as particularly is the case in its northern part where the Fairway Ridge and Chesterfield Plateau are covered by coral reefs and small atoll islands (Fig. 2.6). Its western margin is characterized by local graben and horst structures, and elongate basins such as the Middleton and Lord Howe Basins which separate the Lord Howe Rise from the Dampier Ridge further west. The complex physiography is even more accentuated by a chain of seamounts along the western flank of the Lord Howe Rise, which trend northward from the volcanic Lord Howe Island.

From the two DSDP holes 207 and 208 (Fig. 2.9; Burns et al., 1973) and extensive seismic coverage, the stratigraphy of the sediment cover is fairly well known. Based on the most recent surveys, it has been summarized by Willcox et al. (1981). Since seafloor spreading in the Tasman basin began 80 m.y. ago (Hayes and Ringis, 1973), a pronounced regional unconformity over the Lord Howe Rise which lies within the Cretaceous has been correlated with this event and termed breakup-unconformity. It ties in with the base of shallow-marine silts and clays of Paleocene to Maastrichtian age intersected in DSDP hole 207. The pre-breakup formations below this unconformity are folded and faulted and generally present a horst-and-graben structure which, however, is erosionally planed. The formations are of varied acoustic characteristics and appear to represent crystalline basement, metasediments of Paleozoic to Mesozoic age and intrusives. Local grabens, particularly on the western side of the Lord Howe Rise, may be associated with a pre-breakup rift stage, and are filled with up to 3000 m of continental (deltaic) to shallow-marine sediments of probably (early) Cretaceous age. Their tops show marked erosional truncation, and at the base reflectors lap onto basement. On the eastern boundary of the rise, marked by a major fault which apparently separates continental from oceanic basement, a wedge of sediments at least 2000 m thick extends across the eastern margin of the Lord Howe Rise and beneath the New Caledonia Basin. It probably consists of Cretaceous clastic sediments that were deposited on the ancient continental margin of the Australian supercontinent, before its breakup and the opening of the Tasman Basin. Its correlation with pre-breakup sequences along the western Lord Howe Rise implies that the New Caledonia Basin is older than the Tasman Sea Basin. The difference in water depth between the two basins, incidentally, is mainly the result of different sediment fill.

Also of immediate pre-breakup association are the rhyolite flows, breccias and pumiceous lapilli tuffs drilled in DSDP hole 207, which have been

K–Ar dated as 94 m.y. old (McDougall and van der Lingen, 1974). They were erupted at or near sea level, and are overlain by late Cretaceous sand, silt and clay of a shallow-marine environment with restricted (non-oceanic) circulation. During this earliest post-breakup period, rather strong currents brought foreign, plutonic and metamorphic material to the site, and also reworked the underlying volcanics. Gradual subsidence followed, and open oceanic conditions began in the middle Paleocene with the deposition of mainly carbonate oozes. The present, upper bathyal water depth was reached in the early Eocene, after which the area remained rather stable. However, two breaks of regional significance as found elsewhere in the SW Pacific occur also here: one between the Paleocene and Eocene, the other separating the late middle Eocene from the early middle Miocene. Thus at this site the largest time span is found in the so-called Oligocene unconformity; here the late Eocene, the entire Oligocene and the early Miocene are missing.

This post-breakup sequence, which in DSDP hole 207 has a total thickness of 357 m, can be followed in seismic sections across the whole rise. In some sections there is evidence of probably reef buildups on the breakup-unconformity of late Cretaceous age, and in the shallow-marine sequence immediately above it. For the sequence as a whole, there are considerable thickness variations which probably are due to the variable time span of the Oligocene unconformity, and non-deposition or continuous erosion at active basement highs, and other factors. In the southern part of the rise, refraction measurements have indicated thicknesses of 950–1020 m for the section below the Oligocene unconformity, and 260–400 m for the section above it (Bentz, 1974). In DSDP hole 208 on the northern rise plateau (Figs. 2.6 and 2.9), where most of the Paleogene is absent due to the presence of two sedimentary breaks (i.e., between the middle Paleocene and early middle Eocene, and between the early middle Eocene and the late Oligocene), the thickness below the upper break is 106 m, and above it 488 m. In addition, subsidence has occurred considerably earlier here, since throughout the sequence penetrated—i.e., since at least the Maastrichtian—normal oceanic conditions have prevailed with the deposition of siliceous, nannofossil chalk in Late Cretaceous to middle Eocene times, and of foraminiferal nannofossil ooze in the late Oligocene to late Pleistocene.

Throughout Neogene times, furthermore, there has been widespread volcanism on the Lord Howe Rise. Lord Howe Island and Ball's Pyramid are the only manifestations above water, but they form the southernmost part of the extended Lord Howe Island Seamount Chain which is typically a chain of volcanic features. The rocks of Lord Howe Island have been described by Game (1970) as typical alkali basalts formed in three major eruptive periods. The youngest was radiometrically dated as middle Pliocene (7·7 m.y.), while volcanism probably began as early as mid-Tertiary and continued for 20–25 m.y. with no major change in its character. A big, seismically recorded extrusive feature on the eastern margin of Lord Howe Rise was sampled by Launay *et al.* (1977), who recovered olivine basalt and gabbro, besides hyaloclastic breccias and biomicrites. Similar volcanic piercement structures were crossed in great numbers by seismic lines on the southern Lord Howe Rise (Bentz, 1974). Their particular features suggest that eruptions, which all occurred in relatively deep water and must have been explosive, occurred repeatedly and until very recent time—volcanism indeed may be still active.

2.5 The New Hebrides Island Arc, North Fiji Basin and Fiji Islands Region

2.5.1 The New Hebrides arc and North Fiji Basin

East of the New Hebrides Basin and the New Hebrides Trench, this island arc, which in the average is 200 km wide, stretches for a distance of about 1400 km virtually straight south-southeast to north-northwest (Figs. 2.6 and 2.10). Although it is broadly similar to the Solomons arc in its rock contents, stratigraphy and structural evolution, there are important differences between the two arcs, as has been highlighted by Katz (1982b).

Basement formations are only known in the south of Pentecost Island, where they occur as tectonically isolated blocks of serpentinized periodotites and basaltic lavas metamorphosed to greenschist and amphibolite facies, which have been intruded by gabbro (Mallick and Neef, 1974). The metamorphic age has been dated as 35 m.y., and the gabbro intrusion as 28 m.y. The emplacement at their present location, however, is the result of block-faulting and diapiric intrusions in the latest Miocene and Pliocene (Fig. 2.11). From its general geochemistry and petrology, this basement complex seemingly represents an uplifted portion of oceanic crust.

Sediments and volcanics in the New Hebrides exclusively range from early Miocene to Recent, except for a red mudstone formation in the NW of Malekula, which is of unknown age and could be older (Mitchell, 1971). It may compare with the red, abyssal clay of latest Oligocene to Miocene age in DSDP hole 286 immediately west of the trench at about this same latitude (p. 26). Such correlation, however, would introduce serious tectonic problems, although it is not impossible in view of the likely presence, in this area, of local obduction of the western plate (Daniel and Katz, 1981). The marked structural separation of the red mudstone area from the rest of Malekula Island could indeed hint at such a possibility.

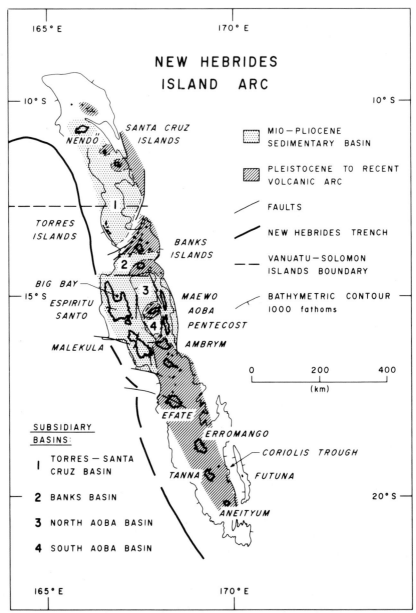

Fig. 2.10. Structural map of the New Hebrides Island arc. [After Katz and Daniel (1981), simplified.]

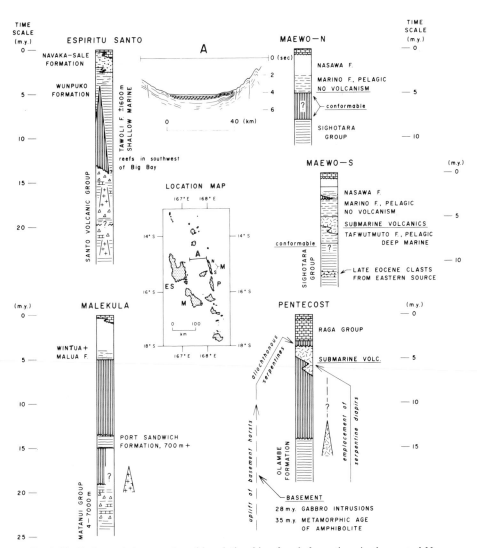

Fig. 2.11. Facies and time-stratigraphic relationship of rock formations in the central New Hebrides. For sources and explanation, see text. Vertical hatching: stratigraphic gap. [From Katz (1982).]

The distribution of thick, Mio–Pliocene sediments is restricted to the central and northern part of the New Hebrides island arc (Fig. 2.10). Here this major sedimentary basin has been penetrated to a variable extent by Plio–Pleistocene to Recent volcanic rocks of both intrusive and extrusive character, the latter including lavas and pyroclastics. South of about 16°30′,

Fig. 2.12. Early to middle(?) Miocene volcaniclastic sequence on Espiritu Santo Island, Peamaeto River west of Big Bay. (Above) Massive-bedded, strongly indurated greywacke sandstone. (Below) Alternating sequence of greywacke sandstone and dark-grey to blackish siltstone and splintery shale.

the whole island arc is built up exclusively by these younger volcanic formations (Katz and Daniel, 1981), and may well represent a distinct—i.e., younger—segment that only more recently has been amalgamated to form the present-day, more extensive island arc (Katz, in press). Thus, contrary to the central and northern part of the New Hebrides island arc, this younger arc segment in the south would entirely postdate the opening of the South Fiji Basin (Larue *et al.*, 1980).

In the central and northern part, the subdivision of the New Hebrides arc into separate geological entities from west to east [Mitchell and Warden (1971); this concept has been taken up again and recently expanded by Carney and Macfarlane (1982), who proposed even four such entities] is artificial. As derived from recent, geologic–geophysical information in the sea-covered areas between the islands (Katz, 1981a, 1982b), it becomes clear that the so-called western and eastern belts are part of one and the same continuous basin—though including marked facies and thickness changes—while the "Central Chain" and "Marginal Province" are younger features which are superimposed on the latter, as well as extending further south.

In this major basin of the New Hebrides island arc (which from north to south has been subdivided into four sub-basins, Fig. 2.10; Katz, 1981a), the early to middle Miocene is characterized by volcaniclastic sediments (Fig. 2.12), volcanic lavas and flow breccias of calc–alkaline to island arc tholeiitic affinities (Mitchell and Warden, 1971; Carney and Macfarlene, 1982). This volcanic assemblage is associated with an organic calcareous rock assemblage (Fig. 2.13), mainly of algal–foraminiferal reef limestones and derived calcirudites and arenites, as well as pelagic calcilutites or marls. Since basement is nowhere exposed under these Miocene series, only minimum thicknesses are known. They appear to be 4000–6000 m in the west (Robinson, 1969; Mitchell and Warden, 1971; Mallick and Greenbaum, 1977), but less than 1000 m in the east (Mallick and Neef, 1974; Carney, in press).

In the North Aoba Basin between the western and eastern island chains (Figs. 2.10 and 2.14), preliminary results from a recent ORSTOM-CCOP/ SOPAC refraction survey using OBS (ocean-bottom seismographs) suggest a total sediment thickness of $3 \cdot 9$ km, subdivided into three layers of $0 \cdot 1$ km (at a velocity of $1 \cdot 7$ km/sec), $1 \cdot 4$ km (at $2 \cdot 0$ km/sec) and $2 \cdot 4$ km (at $3 \cdot 4$ km/sec), respectively; underlying basement showed a velocity of $5 \cdot 8$ km/sec (J. Récy, personal communication). The $2 \cdot 4$ km thick layer at the base of this sedimentary sequence is tentatively correlated here with the early to middle Miocene. Thus a regional decrease in sediment thickness is indicated across the basin, from west to east. This goes parallel with an important, regional facies change: while the western facies is dominated by thick lava and volcanic rudite sequences of an open-marine archipelago,

Fig. 2.13. Cliff of white, vaguely bedded reefal limestone of early to middle Miocene age. Petawata River, west side of Cumberland Peninsula, Espiritu Santo Island.

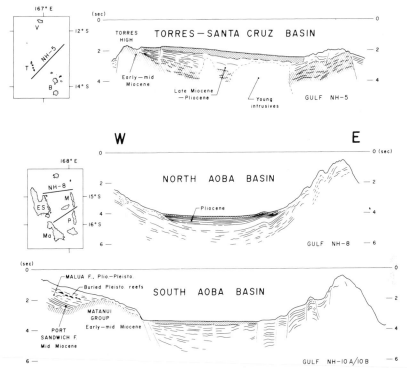

Fig. 2.14. Representative cross-sections through the median basins of the New Hebrides Island arc, from north to south: Torres–Santa Cruz Basin, North Aoba Basin, South Aoba Basin. Dotted layer in all three basins: Pleistocene to Recent. From Gulf Oil Company ("GULFREX") seismic records on open file at the Vanuatu Geological Survey, Port Vila. Name of islands in location insets are: Vanikolo (V), Torres Island (T), Banks Island (B), Espiritu Santo (ES), Malekula (Ma), Maewo (M) and Pentecost (P).

i.e., active volcanic island arc setting, with a range of derived products deposited in mixed environments from sub-littoral to bathyal (Jones, 1967; Mirchell, 1970), the eastern facies is generally fine-grained, having more distal characteristics with respect to the volcanic and sediment source, and is deposited in more generally deep-water environments; primary volcanics, such as lavas etc., are relatively rare in the east (Mallick and Neef, 1974). Considering the respective derivation of sediments, the western facies is coextensive with the source of its deposits (both volcanically derived and organic carbonates, which latter are controlled by, and directly related to, the presence of volcanic edifices), and thus is derived from within its area of deposition. The eastern facies, on the other hand, is of mixed derivation from both the western volcanic source, and an eastern terrigenous source: it partly represents a deep-sea fan environment on a west-dipping paleo-slope (Neef, 1982; Carney, in press), and includes conglomerates containing late

Eocene calcarenite cobbles (Coleman, 1969) and metabasalts and tholeiitic
dacites and rhyolites which seemingly are derived from an ancient landmass
lying east thereof, possibly Fiji (Carney and Macfarlane, 1978).

In the series of the western facies, local unconformities are frequent and
may be of both tectonic and volcanic origin. A more regional tectonic
unconformity is found on Malekula Island between the early Miocene Mat-
anui Group and the middle Miocene Port Sandwich Formation (Mitchell,
1966, 1971); this unconformity is also seen offshore on seismic profiles east
of Malekula (Katz, 1981a) (Fig. 2.14). Widespread intrusions of basalt–
andesite and gabbro to microdiorite bodies, and a major diastrophic event
with strong deformation of the rocks occurred on Espiritu Santo and
Malekula Islands around or immediately after middle Miocene times,
followed by general uplift and erosion (Fig. 2.11). At this time, too, volcanic
activity seems to have ceased.

Eastwards across the median basins (North and South Aoba Basins; Katz,
1981a, 1982), deformation decreases and any larger unconformities disappear. Although deposition of the early to middle Miocene Olambe
Formation on Pentecost Island (Mallick and Neef, 1974) still seems to be
followed by an extended hiatus (Fig. 2.11), there does not seem to be any
major, structural unconformity about it. Also on Maewo Island further
north, there is a conformable relationship—and possibly continuous sedimentation—between the early to middle Miocene Sighotara Group and the
late Miocene Tafwutmuto Formation, which consists of pelagic, deep
marine mudstone deposited under 2000–3000 m of water (Paltech, 1979).

In the west, uplift with further block-faulting, which occurred from the
middle Miocene onwards, caused emergence and erosion which may have
lasted for some 8–9 m.y. or more in some places (Fig. 2.11). As a result,
subsidiary basins formed which during the ensuing late Miocene to Pliocene
transgression were filled with strongly diachronous, terrigenous calcarenites, marls and pelagic mudstones, regionally with a pronounced
unconformity but locally conformable above the underlying, middle and
early Miocene deposits. While this transgression advanced from north to
south, or perhaps NE to SW, the late Miocene to Pliocene sediments in the
west (Tawoli Formation and correlatives) are restricted to several small
basins and are generally shallow-marine. The Tawoli Formation in the Big
Bay–Jordan River area (Mallick and Greenbaum, 1977) (Figs. 2.10, 2.11
and 2.15) is up to 1600 m thick, while corresponding Pliocene deposits in the
NW and SW of Malekula (Mitchell, 1966, 1971), and offshore NE Malekula
(Katz, 1981a) are only a few hundred meters thick at the most. There are no
contemporaneous deposits in the South Aoba Basin (Katz, 1981a, 1982) and
on Pentecost Island (Mallick and Neef, 1974), but they do occur to the north
of it on Maewo Island (Tafwutmuto and Marino Formations, both pelagic

Fig. 2.15. Soft-sediment deformation in thin-bedded, hard calcarenites of the late Miocene to Pliocene Tawoli Formation, west side of Big Bay, Espiritu Santo Island. Massive bed at base is compact, silty mudstone.

and deep-marine; Carney, in press). From there they extend continuously across the North Aoba Basin, onlapping westwards over earlier Miocene formations (Figs. 2.11 and 2.14).

In the central part of the New Hebrides island arc, there is thus a virtually conformable—if not continuous—relationship between all the Miocene and Pliocene formations in the east, as opposed to the presence of a pronounced regional, erosional unconformity which in the west occurs between the middle Miocene and the late Miocene to Pliocene. Sediments above this unconformity are shallow-marine in the west, whereas corresponding sediments in the east are deep-marine; in a northward direction, sedimentation is more continuous and generally of greater thickness (Katz, 1982). Active volcanism during the early to middle Miocene was concentrated in the western area, but apparently was non-existent after the middle Miocene. In turn, there is an intense, submarine volcanic phase in the eastern area from 7 to 4 m.y., i.e., around the Mio–Pliocene boundary; it is exposed on Pentecost and the southern part of Maewo Island, but does not extend to northern Maewo (Mallick and Neef, 1974; Carney, in press) (Fig. 2.11). This volcanism is transitional calc–alkaline to tholeiitic and may herald the more widespread island arc volcanism of Plio–Pleistocene to Recent age, which has developed all along the modern, New Hebrides island arc. In northern

Maewo where no such volcanic activity took place, the Pliocene Marino and Nasawa Formations directly overlie the early to middle Miocene Sighotara Group (Carney, in press) (Fig. 2.11). In Pleistocene to Recent times, reef limestones have formed a great number of plateaux and terraces on most of the recently uplifted islands, both in the east and west (Jouannic *et al.*, 1982); because of recent tilting of some of the islands, some reefs are now submerged as is found east of Malekula (Katz, 1981a). In the deeply subsided central parts of the median basins, a transparent layer some 0·2 sec thick (150–180 m) probably represents Quaternary ash (Katz, 1982).

The same Mio–Pliocene series can be followed throughout the northern part of the New Hebrides island arc, where they form a large, gentle syncline (Banks and Torres-Santa Cruz Basins; Katz, 1982) (Figs. 2.10 and 2.14). Early Miocene is known from Torres Islands (Greenbaum *et al.*, 1975) and from Nendo in Santa Cruz (Craig, 1975), both on the western high flank of the Torres–Santa Cruz basin. On seismic sections across the basin, this early Miocene sequence is overlain, with only a slight unconformity, by a younger series here interpreted as late Miocene to Pliocene that occupies the whole central part of the basin (Fig. 2.14). It is 1·1 sec thick, as against 0·8 in the Banks Basin to the south, 0·4 sec in the North Aoba Basin (or up to 0·7 sec if the transparent layer under the strong reflector band, Fig. 2.14, is included), and nil in the South Aoba Basin (Katz, 1981a, 1982). This corresponds to the regional, southward thinning also observed between Big Bay–Espiritu Santo and Malekula Islands (Mallick and Greenbaum, 1977; Mitchell, 1966, 1971).

In Plio–Pleistocene to Recent times, calc–alkaline to tholeiitic island arc volcanism (Colley and Warden, 1974; Gorton, 1977) has affected large parts of the New Hebrides. In the north it is centered along the central and eastern part of the island arc, including intrusive stocks over wide areas underneath the Torres–Santa Cruz Basin (Ravenne *et al.*, 1977a; Katz, 1982) (Fig. 2.14) and subaerial and submarine extrusives on several of the islands and in the back-arc rift zone (Katz and Daniel, 1981). South of about 16°30′, the entire island arc appears to be formed exclusively by a variety of products of this magmatic phase, including relatively thick, volcano–sedimentary sequences such as the layered, marine tuffs and pyroclastics of the Efate Pumice Formation (Ash *et al.*, 1978). Its exposed thickness (with no base seen) on Efate Island is about 500 m, while offshore it has been found covering wide areas to the N, SW and SE of Efate, with a thickness of 1·0–1·4 sec reflection time. Thus its thickness, assuming an average minimum velocity of 1·7–1·8 km/sec, could be 1000–1200 m or more. A similar, well layered sedimentary formation has been mapped in a continuous belt to the west of the three southern islands of Erromango, Tanna and Aneityum (Fig. 2.10), with a thickness of 0·5–0·7 sec, or 500–600 m (Katz and Daniel, 1981). It may well be contiguous with the Efate Pumice Formation. The age of the formation on Efate Island—from planktonic foraminifera and a whole rock K–Ar determination of 1·58 ± 0·05 m.y., of a rhyodacite block in a pumice breccia (Ash *et al.*, 1978)—is late Pliocene to Pleistocene. Of probably late Pliocene age are also the foraminiferal tuffs exposed on Futuna Island (Carney and Macfarlane, 1979), while sediment samples dredged from the western and eastern scarps of the Coriolis Trough—on the latter of which lies Futuna Island (Fig. 2.10)—are middle to late Pliocene, and those from within the trough Pleistocene (Dugas *et al.*, 1977).

These Plio–Pleistocene sediments in the east of the southern New Hebrides arc cover the entire slope and adjacent abyssal plain about 3000 m deep, from 15° to 20°S (Katz and Daniel, 1981). South of about 16°S they form a kind of extensive sediment apron apparently associated with and derived from the island arc volcanism. Here a 4·5 m piston core obtained at 16°28·8'S and 169°31·4'E, in 3040 m of water, recovered predominantly muddy sand and silt, most of which is dark grey ash material (R/V *Kana Keoki* cruise report No. 74, KK82-03-16 Leg 03, CCOP/SOPAC 11th Session, Wellington, November 1982). On seismic sections the sediments are generally well layered and probably are volcaniclastic turbidites, blanketing a relatively smooth basement surface and burying all but some local, high basement peaks and ridges. Their thickness, which gradually decreases to the east, is locally over 0·6 sec reflection time, but generally is 0·5–0·2 sec (Luyendyk *et al.*, 1974). Between 200 and 300 km east of the island arc, they onlap onto a shallower, rugged basement surface almost devoid of sediment.

The young age of the *North Fiji Basin* which extends between the New Hebrides island arc and Fiji (Chase, 1971; Falvey, 1975) is reflected by this irregular, rough basement surface with a sediment cover that over wide areas is not more than a thin veneer (0·15–0·2 sec reflection time at the most, increasing very locally to 0·3 sec in flat floored troughs containing ponded sediments; R/V *Kana Keoki* cruise report No. 74, KK82-03-16 Leg 03, CCOP/SOPAC 11th Session, Wellington, November 1982), while lacking altogether in others. The area in general is governed by a pelagic sedimentary regime, with sediments often heavily bioturbated. This is strongly contrasting with the volcanic ash series along the western basin margin, where sediments are derived from turbidity currents, and are not bioturbated. The formation of the North Fiji Basin, its precise mechanism and timing, however, are still little understood. Recent studies (Falvey, 1978) suggest Plio–Pleistocene growth by r–r–r triple-junction development.

2.5.2   The Fiji Islands region

The large island group of Fiji, with the main islands of Viti Levu and Vanua
Levu each about 150 km across (Fig. 2.16), reveals a geologic history dating
back to the Eocene. Rocks include volcanic and sedimentary (epiclastic and
carbonates), plutonic, low-grade regional metamorphic and medium-grade
contact-metamorphic types. Based on the regional change in magma composition of volcanic rocks about 5 m.y. ago (from andesitic to shoshonitic
and basaltic), Gill (1976a) has postulated that Fiji shifted from an original
island arc setting into a position of subduction-removed oceanic islands, at
about the Miocene–Pliocene boundary.

Due to various orogenic and tectonic movements as well as large-scale,
repeated magmatic activity, there are no larger and long-lived, well preserved sedimentary basins recognizable. But the two main islands give good
evidence of the margins or parts of basins, which mainly extend out into

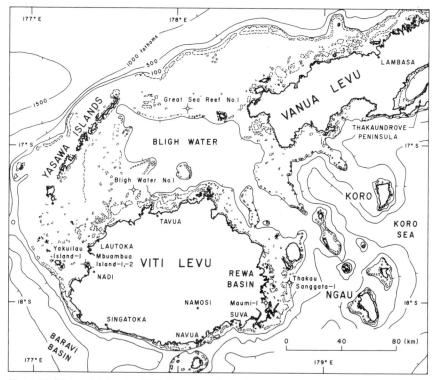

Fig. 2.16. The main islands of Fiji. Shown are the main reef-infested areas around the islands,
and locations and names of oil exploratory wells drilled between 1980 and 1982. [After Eden
and Smith (1983).]

offshore areas. It is possible that in some of these, sedimentation was continuous since the Miocene, or even the Eocene. The following compilation of geologic history and stratigraphy in Fiji is largely based on the recent presentation by Rodda (1982).

The oldest known rocks are Eocene volcanics near Nadi* in the west of Viti Levu, which are mainly basic andesite, and tholeiitic basalt to andesite probably with large volumes of dacite in southern Viti Levu. Limestones associated with these rocks east of Nadi contain larger foraminifera indicating Tertiary b, late Eocene (Cole, 1960). Eocene fossils have also been found in the south of the island. Intrusion of tonalite in the Oligocene (34 m.y.; McDougall, 1963) is an expression of the first orogenic episode evidenced in Viti Levu, which is termed Wainimala Orogeny.

The name *Wainimala Group* has been given to these oldest rocks of Eocene age, but likewise is extended to include a thick sequence of late Oligocene to middle Miocene rocks. These are primary to reworked volcaniclastics (basaltic to dacitic) and pillow lava flows and breccias, but also sandstones which near the Oligocene intrusives contain tonalite detritus. The outcrop area NW of Suva to west of Navua (SE Viti Levu) contains mainly tuffs. In general the strata become finer-grained towards the north, and may continue into Bligh Water where the well Bligh Water No. 1 (Fig. 2.16) reached Oligocene at total depth of 2743 m (Katz, 1981b; Eden and Smith, 1983). Also within these younger Wainimala Group beds, limestone occurs at many stratigraphic levels but is most common near the top where the volcanic rocks grade upwards into sandstone and mudstone (Rodda, 1967). Foraminifera from the upper limestones give an age range of Tertiary e and f, latest Oligocene to early and middle Miocene. The total thickness of the Wainimala Group is hard to establish, and is certainly very variable; some 10 km has been estimated by Rodda (1975).

During a second and major orogenic phase—the important *Colo Orogeny*, which lasted over 5 m.y. of the latest middle Miocene to early late Miocene when folding, faulting and intrusion of plutonics occurred—the *Colo Plutonic Suite* was emplaced between about 11 and 8 m.y. ago (Rodda *et al.*, 1967). The Colo stocks range from olivine gabbro to biotite tonalite and trondhjemite (Rodda, 1967), and mainly intrude anticlines at or near their crests, thus probably are synorogenic. Near their contacts, the Wainimala Group strata locally develop into phyllites and augengneiss; metamorphism in the greenschist facies is widespread. Many small deposits of sulfide minerals and gold are associated with these stocks.

Apparently age equivalent, in parts, with the younger Wainimala rocks are the *Sigatoka Sedimentary Group* and the *Savura Volcanic Group*, which locally overlie (presumed Eocene) Wainimala rocks unconformably; how-

---

* For the spelling and pronunciation of Fijian names, see Rodda (1984).

ever, Sigatoka sedimentary strata also grade laterally into typical Wainimala rocks, while the Savura Volcanic Group has close petrological and geochemical affinities with the upper Wainimala rocks in central to eastern Viti Levu. Its most common rock is andesitic, but overall compositions range from basalt to rhyolite (Rodda, 1967). Apart from volcanic flows with pillow lava and breccia, most volcaniclastic rock types are represented, from polymict conglomerate and breccia to mudstone. Rocks of the Sigatoka Group are predominantly sedimentary, mostly sandstone and mudstone but including large and small masses of limestone. A large outcrop in a highway cut near Yako village south of Nadi consists of excellently bedded, partly platy–flaggy, fine-grained silt–sandstone which is highly calcareous, hard and dense; colours are mainly green to purplish grey. Fossil ages assigned to the Sigatoka Sedimentary Group are Tertiary e, upper e and f (Rodda, 1967).

After uplift and erosion, widespread sedimentation took place in several basins during the late Miocene and continued into the Pliocene. Andesitic volcanism, predominantly from three main centres in SE, NE and central–western Viti Levu supplied large volumes of volcaniclastic material into these sedimentary basins. Basal sediments also contain detritus derived from the Colo intrusives. The *Navosa Sedimentary Group* in central–west Viti Levu is somewhat similar to the Sigatoka Group rocks, but is distinguished by a basal conglomerate with abundant plutonic (Colo) detritus, whereas Sigatoka conglomerate has none. Also, limestone included in the Navosa Group is brown, impure and well bedded, whereas Sigatoka (and Wainimala) limestones are white to light grey, massive to crystalline (Rodda, 1967). In the SE, the *Medrausucu Group* comprises lavas and autobrecciated agglomerates (the Namosi Andesite with disseminated porphyry copper mineralization) which laterally grade into conglomerates, greywackes, mudstones and marls. The more distal facies towards Suva include, in upwards sequence, the Veisari Sandstone, Lami Limestone and Suva Marl, which contain particularly rich faunas with planktonic foraminifera indicating zones N.17, N.18, and N.19 (late Miocene to early Pliocene). Potassium–argon dating of Namosi Andesite has given concordant ages of $5 \cdot 7 \pm 0 \cdot 1$ m.y. (Gill and McDougall, 1973). The change from the rather deep-water, low-carbonate Veisari Sandstone to the shallow-water, partly reef and larger foraminifera Lami Limestone has been attributed by Rodda (1982) to an event associated with the Tortonian–Messinian eustatic sea level change. Here there is an unconformity which separates strata of the lower subzone of NN.11 from those of upper NN.12. This regression was followed by a new transgression, indicated by the deep-water Suva Marl which contains about equal quantities of terrigenous silt and carbonate, besides up to 5% tuff of various kinds, some of them crystal tuff

Fig. 2.17. Late Miocene, volcaniclastic siltstones of the Nadi Sedimentary Group, NE of the town of Nadi (see map, Fig. 2.16).

containing biotite. The age of the Suva Marl corresponds to nannoplankton zones NN.13/14 and NN.15 (Rodda, 1982).

In the west of Viti Levu, on land and offshore around Nadi Bay, the *Nadi Sedimentary Group* (Fig. 2.17) was deposited in the late Miocene to early Pliocene, resting unconformably on Wainimala and Sigatoka Group rocks. The sediments of the Nadi Group form a rather monotonous, buff grey and well-bedded, marine sequence of mainly andesitic–dacitic crystal–volcanic sandstones but with a clear admixture, in certain horizons, of more acidic, plutonic detritus which in places is rather rich in quartz. The rocks are fine-grained to coarse-grained and generally very calcareous, with variable contents of muddy matrix. Much carbonaceous plant material and coaly fragments are concentrated in lenses and laminae of current-bedded sequences. Intercalations of coarse, polymict conglomerate contain pebbles and boulders of basaltic to dacitic volcanics and sediments derived from the Wainimala and Sigatoka Group rocks, as well as from the Colo intrusives. The thickness is at least several hundred metres on land, and probably increases offshore to the west to well over 1000 m.

To the north the Nadi Group is unconformably overlain by the shoshonitic *Koroimavua Volcanic Group* near Lautoka, which includes monzonite intrusions and the reworked epiclastic Vuda Beds with conglomerate, sandstone as well as limestone (near Vuda Point on the coast southwest of

Lautoka). K–Ar ages of biotite from lava flows of the Sabeto Volcanics, which are part of the Koroimavua Group, are 5·35 ± 0·1 m.y., while biotite from a nearby, subvolcanic monzonite plug yielded an age of 4·9 ± 0·1 m.y. (Gill and McDougall, 1973). These rocks, therefore, are probably age equivalent with some of the Nadi Group sediments. Of similar age are rocks in the Yasawa Islands to the west, where the 1000 m thick Koromasoli Sandstone on Waya, which laterally grades into volcanic breccia, has been correlated with the Nadi Sedimentary Group (Wood, 1980). Strata of sandstone grade or finer, partly certainly epiclastic but partly dacitic tuff, are also found on Naviti Island and the four small islands immediately south, with a thickness of at least 1200 m (Rodda, 1982). Of older, middle Miocene age, however, are the massive micrite and underlying, muddy calcarenite with coral fragments and large foraminifera on Sawa-i-Lau Island, interpreted as a fore-reef deposit (Lindner, 1975).

A very extensive area in northern and east Viti Levu is covered by the *Ba Volcanic Group* of essentially Pliocene age. Four main basaltic centres are known, which were largely submarine, with rocks ranging from olivine and augite basalt to feldspar basalt; trachybasalt and intrusions of trachyte and micro-granodiorite occur in some places (Rodda, 1967). The best known is undoubtedly the Tavua volcanic centre because of its associated Au–Ag–Te mineralization (Emperor Gold Mine). Radiometric dating shows that the Tavua volcanism probably was confined to the interval between about 5 and 4 m.y. (Gill and McDougall, 1973). Sediments which are associated with the Ba Volcanic Group occur well away from the volcanic centres, and include strata with no basaltic detritus, although most are epiclastic basaltic sandstone. Nearly 1000 m of sediments occur in the center of Viti Levu, and thick sediments are also found in the east, i.e., in the Rewa Basin north of Suva (Fig. 2.16). The sediments of the Rewa Basin are defined as *Verata Sedimentary Group* which in the north is gradational with the Ba Group; also the southern boundary with the Medrausucu Group is partly gradational. The Verata sediments are mainly flat-lying and consist of well-bedded, muddy sandstone, with calcareous silt to fine sandstone or marl in its top portion (Rodda, 1982).

While in several areas of Viti Levu strata belonging to the planktonic zone N.19 are well established, and the upper subzone of NN.15 is represented near Suva, no faunas or floras indicating younger zones (NN.16, N.20 or N.21) have been found. A general emergence due to prolonged tectonic doming of the island has interrupted further sedimentation in younger basins. However, in areas marginal to Viti Levu, sedimentation has undoubtedly continued. Thus in the west there are young marine beds which unconformably overlie strata of the Nadi Sedimentary Group. These *Meigunyah beds* have not yielded foraminifera diagnostic of any zone, and may be Pleistocene to Recent. Above sea level near Nadi, their thickness is perhaps 40 m at the most, but a considerable thickness has been found in drillholes (Rodda, 1982). Presumably they are deposits of an older *Nadi Bay*, and it is probable that within the present bay there is a continuous sequence from the Meigunyah beds up into present day sediments.

Similarly, the *Cuvu Sedimentary Group* in the SW (west of Sigatoka), which consists of marl and limestone deposited on a Pliocene erosion surface (Rodda, 1967, 1982), may be continuous with Recent sediments being deposited offshore. On land the Volivoli Limestone which forms the top portion of the Cuvu Group (Rodda, 1982) overlaps older formations inland, and probably is the result of a young transgression. The Cuvu Group beds are not folded but gently dip seaward, and thus would correlate with the youngest sediments in the offshore *Baravi Basin* (Larue *et al.*, 1980). This basin, with a NW-trending axis about parallel to the coast of SW Viti Levu (Fig. 2.16), is some 80 km long and 30 km wide. Gravity anomalies suggest a total sediment thickness of about 3 km, while recent seismic surveys (R/V *Kana Keoki* and R/V *Lee*, April 1982; CCOP/SOPAC 11th Session, Wellington, November 1982) show sediments up to 2·0 sec reflection time, i.e., probably in excess of 2500 m thick. The age of these sediments most likely is (?late) Miocene to Pliocene. Towards the NW, the Baravi Basin is terminated by a ridge and trough, the latter over 3100 m deep, which seem to be in line with the Yasawa Islands trend or Yasawa Zone (Wood, 1980), interpreted as a transform fault which offsets the Fiji Platform against the North Fiji Basin. The Baravi Basin, therefore, would originally, i.e., during the Miocene, have been a direct continuation of the North and South Aoba Basins of the New Hebrides (Larue *et al.*, 1980).

Bligh Water to the north of Viti Levu includes other sedimentary basins which are marginal to the updoming main island, and thus represent areas which partly may have been subsiding more or less continuously since the Miocene, or even since the Eocene. Drilling in the Bligh Water areas has confirmed sequences down to the early Tertiary, and the existence of sedimentary thicknesses in excess of what seismic surveys had indicated (Katz, 1981b).

To the northeast of Viti Levu, the large island of *Vanua Levu* (Rickard, 1966) mainly consists of the products of a long and complex period of submarine tholeiitic volcanism, which range from basaltic to dacitic with rare rhyolite. This *Macuadrove Supergroup* is essentially of Miocene age (Hindle, 1976). A major change occurred in the Pliocene, when part of the area was uplifted and subaerial eruption of alkalic lavas (*Bua Volcanic Group*; Coulson, 1971) ranging from hawaiitic basalt to trachyte took place, which formed the Seatura shield volcano. The beginning of these subaerial eruptions is no younger than 3·4 m.y. (Rodda, 1982). Small, intervolcanic

c

sedimentary basins formed in several places. However, since their strata have remained essentially subhorizontal, their thickness is unknown. The biggest of these basins occurs in the lower reaches of the Dreketi and Sarwaga Rivers (the Dreketi Basin), and includes lava flows, reworked lapillistone and tuff, but also finer-grained grit, siltstone and mudstone with planktonic foraminifera indicating zones N.17 to N.19. Much of the basin was covered by the Bua Basalt from the Seatura Volcano during the middle Pliocene. Another basin in the Labasa area contains epiclastic volcanic grit, sandstone and siltstone which range in age from zone N.17 or N.18 to zone N.19 and early N.20. Localities in this area are paratype localities for zones N.18, N.19 and N.20 of Blow (1969). In SE Vanua Levu, a sequence consisting of conglomerate, marl and limestone is exposed on the south coast of the Cakaudrove Peninsula (Woodrow, 1976). Samples from the marls indicate zones N.18, N.19, N.21 and N.22 (Rodda. 1982). Thus the main emergence of this part of Vanua Levu appears to have occurred in the Pleistocene.

The islands of the *Koro Sea* to the south of Vanua Levu (and east of Viti Levu; Fig. 2.16) are mostly single volcanoes of alkali basalt, generally subaerial but including, for a short distance above sea level, pillow basalts. Their age is late Miocene and Pliocene, possibly ranging into Pleistocene (Rodda, 1975). The islands of Koro and Taveuni are the youngest volcanic islands in Fiji, with several phases of activity which in Tavenui continued up to less than 2000 years ago (Rodda, 1975). Between these islands of the Koro Sea, deep basins (2000 to nearly 3000 m deep) contain relatively thick,

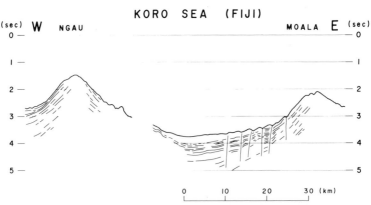

Fig. 2.18. Volcanic buildups in the Koro Sea east of Viti Levu, Fiji (see Fig. 2.16). Between the volcanoes, deeply subsiding basins filled with thick sediments probably mainly of locally derived, volcanogenic detritus. From Mobil Oil Corporation seismic line 72-221. Vertical scale: seconds reflection time.

turbiditic sedimentary series probably all derived from the various volcanic sources. Single-channel seismic reflection records (Fig. 2.18) show thicknesses of up to over 1 sec reflection time. The age of these volcaniclastic sediments therefore is mainly Plio–Pleistocene to recent; only in their basal portions may they date back to the late Miocene.

East of the Koro Sea, the islands of the *Lau Group* sit on top of the northern part of the Lau–Colville Ridge, which is supposed to be an extinct island arc, or remnant arc (Karig, 1970; Gill, 1976b). These islands are mainly atoll reefs, but some have older limestones and volcanic rocks exposed (Ladd and Hoffmeister, 1945). The correct stratigraphy and correlation of these limestones and volcanics, however, has not been easy to establish. As Rodda (1982) has noted, there are various complications with regard to the originally conceived, fairly simple stratigraphy. Recent work (Woodhall, in press) has shown at least three periods of volcanism and four of limestone formation; in some cases, deposition of limestone may have been continuous during a volcanic episode.

The oldest volcanic rocks are the Lau Volcanics (Ladd and Hoffmeister 1945), which are predominantly basaltic andesites to dacites and 9·0–6·4 m.y. old (late Miocene; Gill, 1976b). The volcanics exposed on Nayau and Yacata islands (12·45 m.y. and 14·55 m.y., respectively; Whelan *et al.*, in press), which by Cole *et al.* (in press) have been included in their Lau Volcanic Group, may actually belong to a separate, and earlier volcanic phase (see below). The Korobasaga Volcanic Group, 4·5–2·5 m.y. old, erupted in a more submarine environment and consists mostly of tholeiitic basalt (Cole *et al.*, in press). Alkali basalts of the Mago Volcanic Group are <2·5 m.y. old and may reach into Pleistocene to Holocene times (Rodda, 1982). They may be part of the Taveuni and Koro Sea province of alkali basalt volcanism. The three chemically distinct phases of volcanism (Cole *et al.*, in press) may reflect the changing tectonic environments of the area, which gradually shifted from an island arc to a rifting, oceanic-type extensional stage (Gill, 1976a,b).

On paleontological evidence, based mostly on larger foraminifera, the most extensive limestones in the Lau Islands belong to the Tertiary f stage, which spans the middle and late Miocene. However, by far the most limestones probably are of late Miocene, or even early Pliocene age (N17 and N18/N19; Woodhall, in press). Pliocene limestones are known on at least three islands, and a Pleistocene coralgal reef occurs on Fulaga in the south of the group, where it is uplifted to at least 27 m above sea level (Rodda, 1982). Undoubted middle Miocene limestone is known from a single locality on northern Vanua Balavu, which is the "Futuna Limestone" type section of Ladd and Hoffmeister (1945). This limestone contains

planktonic foraminifera of the N13/N14 zones (Woodhall, in press). It is
thus older than all radiometrically dated rocks of the Lau Volcanic Group,
with the exception of those on Nayau and Yacata islands. However, the
Futuna Limestone—15 m of bedded foraminiferal limestone—contains
intervals with volcanic detritus including a 1–2 m thick bed that includes
clasts of pebble size (Woodhall, in press). Volcanic rocks of middle Miocene
age, or older, were thus being eroded in an adjacent area: these may be part
of the same complex as the volcanics exposed on Nayau and Yacata islands.
On the other hand, Woodhall (in press) has interpreted the contact between
Futuna Limestone and the overlying Korogara Limestone as a disconfor-
mity representing a depositional break, which may have been coincident
with the outbreak of late Miocene volcanism—i.e. that of the Lau Volcanic
Group. In this light, not three but four major phases of volcanism took place
on the Lau Ridge. The earliest which is no younger than middle Miocene,
may correlate with an extensive phase that probably mainly occurred further
south along the Lau–Colville Ridge: it was the source of the thick middle
Miocene, volcaniclastic sequence drilled in DSDP holes 205 and 285 in the
South Fiji Basin (Fig. 2.19), which is thought to be derived from the
Lau–Colville Ridge (Packham and Terrill, 1975).

   From the detailed descriptions of Ladd and Hoffmeister (1945), it follows
that—while some of the limestones in the Lau Islands are tuffaceous—most

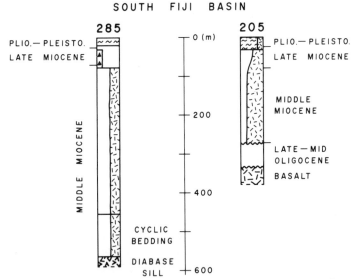

Fig. 2.19. Stratigraphic columns of DSDP holes 205 and 285 in the South Fiji Basin. For
location see Fig. 2.7, and legend in Fig. 2.9.

limestones quite generally are very pure, massive carbonate rocks which often are recrystallized. True coral reef rocks do occur in a few places, but by far the most common rock types in the Lau Islands are algal calcilutites, which are hard and compact rocks of red, buff, cream to white colours. The organic remains are embedded in a calcareous mud or paste, and may in some varieties mainly be foraminifera instead of algae. In general, calcareous algae and foraminifera are the main rock formers in most limestones. Their greatest thickness is found on the island of Vatu Vara, which is wholly limestone from sea level to its summit, 324 m high (Rodda, 1982).

2.6 Southeastern Sector from the South Fiji Basin across Lau and Tonga Ridges to the SW Pacific Basin

2.6.1 The South Fiji basin

This large basin, which lies south of Fiji and west of the Lau–Colville Ridge, extends from about 22° to 34°S (Fig. 2.7). Its triangular basin floor is between 4000 and 4700 m deep and covers an area of 800 000 km^2. A Central Ridge separates the much larger, northern Minerva Abyssal Plain from the southern Kupe Abyssal Plain (Packham and Terrill, 1975). In the south it borders against the continental margin of the North Island, New Zealand.

The sedimentary sequence in the Minerva Abyssal Plain has been sampled in the two DSDP holes 205 and 285 (Fig. 2.19; Burns et al., 1973; Andrews et al., 1975). While hole 285 terminated in an intrusive diabase sill in the middle Miocene, hole 205 reached what is considered oceanic basement in the form of a finely crystalline, vesicular basalt flow which was extruded while late middle Oligocene to late Oligocene nannofossil ooze was being deposited. This agrees with an Oligocene age of the basin as postulated by Weissel and Watts (1975) and Watts et al. (1977) from magnetic anomaly lineations. Consistent with all magnetic anomaly data, most of the basin was formed between 33 and 26 m.y. ago (Davey, 1982). Older crust, and thus an older sediment sequence, probably of Eocene age, may occur in the northwest of the basin; as has been suggested (Lapouille, 1978; Packham and Terrill, 1975), the South Fiji Basin and the southern New Hebrides Basin may be continuous , i.e., part of the same body of sea floor which is getting successively older to the north, i.e., towards DSDP hole 286 (Weissel et al., 1982b).

The late Oligocene sequence in hole 205 is 61 m thick. Intermittent showers of intermediate to acidic volcanic ash contributed to the otherwise biogenic sedimentation of nannofossil ooze, deposited in a quiet deep-sea environment close to the depth of total foram solution. In the early Miocene there was non-deposition or very slight submarine erosion. From the early middle Miocene onwards, a closely similar succession was found in both holes 205 and 285. It is characterized by thick, middle Miocene volcaniclastics with varying amounts of biogenic ooze. The volcanic components mainly consist of glass shard ash, but occasional conglomeratic intercalations contain volcanic rock pebbles and pumice fragments. In hole 285 the coarse-grained to fine-grained, microcross-laminated sand–silt cyclic deposits give clear evidence of reworking, i.e., redeposition of this clastic sediment, due to ocean bottom currents. Upwards in the section the volcanic components gradually diminish and become very subordinate among the nannofossil oozes of the late Miocene. Tectonic movements during the middle and late Miocene resulted in local exposure and the reworking of older material, with admixture of shallow-water, benthonic foraminifera. In Pliocene times, both areas subsided to below nannofossil solution depth (which is deeper than foram solution depth), probably as a result of lithospheric cooling. "Abyssal red clay," i.e., dark reddish-brown, iron-oxide clay rich in glass shard ash, was deposited. Occasional pumice fragments and some layers of nearly pure vitric ash occur in hole 205.

The thickness of late Miocene to Recent sediments is nearly equal at both sites: 37 m of late Miocene in hole 205 and 38 m in hole 285, and 30 m of Pliocene to Recent sediments in hole 205 as against 22 m in hole 285. However, there are great differences in the middle Miocene: 209 m at site 205 but 505 m at site 285. Thus while sedimentation rates in both areas are highest in the middle Miocene and decline markedly upwards—probably mainly as a result of the decrease of volcanic debris—they are twice as high, in the middle Miocene, at site 285 than at site 205. From the sediment accumulation curves of the two sites, Packham and Terrill (1975) have calculated the following rates of sedimentation for the sites 205 and 285, respectively: middle Miocene, 40 and 100 B; late Miocene, 6 and 7 B; Pliocene to Recent, 6 and 4 B. From their detailed study of sedimentary structure and composition, particularly in the middle Miocene volcaniclastics, they have further come to the conclusion that the source of the detrital material must have been in the Lau–Colville Ridge to the east, from where transport occurred by turbidity current across the gently westward sloping, abyssal plain. The fact that the thicker sediment sequence was found in hole 285, i.e., farther away from the Lau Ridge than hole 205, is of only local significance. Davey (1982) has shown that regionally the sediment thickness in the South Fiji Basin increases from west to east, with the greatest thickness (over 1 sec reflection time) all along the base of the Lau–Colville Ridge. Also, the thickness is generally greater in the southern (Kupe

Abyssal Plain) than in the northern basin, probably because of considerable addition of volcanic and terrigenous material from New Zealand.

2.6.2   The Lau Ridge and Lau Basin

These structural units, which have long been regarded as classic examples for the concept of remnant arc and interarc basin (Karig, 1970), are still subject to considerable controversy with regard to their geologic history and tectonic significance (Katz, 1977, 1978a). Factual data are sparse and tenuous, and interpretations therefore ambiguous. The stratigraphy of the *Lau Ridge* is only known from outcrops in the very small and low relief Lau Island i the northern part of the ridge, which has been summarized previously (pp. 57–59). It should be noted that sedimentary rocks are exclusively shallow-water limestones of middle to late Miocene and younger age, which have been deposited in alternating sequence with, and/or during, periods of volcanic eruptions. The earliest volcanism for which there is evidence, if only indirectly, is middle Miocene: its detritus is incorporated in the late middle Miocene Futuna Limestone on Vanua Balavu (Rodda, 1982), while thick volcaniclastic sequences apparently derived by turbidity currents from the Lau–Colville Ridge form the middle Miocene interval in the sedimentary series of the South Fiji Basin (p. 60; Packham and Terrill, 1975). The basement of the Lau–Colville Ridge, however, is probably still older. Seismic sections between 25° and 31°S give clear evidence of South Fiji Basin sediments onlapping onto the Colville Ridge basement [Mobil Oil Corporation line 72-188/189; profile K in Davey (1982); and profiles L, M, N, O in Packham and Terrill (1975)], which thus pre-dates the middle Miocene at least, and possibly is older than late Oligocene which is the age of opening of the South Fiji Basin (Davey, 1982), and of its oldest sediments (Andrews *et al.*, 1975).

The *Lau Basin* is supposed to be a young marginal basin formed by rifting apart of the former Lau–Tonga Ridge. On the basis of various geophysical and geochemical considerations, this process of rifting and crustal genesis would have begun, according to various authors, between 3 and 5 m.y. ago (Karig, 1970; Sclater *et al.*, 1972; Hawkins, 1974, 1976; Gill, 1976b; Weissel, 1977; Cherkis, 1979; Malahoff *et al.*, 1982b). Sediments were thought to be very thin and sporadic, and mainly of volcanic origin derived from either the Lau Ridge or Tonga Ridge.

The drilling of DSDP hole 203 in the south–central Lau Basin (Figs. 2.20 and 2.21) (Burns *et al.*, 1973) seemed to prove this point. The results of this well, which bottomed in the Pliocene, have indeed been referred to not

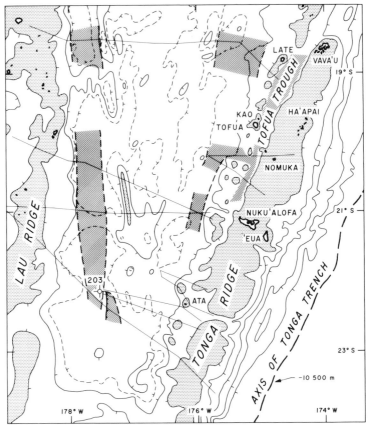

Fig. 2.20. Sediment distribution (shaded areas) in the southern Lau Basin between Lau Ridge and Tonga Ridge. Bathymetric contour interval 1000 m for solid lines, 500 m for broken lines [contours after Eade (1971), modified from existing seismic records]. Dotted areas: above 1000 m deep. Thin lines are seismic tracks; 203 is site of DSDP hole.

infrequently as if the 409 m section that was penetrated would represent virtually the entire sediment sequence that existed in the basin, and would consist entirely or mainly of ash and volcanic sand (Hawkins, 1974; Gill, 1976b; cf. Katz, 1976a, 1978a). This is in spite of the fact that only five widely spaced cores of a combined thickness of not more than 20·3 m were recovered (Fig. 2.21), of which considerable parts consisted of pure nanno ooze with no volcanic components. In other parts, the nanno ooze contains some 5–15% of volcanic glass, while occasional real ash layers are intercalated along sharp boundaries. Such ash is partly fairly coarse, i.e., of silt to sand size, while even some round cobbles of basalt occurred in the section near the base of the hole; volcanic debris, therefore, appears to be mainly of

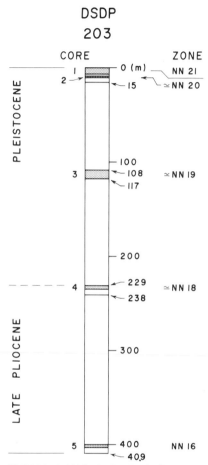

Fig. 2.21. Well section of DSDP hole 203 in the Lau Basin (for location see Fig. 2.20), showing drill cores obtained and their respective nannofossil zonations. [From Burns *et al.* (1973).]

local derivation and within well defined intervals. It has been argued, too, that the relatively high sedimentation rates (120 B on the basis of the Plio–Pleistocene boundary, and 80–120 B for zone N.21, but 50 B in the latter part of the Pleistocene—all figures uncorrected for compaction) suggest an appreciable volcanic component over and above purely biogenic sedimentation, also in the uncored parts of the section. Again, this may be true for some intervals but not necessarily for the section as a whole. In summary, there is still very incomplete knowledge of the 409 m thick section that was drilled through, as less than 5% of it has been recovered and actually seen. In particular, the nearly 300 m of mostly Pliocene strata below

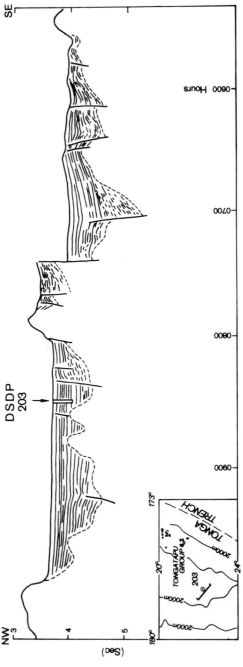

Fig. 2.22. Tracing of seismic profile (Mobil Oil Corporation line 72-195; for location see Fig. 2.20) through DSDP site 203 in the southern Lau Basin. Vertical scale in seconds reflection time, horizontal scale in hours shiptime. [From Katz (1976a).]

core 3 (which corresponds to the NN.19 zone, early Pleistocene) have only been substantiated by two cores of 2·4 and 1·5 m recovery, respectively (Fig. 2.21).

From this very limited information, it can only be said that the principal lithologies probably throughout the section are calcareous biogenic ooze and volcanic ash—in which distribution and proportion, however, is unknown. Also, the total thickness drilled is not nearly that of the whole sediment sequence in the basin. Basement was not reached, and a Mobil Oil Corporation seismic line which passes nearby (Fig. 2.22) (Katz, 1976a) shows this sequence to reach a thickness of 0·94 sec reflection time about 10 km west of the drill site. Sonic velocities measured on the well core 3 (108–117 m) range from 1·54 to 1·56 km/sec, and on cores 4 and 5 (230 and 400 m) from 1·66 to 2·24 km/sec; on the Mobil seismic line a sonobuoy refraction recording near the drill site, from about 0·3 sec below sea bottom, indicated a velocity of 2·47 km/sec. Thus the average velocity for the entire sequence would probably be not less than 2·0–2·5 km/sec, or, conservatively estimated, perhaps about 2·2 km/sec. This gives a total sediment thickness of at least 1000 m. Considering the structural attitude of the reflections across this local basin (Fig. 2.22), the base of the sequence would correspond to a stratigraphic level about 600 m below core 5 of DSDP hole 203. Since this core is low in the NN.16 zone, i.e., in the lowest part of the late Pliocene (Edwards, 1973) or about 3–3·5 m.y. old, an extrapolation backwards in time, assuming similar sedimentation rates as found in the well section, say about 100 B, would add another 6 m.y., and thus give an age at the base of the sequence of 9 m.y., i.e., Tortonian, middle late Miocene.

On the other hand, one might argue that sedimentation rates below the well section are appreciably greater, as the section might represent a rapidly poured-in, volcaniclastic wedge of coarse debris derived from nearby, local volcanic sources that were particularly active during the initial stages of basin formation. Such coarse, igneous–volcanic breccias and conglomerates were found in one or two places along the IPOD-DSDP Mariana transect, which is by far the best known of similar island arc–back-arc areas anywhere in the world (Kroenke et al., 1980; Hussong et al., 1981). Indeed, at site 451 volcaniclastics 865 m thick were deposited during the late Miocene at rates of 400 B (Kroenke et al., 1980). However, this site is on the West Mariana Ridge, which at the time was an active volcanic island arc with subaerial volcanic islands, from where the clastic wedge of site 451 was mainly derived. Thus its tectonic and paleogeographic setting does not compare with site 203 in the Lau Basin.

In the actively spreading Mariana Trough between the West Mariana and Mariana ridges, which most closely corresponds to the Lau Basin, the holes 453, 454 and 456 all yielded turbiditic sequences of mainly silt-size volcaniclastic sediments that contain considerable hemipelagic and biogenic components (Hussong, Uyeda et al., 1981). These sequences thus compare with the one drilled in the Lau Basin. Also, their average sedimentation rate is of the order of 60–160 B, thus similar to the 100 B in the Lau Basin. The fine-grained volcaniclastics in the Mariana Trough directly overlie basaltic basement, except at site 453 where in the bottom of the hole 150 m of hydrothermally altered, coarse gabbrometabasalt polymict breccia and metavolcanic breccias were drilled. Geochemically these rocks are arc-related, and it is uncertain whether they are representative of the upper crust at this site, or whether there are back-arc basalts at greater depth. From seismic evidence it has been suggested that they are not part of the crust here, but may have been tectonically displaced from the West Mariana Ridge during initial rifting of the Mariana Trough. Their interval velocity, as computed from multichannel seismic reflection data, is 5·3 km/sec, which is in fairly good agreement with measured velocities of breccia samples which gave an average of 4·75 km/sec (but much higher velocities were measured on some of the gabbro and tectonized serpentinite samples in the bottom of the hole). These velocities strongly contrast with those of the sediments above, which range from 1·54 to 2·19 km/sec.

It is concluded that the polymict and altered, volcanic–igneous breccia in hole 453 is not a normal part of the sedimentary sequence at this site. The sediments at site 203 in the Lau Basin, on the other hand, appear to form a continuous sequence from the well section further down, as is shown by the evenly bedded, strong reflectors that occur throughout the sequence [Fig. 2.22; this is in marked contrast to the seismic character displayed in the small pond around site 453 (Hussong et al., 1981)].

From the seismic picture in the Lau Basin, therefore, as well as the comparison with data from the Mariana transect, there is nothing to suggest that the lower part of the Lau Basin section would consist of sediments appreciably different from those drilled in hole 203. In particular, any argument for a much higher sedimentation rate (a rate of 100 B, after all, is already quite high) would be purely speculative. From all the evidence available, the age of the base of this sequence is unlikely to be much younger than calculated above, i.e., about 9 m.y.

It could in fact be considerably older, since both an increase in seismic velocity below the well section and greater compaction of the sediments would tend to increase total thickness as well as age span of the sequence. The average seismic velocity as used above is most probably a minimum, and a velocity of only 0·1 km/sec higher would increase the total thickness to 1081 m, or to about 680 m below core 5 instead of 600 m. Also, by extrapolating the same sedimentation rate uncorrected for compaction, an error in the age calculation is introduced which may be considerable.

Depending on these factors, the age of the base of the sequence could easily be pushed back to as far as 12–13 m.y., which is the boundary between middle and late Miocene.

This of course has serious implications with regard to the age of the Lau Basin. Gill's postulate (1976b) that it has opened only after the Lau Volcanics erupted on the Lau Ridge (6·4–9·0 m.y.) appears to contradict the above evidence (Katz, 1978a). Indeed, at least in its southern part the Lau Basin must be older. The young sediments in the near-continuous trough in the west of the basin (Fig. 2.20) thicken southward and attain at least 1000 m at about 22°S. Here their base is of late Miocene age, perhaps of the earliest part of late Miocene. In addition, seismic evidence of Mobil Oil line 72-195 (Fig. 2.22) suggests the presence of an older sediment series, of different seismic character and slightly unconformable, which locally occurs underneath. It onlaps onto older basement highs and thickens into fault angle depressions, as shown between 0630 and 0700 hours of profile Fig. 2.22. At this place the maximum thickness of this older series is 0·7 sec (1·3 sec total sediment thickness). Assuming a somewhat higher seismic velocity because of greater age and compaction, say 2·5–3·0 km/sec, it could amount to 900–1000 m. The age of this series—since it rests unconformably below the late Miocene–Pliocene sequence—may be early Miocene to Oligocene–Eocene, and thus correlate with the older sediment series of the Tonga Ridge (see below).

Important tectonic deformation occurred prior to, and/or contemporaneous with, deposition of this older sediment series, as is indicated by its distribution around fault angle depressions (Fig. 2.22, 0700 hours). While apparently wedging out against basement highs, it may have been affected by some erosion after its deposition, but prior to the overlying, late Miocene–Pliocene sequence. This would suggest a second tectonic phase of about middle Miocene age. Further deformation took place in Pleistocene to Recent time, with block faulting and tilting affecting also the youngest sediments. Piercement structures of volcanic–igneous rocks apparently are associated with this young faulting, and thus also penetrate and displace older basement. Dredge samples investigated by Hawkins (1974, 1976, 1977), which are mainly subalkaline, tholeiitic basalts similar to oceanic ridge basalt, all originate from high-lying ridges, fault scarps and seamounts, and may in fact represent such younger intrusions and not the original, pre-sedimentary basement.

This question of possibly two different "basements" is most important, but remains unresolved. Also unresolved is the age of the oldest sediments in the Lau Basin. In a general sense, thickness and age of sediments—which mainly occur in two narrow belts in the east and west of the basin, which converge southwards (Fig. 2.20)—increase to the south, and it may well be that the thickest sediments including the oldest series occur in the big and centrally located depression around latitude 23°S (Fig. 2.20). No data are available from that area. From all this it becomes obvious, however, that the tectonic evolution of the Lau Basin is more complex than generally envisaged; so far it remains engimatic (Katz, 1977).

2.6.3   The Tonga ridge

As the easternmost of the active volcanic island arcs in the SW Pacific, this major structural feature which to the south continues into the Kermadec– New Zealand system (Katz, 1974), lies immediately adjacent to the Pacific plate, i.e., the large abyssal plain which extends east of the Tonga– Kermadec Trench (Figs. 2.7 and 2.20). From outcrops on the islands, and land and marine seismic profiles from oil companies and several wells drilled on the main island of Tongatapu, the stratigraphy has been fairly well established (for a summary see Katz, 1976a). However, the oldest part of the sequence and its relationship with igneous basement is only exposed on the steep eastern side of 'Eau Island (Hoffmeister, 1932; Stearns, 1971), and is partly of very difficult access, and/or heavily overgrown by tropical bush.

The basal igneous sequence which is exposed mainly in the lower cliffs of the east coast consists of lava, agglomerate, conglomerate and tuff which compositionally include, according to Ewart and Bryan (1972), high-alumina tholeiite, olivine basalt, basaltic andesite, quartz gabbro and dacite tuff. Radiometric dating ($^{40}Ar/^{39}Ar$) of plagioclases from this series has given ages of 46·6 m.y. for quartz gabbro, and 46·1 m.y. for tholeiitic basalt (Ewart *et al.*, 1977). The sequence is cut by acid andesite, and the whole complex is unconformably overlain by late Eocene, calcareous sediments.

Petrological and geochemical characteristics suggest that the 'Eua igneous suite represents a very early stage of island arc evolution, being decidedly less fractionated (i.e., more "primitive") than the more recent Tongan volcanic rocks. Ewart and Bryan (1972) conclude that the 'Euan volcanics may form the topmost part of an ophiolite complex; this is supported by the occurrence of peridotite and dunite low down at the inner flank of the Tonga Trench (Fisher and Engel, 1969). Thus the Tonga ridge structure may initially have formed directly on oceanic crust, perhaps through fracture zone volcanism along an extensive transform fault system which was the precursor of the present-day Tonga Trench (Stearns, 1971; Kroenke and Tongilava, 1975; Hilde *et al.*, 1977). Volcanic growth and/or uplift led finally to the conditions which allowed deposition of the late Eocene limestone series.

The sedimentary sequence as exposed in the southern Tonga islands (mainly 'Eua) is shown in Fig. 2.23. Although many refinements have been

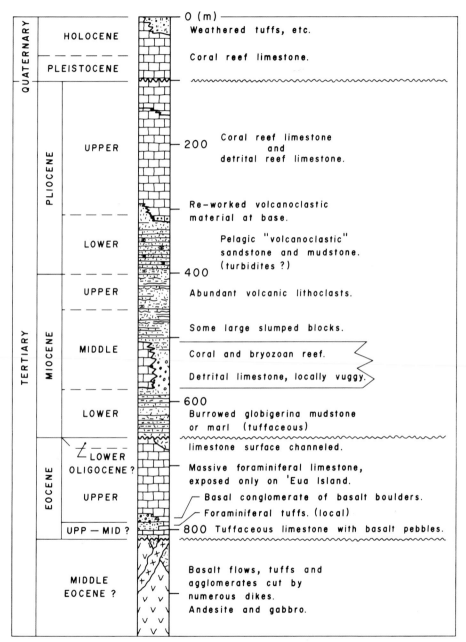

Fig. 2.23. Stratigraphic sequence of the Tonga Ridge as revealed from outcrops. [After Mulder and Nieuwenhuizen (1971).]

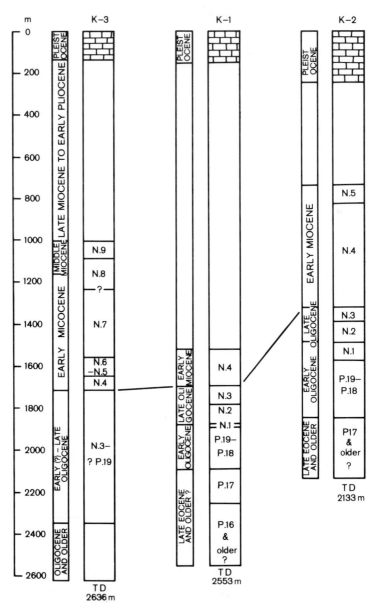

Fig. 2.24. Biostratigraphic columns of wells Kumimonu-1, -2 and -3 on Tongatapu. For location see Fig. 2.27. (From W. M. Barney and H. R. Warters, with kind permission.)

added in more recent time, particularly in the course of oil exploration activities by Webb Tonga, Inc. (Katz, 1978b, 1979), this column still appears basically correct. However, marine seismic and well data (Katz, 1974, 1976a) have shown that westwards the section considerably increases in thickness. Thus the two wells drilled by Shell in Tongatapu in 1971, at TD 1686 m, bottomed in the upper part of the early Miocene, after penetrating over 1500 m of a rather monotonous series of marine volcaniclastic sediments of early Pliocene and Miocene age, which were deposited in outer shelf and bathyal environments (Tonga Shell, 1972; Katz, 1976a). From the three Kumimonu wells drilled by Webb Tonga, Inc., in 1978, the total thickness of the Mio–Pliocene volcaniclastic section underneath Tongatapu is between 1070 m and 1580 m (Fig. 2.24). According to Robertson Research, depositional environments were shallowing in the lower part of the sequence, being generally sublittoral in the early Miocene. No limestones were found in the early Tertiary. Sediments older than Miocene continue to be mainly volcaniclastic, with a total Oligocene to Eocene thickness, in the three wells, of 679 to 922 m. This again is very much thicker than on 'Eua. It is noteworthy, too, that seismic interpretation which immediately south of Tongatapu had predicted over 4 km of sediments, including early Tertiary (Katz, 1974) (Fig. 2.25), was basically confirmed by the well Kumimonu-3 which on a local high structure near Nuku'alofa found Oligocene and/or older sediments from 1714 m down to TD at 2636 m (Katz, 1979). On the other hand, the regional extent of an Eocene limestone formation as originally suggested from the seismic correlation (Fig. 2.25), has proven incorrect. Equally unsupported by the well sections is the more speculative, stratigraphic–lithologic subdivision suggested by Kroenke and Tongilava (1975) in their interpretation of the same seismic line. Indeed, no carbonate platform or reef buildups were found underneath Tongatapu, nor any other, calcareous formation. Thus it appears that limestone developments as shown on 'Eua may be limited to the area east of Tongatapu, which apparently was relatively high through most of its history.

While the detailed stratigraphic relationships at the base of the thick limestone complex in 'Eua are still not very clear—there are obviously some differences between various outcrops in the "Nine Gulches" (Tele-a-hiva; Stearns, 1971)—it is evident that the massive, dense foraminiferal limestone which forms the high backbone of 'Eua, with cliffs over 100 m high falling down to the east (Fig. 2.26), is overlying the richly fossiliferous, tuffaceous limestone of Stearns (1971). This has been found 6–8 m thick by the writer, consisting of well-bedded and coarse-grained soft tuffaceous calcarenite resting on lava flows. According to Ladd (1970), the fauna includes foraminifera, discoasters, corals, hydrozoans, brachiopods, bryozoans, annelids, crinoids, echinoids, ostracodes, barnacles, decapod crustaceans,

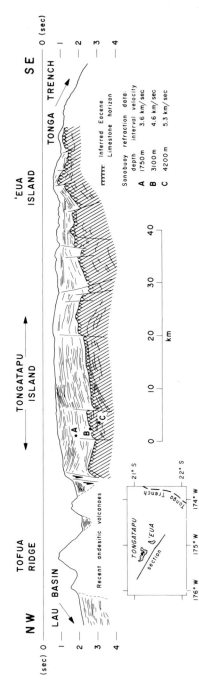

Fig. 2.25. Tracing of seismic section across Tonga Ridge south of Tongatapu. Interval above hatched portion is interpreted as Miocene and younger. Vertical scale in seconds reflection time. [After Katz (1974).]

Fig. 2.26. View south along the eastern side of 'Eua Island, Tonga. The high cliffs and ridge in the centre-right background, up to 300 m above sea level, are formed by massive, late Eocene limestone.

molluscs, shark teeth, otoliths, and spores and other plant microfossils; with the exception of the larger foraminifera which indicate shallow warm waters in a sheltered situation, most other forms suggest deposition at a depth of not less than 200 m. However, Mulder and Nieuwenhuizen (1971) have pointed to the fact that a considerable relief must have existed before deposition of the Eocene sediments: locally there are coarse conglomerates and boulder beds up to 25 m thick, with mainly volcanic debris, often angular, but also including large limestone boulders which contain the same foraminifera as are found in the matrix. This must indicate contemporaneous tectonic unrest, probably further uplift. At other places, the younger massive, thick limestones rest directly on basement volcanics.

These rocks of the main limestone complex, which are extremely hard and ringing when struck with a hammer, have been described by Mulder and Nieuwenhuizen (1971) as "some 100 m of massive to coarsely-bedded cream-coloured limestones, mostly bioclastic, algal, lime wackestones, often rich in larger foraminifera but not forming foraminiferal packstones (lumachelles)". Most rocks are free from coral or coral debris, and while deposition mainly occurred in a shallow marine and protected (lagoonal) environment, some large-scale foresetting was observed by Mulder in the northern part of the island, indicating more agitated sedimentary conditions. Further uplift is suggested by the absence of Oligocene rocks—whereas several hundred meters of marine Oligocene sediments were found in the Kumimonu wells on Tongatapu Island further west (Fig. 2.24). Thus the late Eocene limestone surface of 'Eua—exposing a large number of sinkholes on the high plateau, which seemingly are related to fracture and fault systems—probably already developed a considerable relief during Oligocene time. With the following Mio–Pliocene transgression, locally accentuated submarine slope conditions were established; submarine channels in the late Eocene limestones were filled with bathyal, lithic sandstones of Miocene and Pliocene age (Mulder and Nieuwenhuizen, 1971).

The Miocene to early Pliocene succession generally consists of a lithic, volcaniclastic sand–siltstone series with subordinate mudstone intercalations. All the sediments contain an abundant, planktonic fauna (*Orbulina*; Mulder and Nieuwenhuizen, 1971). In 'Eua, the marked age variations of the base of the succession suggests transgression onto an appreciable relief. According to Mulder and Nieuwenhuizen's detailed petrographic examinations, the formation represents a mainly detrital, epiclastic sedimentary sequence, including turbidites. Indications of an admixture of contemporaneous, pyroclastic material were found only locally. Also in the Kumifonua well sections on Tongatapu (Tonga Shell, 1972), the sediments were mainly detrital with volcanic lithoclasts comprising acidic and basic andesite and basalt. Evidence for intra-Miocene volcanism, however, was found in an agglomerate cored at the bottom of Kumifonua-2 well, which shows flow structures and hydrothermal alteration.

Similar Miocene, volcaniclastic sediments occur on various small islands of the Nomuka Group about 100 km north of Tongatapu (Fig. 2.27). As described by Mulder and Nieuwenhuizen (1971), they are mainly pelagic and deep-marine, rich in planktonic foraminifera, and often represent submarine canyon and fan deposits; typical turbidites as occur on Nomuka-iti exhibit features of internal slumping, load casts, convolute and graded bedding. Very fine-grained, silty mudstone to claystone graded cycles—each cycle only about 1·5–2 cm thick—were seen on Fonoifua and Tanoa, indicating rather distal and relatively quiet, deep-marine deposition where, however, occasional high-energy bottom currents scoured the pelagic mud deposits. Also, there is a 3-cm layer of a poorly sorted pyroclastic intercalation with fresh-looking, somewhat vesicular andesitic fragments (lapilli), thus indicating contemporaneous volcanism. Much coarser grained, lithic sandstones are found on Kelefesia and Tonumea; they contain abundant igneous lithoclasts, which locally are up to 30 cm and clearly rolled. Together with lenticular, grey mudstone intercalations, some graded bedding and foresetting is noted; scour and fill structures, channelling, strong wedging and internal slumping can also be observed. The sedimentary

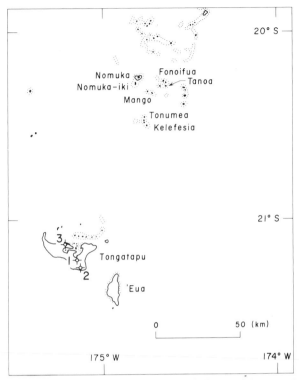

Fig. 2.27. The southern Tongan Islands: Tongatapu and 'Eua, and the Nomuka Group. 1, 2 and 3 on Tongatapu: well locations of Kumimonu-1, -2 and -3 drilled by Webb Tonga, Inc., in 1978. Kumimonu-3 is located in the town of Nuku'alofa. The earlier wells Kumifonua-1 and -2 drilled by Shell in 1971 are located 2·5 and 3 km, respectively, to the SE (No. 1) and WSW (No. 2) of Kumimonu-3.

structures indicate a possible western source of the detrital material. On Tonumea, locally calcareous siltstone and mudstone become equally important as lithic sandstone.

Shallow-water deposits of Miocene age are found on the islands of Mango and Nomuka. On Mango about 100 m of unbedded, coarse rudites contain angular and subrounded, unsorted pebbles and fragments of igneous rocks and subordinate limestones, as well as large, rounded boulders of mudstone or siltstone. The igneous fragments are up to 50 cm across and represent a great variety of rocks, but are mostly of andesitic or basaltic composition; some pebbles of quartz gabbro also occur. These rock types could well indicate derivation from a volcanic basement, such as exposed on 'Eua. However, Mulder and Nieuwenhuizen (1971) also mention one find of a basaltic pebble with centimetre-size inclusions of a shallow-water lime

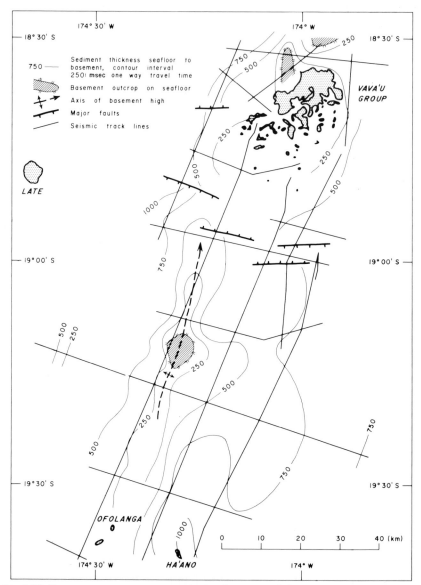

Fig. 2.28. Basement contour map over the northern part of the central Tongan Platform, between Ha'apai and Vava'u Islands. [From Mulder and Nieuwenhuizen (1971)].

wackestone containing *Lepidocyclina* fragments, which indicate a late Oligocene to middle Miocene age. Thus contemporaneous Miocene volcanism is indicated here. The limestone fragments which locally reach boulder size—1·5 × 0·75 m according to Mulder and Nieuwenhuizen (1971), while Lister (1891) mentions the largest boulder he has seen as being 3 × 2 × 1 m—are bioclastic coralline and bryozoan limestones with larger foraminifera indicating an early middle Miocene age. Apparently, they are derived from a nearby shelf or reef fringing a volcanic island, which may have been active during Miocene times. As Lister (1891) has noted, the source of these rudites must have been only a short distance south of Mango.

Of similar age and type as the limestone fragments of Mango are the massively bedded, dark-yellow bioclastic lime wackestones and packstones which, in about a horizontal position, form the highest hills (nearly 50 m high) of Nomuka. They are highly foraminiferal (mainly *Cycloclypeus*) and quite different from the Quaternary raised coral limestones. As indicated by the larger foraminifera, they are of early middle Miocene to middle Miocene age. In analogy to evidence from Mango, the occurrence of these shallow marine limestones suggests that Nomuka is underlain by an extinct, Miocene volcano, or alternatively by older volcanic basement. The Ano Ava lagoon, in fact, could represent the remnant of an old caldera or crater. Shallow basement in this area is also suggested from seismic surveys which regionally indicate a decrease of total sediment thickness from south to north, along the Tonga ridge or platform. Thus north of Nomuka, the sediments are generally less than 1 sec thick (one-way seismic travel time, Fig. 2.28), i.e., less than about 2500 m, or indeed less than 1000 m along a ridge on the western side of the platform (which corresponds to the position of Nomuka Island). From this situation and the widespread occurrence of Miocene rocks in the islands of the Nomuka Group, one could assume that from south to north the early Tertiary sediments gradually disappear, i.e., wedge out underneath the Tonga platform.

Further north, even most of the Miocene sediments probably wedge out. Around the islands of the Vava'u Group, seismic data indicate little more than 600 m of sediments, with basement outcropping on the seafloor NW and north of Vava'u (Fig. 2.28). The islands of Vava'u consist nearly exclusively of limestone, which is not older than Pliocene (Mulder and Nieuwenhuizen, 1971). Clearly two formations are distinguished, of which the older, about 250 m thick, is tilted and unconformably overlain by subhorizontal, raised Quaternary reef terraces. Only at one place, in the western cliffs of Hunga Island, has a 1 m thick, persistent intercalation of medium-grained volcaniclastic sediment been observed in these older limestones. According to Mulder and Nieuwenhuizen (1971), the clastic fragments are clearly detrital and represent a variety of igneous rocks, while the

Fig. 2.29. North coast of Vava'u (Matakiniua): 170 m high, southward-tilted platform of Pliocene limestone.

calcareous matrix contains an abundant planktonic fauna (mainly *Orbulina*) as well as larger foraminifera. As described by these authors, the older limestone sequence consists of

relatively thick (±50 m), massively-bedded detrital limestones and local rather soft, chalky bryozoan limestones, alternating with thinner (5–25 m) *Cycloclypeus* packstones. The detrital limestones show large-scale fore-set bedding and contain clearly rolled coral colonies, thick-shelled gastropods and pelecypods. They are interpreted as reef slope deposits, which is in agreement with the moderate water depth conditions indicated by the intercalated *Cycloclypeus* packstones. These latter contain few other foraminifera than *Cycloclypeus*, and no closer age determination than Pliocene or younger was possible.

Thus while this lower limestone sequence, which is beautifully exposed in the 100–200 m high, outward-facing cliffs of Vava'u (Fig. 2.29), mainly represents reef and reef slope deposits, the central part of the main island appears to be underlain by a different facies, as is exposed in the quarry of Neiafu and numerous water wells. It consists of soft, horizontally bedded porous limestone, locally containing coral colonies and well preserved solitary corals, gastropods, pelecypods and foraminifera, mainly *Orbulina* (Mulder and Nieuwenhuizen, 1971). A lagoonal environment is indicated for these limestones. Considering the seismic evidence of shallow basement in this area, Mulder and Nieuwenhuizen (1971) have speculated that the

Pliocene limestones of Vava'u are built on a drowned caldera complex, with the reef and reef slope deposits now exposed in the outward cliffs representing the edges of the caldera, and the lagoonal limestones of Neiafu quarry its central portions. In this way Vava'u would reflect—from both its morphology and facies distribution of its sediments, but disregrading its younger tilt to the S—the original configuration of its volcanic basement.

The younger, Pleistocene to Recent limestones of Vava'u represent various reef and lagoonal formations, which demonstrate intermittent uplift to a maximum of 180 m in the western part of the main island. Correlating the various terrace levels, successively less uplift is observed due east and south. Further south there is subsidence actively taking place: drowned reefs are seen here on seismic sections, indicating that the growth of the eastern barrier reef could obviously not keep pace with submergence (Mulder and Nieuwenhuizen, 1971). The great water depth across the platform, which at 19°25′S (between the Vava'u and Ha'apai Groups) is 500–600 m, obviously is a result of such regional tilting, perhaps combined with cross-faulting as is also observed around 19°00′S (Fig. 2.28).

The entire Ha'apai Group of islands, from 19°30′S to 20°10′S (Fig. 2.20), apparently is built on a submerged, Quaternary (?Plio–Pleistocene) reef platform with many of the islands representing differentially uplifted reef terraces. Such uplift is greatest in the east, where the largest islands occur which are directly above the eastern barrier reef (for the detailed bathymetry, see Eade, 1972). Thus there is evidence here of a young westward tilt across the Tonga platform. In many places, the raised coral limestones are capped by weathered pyroclastics which may be up to 20 m thick. Locally they were deposited below water, as is indicated by the occasional presence of smaller foraminifera and pelecypods (Mulder and Nieuwenhuizen, 1971). These volcanic products probably are derived from Kano and Tofua volcanoes in the west, which have erupted repeatedly in Pleistocene to Recent times.

Quaternary coral limestones also are widely present on Tongatapu and 'Eua Islands in the south, where they unconformably overlie early Pliocene and Miocene volcaniclastics (Fig. 2 in Katz, 1976a). They form various raised terraces around 'Eua (Hoffmeister, 1932), and underlie all of Tongatapu which itself is uplifted in the south but gradually tilting northwards. The thickness of this young reef platform in Tongatapu, from which the 30 000–40 000 inhabitants of the island draw all their fresh water supply, seems to be increasing from north to south: in the three wells Kumifonua-1 and -2 (Tonga Shell) and Kumimonu-3 (Webb Tonga), all drilled near the northern shore around Nuku'alofa (Fig. 2.27), its thickness is between 134 and 143 m; in Kumimonu-1 near Malapo in the south-central part of the island, it is 152 m thick, and 247 m in Kumimonu-2 at Fua'amotu in the SE

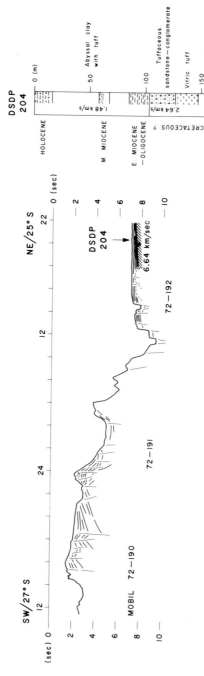

Fig. 2.30. Tracing of seismic profile across Tonga–Kermadec Ridge and Trench at about 26°S, with location and well section of DSDP hole 204. Horizontal scale of profile in hours shiptime, vertical scale in seconds reflection time. From Mobil Oil Corporation's lines 72-190, -191, -192 (for location see Katz, 1974.)

corner of Tongatapu (Fig. 2.24). All figures include the surface sand or soil layer, which generally is no more than 0·5–2·0 m. The rich fertile soil in Tongatapu is derived from young volcanic ash which to a greater or lesser extent is blanketing all the Tongan islands.

This recent volcanism is concentrated west of the Tonga platform, from which it is separated by the Tofua Trough (Fig. 2.20); this is generally about 1700 m deep and filled with up to 1·5 sec thick, probably mainly volcaniclastic sediments. The line of volcanic islands and shoals which build up the Tofua Ridge is part of a 2800 km long, linear belt of Pleistocene to Recent volcanic activity that extends from the North Island, New Zealand along the Kermadec and Tonga Ridges to as far north as Nuia Toputapu and Tafahi Islands south of Samoa, thus over 23½° latitude (Fig. 2.7). The rocks of this volcanic chain in Tonga are dominated by basaltic andesite, with locally developed andesite and dacite; the 1967–1968 eruption of Metis Shoal produced rhyolite. Detailed descriptions of the petrology and geochemistry are given by Bryan *et al.* (1972), Ewart and Bryan (1973), and Ewart *et al.* (1977), where also further references are listed. The history of the volcanoes along the western margin of the Tonga Ridge is one of intense modern activity, with more than 35 eruptions recorded over the 1770–1970 period (Richard, 1962; Bryan *et al.*, 1972). These include both subaerial and submarine eruptions; in fact, since all the volcanoes are structures rising between 1500 and 2000 m up from the sea floor, only a limited number of them emerge with their summit portions above the sea (Davey, 1980; Brothers *et al.*, 1980).

2.6.4 Abyssal plain east of the Tonga Ridge–Trench: DSDP hole 204

The lower slope and trench east of the Tonga Ridge are virtually devoid of sediments (Katz, 1974), but the abyssal plain beyond the trench, below 5000 m deep and forming part of the SW Pacific Basin, is covered by a uniform sequence a few hundred metres thick. Here the only direct information is available from DSDP hole 204, which was drilled 100 km east of the trench axis, at 24°57·27'S and 174°06·69'W (Figs. 2.7 and 2.30) (Burns *et al.*, 1973). At this site a 0·14 sec thick transparent layer is 103 m thick and consists of dark brown, red and greenish-grey abyssal clay and ash, which were deposited below the carbonate compensation depth near a region of active, dominantly andesitic volcanism. Radiolaria ages suggest a range from Quaternary at the top, through middle Miocene in the middle portion (core 3) to early Miocene or Oligocene at the base (cores 4 and 5). However, no planktonic foraminifera and unreworked, calcareous nannoplankton were obtained, while the siliceous microfossils were very rare and partly destroyed. Thus the ages of the cores are questionable. On the other hand, reworked, well preserved late Cretaceous radiolaria occur at the top of the section, and more poorly preserved, late Cretaceous nannofossils and radiolaria near its base.

An abrupt transition, probably representing an unconformity, occurs at this level of 103 m, which is marked by the first of a series of strong seismic reflectors. Lithified, tuffaceous sandstone and conglomerate follow from 103 to 126·5 m; they are barren of fossils except for *Inoceramus* (?) fragments. This interval consists mainly of yellowish brown, granular to pebbly, coarse tuffaceous conglomerate. Sedimentary structures including cross-bedding, graded bedding and contorted lamination indicating soft sediment deformation occur in the finer-grained beds. This and the rounding of clasts in the conglomerates indicate a high-energy environment with submarine transport along a fairly steep paleoslope. Thus a relatively high relief, and proximity to a volcanic source are suggested by this interval.

Along a sharp contact, the basal conglomerate of this unit overlies a layer of dark greenish-grey vitric tuff, which occurs from 126·5 to 147 m. Here there is only a vague stratification, and generally much less evidence of sediment transport than in the overlying unit. The tuff mainly consists of angular shards of devitrified basaltic to andesitic glass. This unit probably formed directly from the extrusive activity of nearby volcanoes (the Louisville Ridge?), with little or no sedimentary transport.

Because of the lack of fossils in the lower two units of the section drilled, their age remains unknown. However, because of the presence of reworked late Cretaceous microfossils in the overlying abyssal clay and ash, it has been concluded that a Cretaceous age seems likely for the sequence below the unconformity at 103 m (Burns *et al.*, 1973).

Basement was not reached in hole 204, nor is it clearly distinguished on the seismic records. However, a sonobuoy refraction profile immediately NW of the drill site (Katz, 1974) has given a velocity of 6·64 km/sec about 0·53 sec below seafloor. This would indicate that oceanic layer 3 underlies these sediments at a shallow depth of about 600 m (Fig. 2.30).

3

Hydrocarbon Potential in the SW Pacific

H. R. KATZ

Pacific Geo Consultants Ltd., Lower Hutt, New Zealand

3.1 Introduction

With the highly complex, tectonic framework that characterizes the SW
Pacific, extreme variations are found in the sedimentary environments and
tectonostratigraphic conditions. Sediment thickness over wide areas is very
limited. Particularly in the young oceanic basins, sediments generally form a
thin veneer only. But even in older basins with a history reaching back into
the early Tertiary, such as the Coral Sea, East Caroline, New Hebrides and
South Fiji basins, the total sediment thickness remains well below 1000 m.
Only in the Pacific basin to the north of the Solomon Islands, i.e., the
Malaita–Ontong Java Province (Katz, 1980) (Fig. 2.2), where the stratigraphic column ranges from middle-late Cretaceous to Plio–Pleistocene, is the
thickness somewhat greater, i.e., from 1000 m to 1500 m.

In these oceanic basins, the sediments are mainly monotonous, pelagic
oozes of low organic content, and occasionally volcanic ashes, Sedimentation rates invariably are very low. It can safely be concluded that conditions
for the generation and maturation as well as accumulation of hydrocarbons
are extremely adverse. In addition, the water depth of generally well over
4000 m would certainly preclude exploitation in the foreseeable future.

The situation is different in the tectonically unstable, orogenic belts of the
island arc active margins. Elongate basins of Tertiary age have accumulated
from 3 to 5 km of sediments, which mainly are volcaniclastic but which
include some pelagites, and in particular various shallow-water, organic

Sedimentation and Mineral Deposits
in the Southwestern Pacific Ocean
ISBN 0-12-195870-1

carbonates. Evidence for the generation of mature hydrocarbons has been found in Tonga and Fiji, while suitable reservoir rocks in the form of limestone reefs and calcarenites and possibly some volcaniclastic turbidites may be present in many places. In addition to stratigraphic traps, structural traps may exist which have formed in the course of active deformation of many of these basins.

The continental fragments of the New Caledonia–Norfolk and Lord Howe Rises, with a varied history that goes back through Mesozoic and into Paleozoic times, may also include areas with a potential for hydrocarbons. In New Caledonia this is mainly associated with late Cretaceous to early Tertiary deposits. Oil and gas shows have been found in several places, particularly also in two exploratory wells drilled in 1954–55 on the Gouaro anticline near Bourail. Thick sediments are found further south along the Norfolk Rise, and also in the New Caledonia and Fairway basins. The passive-margin type rift basins on the western side of Lord Howe Rise, with up to 4 km thick sediments which probably are mainly continental to shallow-marine and of Cretaceous age, may be particularly interesting.

On the whole, petroleum exploration has hardly scratched the SW Pacific region. The following discussion is of necessity very preliminary, general and sketchy. It must be read in conjunction with Chapter 2 on stratigraphy, where many aspects of sedimentation, depositional environments and facies distribution are set out in greater detail.

3.2   The Young Island Arc Basins

3.2.1   General

From the Manus and New Ireland Islands in the NW to the Tonga Islands in the SE, these basins are all associated with active plate margins. With regard to present day subduction regimes they are all on the upper plate, and may thus belong to either the India–Australia or Pacific plate. The lithologies of basin sediments mainly comprise any mixture between the two end members of volcaniclastics (both epi- and pyroclastics) and organic limestones. The latter generally are subordinate and related to reef and reef-associated environments, such as carbonate platforms, fringing and pinnacle reefs; they include back-reef and fore-reef deposits, reef-derived calcarenites, etc. Carbonate platform formations are most widely developed in the NW (New Ireland and partly Solomons), but also occur in Tonga. The volcaniclastics which make up the bulk of sediments range from coarse rudites to grey-wacke sand–siltstones and shales, and are of depositional environments that range from near-shore to bathyal. Slump deposits and turbidite sequences are widely present. Pelagic shales and marls occur occasionally. Particularly in the New Hebrides and Fiji, submarine lava flows are an important component in the overall basin fill. The oldest sediments in these basins are late Eocene in Tonga and Fiji, Oligocene in New Ireland, and early Miocene (perhaps including late Oligocene) in the Solomons and New Hebrides.

Low-grade burial metamorphism with zeolitization of sandstones and recrystallization of limestones has occurred locally. Also, some basinal areas have been affected by strong deformation with folding and thrusting, and by intrusions of andesitic–dioritic rocks, uplift and deep erosion. Large parts, however, have remained intact and relatively little disturbed. In these areas, which mainly are presently located offshore, extensional tectonics with normal faulting are characteristic; locally some folding has occurred, too.

Entrapment of oil and gas may thus be possible on structural grounds, e.g., in fault blocks and anticlines, but also because of lithologic discontinuities particularly around biohermal bodies of carbonate rocks. Reefs and reef-associated limestones have invariably been the principal target for oil exploration in island arcs. Their recognition, i.e., unambiguous definition by seismic surveys, however, appears to be a major problem. Another problem is seen in the availability of porosity and permeability of potential reservoir rocks. Many limestones now exposed are highly recrystallized and dense, though in some places they have acquired secondary porosity through intense fracturing. The volcaniclastic sandstones quite generally are quartz-deficient and texturally and mineralogically immature, thus dirty and tight. No detailed studies have been carried out to determine the extent to which they could form reservoirs. Yet it is known from other areas (e.g., Tertiary basins in California) that volcaniclastic turbidites can indeed be prolific oil producers.

On the basic question of hydrocarbon generation, very little is known. No possible source rock formations have been identified in any of these island arcs. However, favourable areas for the accumulation of source material could have existed in stagnant lagoonal and back-reef environments. The fact that most limestones are mainly of algal composition may enhance their source potential. Carbonaceous, silty mudstones, locally with plant fragments and pyrite which sometimes emit a slightly foetid smell, have been mentioned from some places, e.g., in the New Hebrides (Mitchell, 1966). Also, rapidly buried forereef deposits, and sediments in deep and silled basins with possibly euxinic conditions, may have provided a source for hydrocarbons where sufficient organic material was included. The establishment of anaerobic environments may have been accelerated by the emission of volcanic gases, while the regional, magmatic–volcanic activity suggests that high heatflow regimes existed over much of the basins. Sediments may thus have reached thermal maturity more readily. Indeed, the presence of mature hydrocarbons has been reported from geochemical analyses of active oil and gas seeps in offshore areas of Fiji (Stoen, 1979; Horvitz, 1980). Recent, more detailed investigations of the well-known oil seeps in Tonga have indicated that the oil there is a biodegraded crude derived from a thermally mature source (Sandstrom and Philp, 1984).

3.2.2   Manus–New Ireland basin

Contrary to what is the rule in all other island arc basins in the SW Pacific, this basin saw predominantly limestone deposition in Mio–Pliocene times, with volcaniclastics being subordinate, local and restricted to short time intervals.

These limestones, which after cessation of active volcanism in Eocene–Oligocene time spread across most of the area, are mainly of shallow-water origin representing large fringing reefs deposited over a gradually subsiding, volcanic basement. Thus an extensive reefal platform was built up, many hundred to well over 1000 m thick. It predominantly consists of well-bedded to massive, coralgal biomicrite that is very pure and whitish to cream coloured, often recrystallized. Calcarenite, calcilutite and other fine-grained calcareous sediments also occur.

The western margin of the basin has been uplifted and is at present mainly subaerial, forming the New Ireland, New Hanover and Manus Islands (Fig. 3.1). The present basin is entirely offshore, forming a gentle depression to the NE and N of this chain of islands; it is bordered on the opposite side by the North East Ridge (de Broin *et al.*, 1977), which is situated above the West Melanesian Trench. According to Exon and Tiffin (1982), no compressional features are recognized in this basin; normal faults, however, are widespread. Including the Oligocene Jaulu Volcanics, which in terms of petroleum geology must probably be regarded as basement, the total basin fill is estimated by Exon and Tiffin (1982) to be over 5 km thick. They suggest that the Miocene shelf limestones are up to 2000 m thick, the remainder of the sequence being formed by another 2000 m of younger limestones and volcanogenic sediments. Local unconformities are found in various parts of the sequence.

In New Ireland, the siltstones of the Lossuk River Beds at the base of the limestones have been regarded as potential source rocks, while the Lelet Limestones would contain good reservoir rocks (Ripper, 1969). In spite of this, however, the potential for hydrocarbon accumulation in the offshore basin must be regarded as extremely limited. Its very simple, synclinal structure with its higher, western flank open, and a fairly uniform sediment fill which to a great extent is supposed to consist of shallow-water lime-

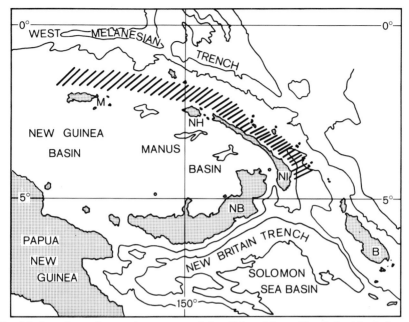

Fig. 3.1. Location and approximate extent of the offshore Manus–New Ireland basin (cross-hatched). Bathymetric contours: 2500 and 5000 m. B, Bougainville Island; M, Manus Island; NB, New Britain; NH, New Hanover; NI, New Ireland.

stones, represent a scenario that obviously is not very attractive. It seems that the mere fact of a comparatively thick (though uniform) sediment sequence is not a sufficiently valid argument for postulating a petroleum potential, as was invoked by Exon and Tiffin (1982). On the other hand, the basin, which is of considerable dimensions (900–1200 km long and 160 km wide), has only been covered by preliminary seismic surveys of reconnaissance nature, and no samples and drilling data are available. Indeed the results of recent and more detailed work in New Ireland both on land and offshore (Exon *et al.*, in press) tend to upgrade prospects considerably.

3.2.3 Solomon Islands

The geological history and structure of the Mio–Pliocene basins in the Solomons, which are much more complex and varied than the New Ireland basin and thus of considerably greater potential than the latter, have been described in detail by Katz (1980, 1982). Sonobuoy data have indicated a total sediment thickness within the Central Solomons Trough of over 5 km (Vedder *et al.*, 1982). Carbonate platforms, reefal bioherms and cal-

D

carenites are extensively developed on the islands both north and south of
the Central Solomons Trough, in particular on Choiseul and Guadalcanal
Islands. In Guadalcanal they are locally up to 100 km thick (Hackman,
1980). In places, large portions of reefs have broken off from shelf edges and
slumped into deeper parts of the basin (van Deventer, 1971), embedded in
volcaniclastic silt–sand or mudstone. While the volcaniclastics generally are
highly susceptible to chemical alteration and compaction, and thus would
constitute poor reservoir rocks, they may serve as an adequate seal or cap
rock around carbonate bodies, whether these are *in situ* reefs or displaced,
exotic olistoliths. However, some quartz-rich, detrital sediments may also
occur and provide suitable reservoir rocks, particularly where sands are
derived from the erosion of quartz-diorite or basement schist with extensive
quartz veining, as exist in Guadalcanal (Hackman, 1980). Metabasalts in the
basement of Guadalcanal were found to contain up to 40% by volume of
quartz (van Deventer, 1971).

While marginal parts of the original sedimentary basin have been uplifted
in the present islands and deeply eroded, large tracts have been preserved in
the downfaulted Central Solomons Trough and the western Bougainville
shelf. Structural traps may exist in fault blocks and anticlines particularly
along the edge of the Central Solomons Trough, as shown by Katz (1980).
However, water depth is considerable, i.e., up to 1000 m. Prospects in
shallower water are particularly seen near the north coast of Guadalcanal
and on the western Bougainville shelf, perhaps also in Manning and
Bougainville Straits. The only onland prospect that may exist is in the
eastern part of the Guadalcanal Plains. The total length of the presently
prospective basin is 800–900 km, and the average width about 50 km (Fig.
3.2).

To what extent reef limestones exist in subsurface underneath the downfaulted Central Solomons Trough is still an open question. One well
(L'Etoile 1) was drilled on a reef target west of Bougainville Island (Fig. 3.2;
Oceanic Exploration Co., 1975), on the western edge of a deep basin that
was shown by detailed seismic surveys to underlie the shelf platform. As interpreted, the reef was thought analogous to the 1200 m thick, early to middle
Miocene Keriaka Limestone which is exposed on land on the eastern edge of
the same basin (Fig. 3.3). The rationale behind the identification of the
drilling target as a possible reef body was mainly based on the distribution of
seismic interval velocities (Shell Development, 1973). Although magnetic
profiles were also obtained, no meaningful anomalies were observed.
However, this may have been due to the high proportion of volcaniclastic
sediments and hence highly magnetic material in the sedimentary section,
which effectively could blanket out any deeper seated anomalies.

The predicted and drilled sections are shown in Fig. 3.4. The well was

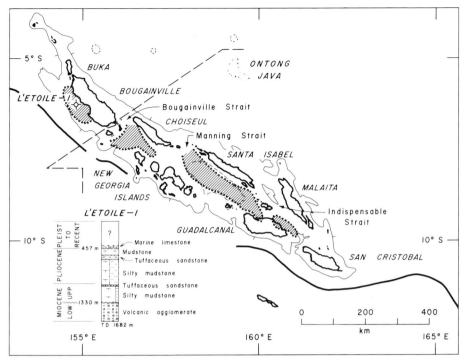

Fig. 3.2. Basin areas with a potential for hydrocarbons in the Solomon Islands, and location of L'Etoile No. 1 well. Broken line: international boundary between Papua New Guinea and the Solomons Republic. Bathymetric contour is 1000 fathoms (800 fathoms for dashed contour south of Santa Isabel). [After Katz (1980).]

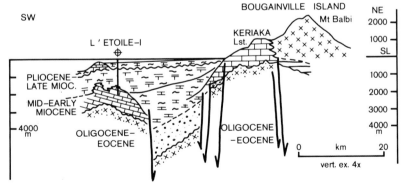

Fig. 3.3. Interpretative profile across the shelf basin west of Bougainville Island, before drilling of L'Etoile No. 1 well. [From Shell Development Australia Pty. Ltd. (1975).]

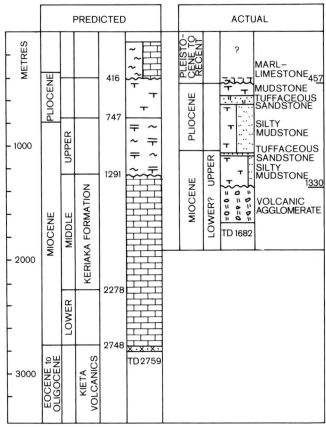

Fig. 3.4. Predicted and actual well sections of L'Etoile No. 1 drilled in 1975 to the west of Bougainville Island (for location see Fig. 3.2). [From Shell Development (Australia) Pty. Ltd. (1975).]

stopped at TD 5515 feet (1682 m), in a section of volcanic agglomerate of probably early Miocene age. All seismic reflectors which might represent the upper boundary of a carbonate reef zone had been penetrated, but with the exception of a thin limestone at 1340 feet (408 m—no samples were obtained above this depth), no carbonate bodies were encountered. The section between this limestone and the basal, volcanic agglomerate consisted entirely of muddy, volcaniclastic silt–sandstones. From foraminiferal studies of sidewall cores the Plio–Pleistocene boundary occurred between 1500 and 1650 feet (457–505 m), and the Pliocene to Miocene contact between 3402 and 3491 feet (1037–1065 m). For the lowermost part of the section, a whole rock K–Ar analysis of rock fragments excluding matrix gave an age of 18·2 ± 7 m.y., i.e., early to middle Miocene (Oceanic Exploration Co., 1975).

This well, which was the first in the SW Pacific island arcs to reach its stated objective, gave evidence that the seismic anomaly on which it was drilled represented a large accumulation of volcanic material rather than the reef complex which had been expected. Thus the eastern and western margins of the sedimentary basin west of Bougainville, in this area at least, have a different geological history. Considering its closeness to the West Bougainville Trench, this may not be surprising. However, the main problem which has remained is with regard to the recognition from seismic surveys of the occurrence or not of reef bodies.

In the easternmost area of the Solomon Islands, very thick sediments are also found in Indispensable Strait. They are flat and undeformed, resting uncomfortably on the folded formations of the Malaita foldbelt. Thus they are not older than Plio–Pleistocene, and probably have no potential for hydrocarbons.

The Cretaceous to Pliocene sediments of Malaita form a continuous and conformable, rather monotonous, pelagic sequence of mainly cream-coloured carbonate mud including siliceous marl and chert, white aphanitic limestone with brown chert nodules, nannofossil chalk and blue-grey and brown calcisiltites (Hughes and Turner, 1977). This is closely similar to what was found in various DSDP drillholes on the Ontong Java Plateau (Andrews et al., 1975) (Fig. 2.2). The total thickness across both areas ranges from 1300 to 1750 m. A slow rate of deposition and quiet conditions of open oceanic circulation have led to well oxygenated environments; bioturbation is very common through most of the sequence. In the Oligocene, however, volcanogenic silt–sand detritus began to appear in northern Santa Isabel, and later also on Malaita (Coleman et al., 1978). Although the predominantly pelagic limestone environment continued, distal turbidites and slump deposits became an important feature in the younger part of the sequence. The seas were rapidly shallowing during the late Miocene and Pliocene. From these observations it is concluded that the Malaita–Ontong Java Province (Katz, 1980) was approaching a position close to a proto-Solomons island arc, whereby sedimentation became more and more influenced by this proximity to a tectonically unstable area. In the late Pliocene the leading edge of the Malaita–Ontong Java Province was highly deformed into a typical foldbelt, indicating that by this time the final collision and amalgamation with the modern Solomons arc was taking place. The tectonic contact thus formed between the two units is clearly visible along a narrow belt of faults and thrusts associated with cold intrusions of ultramafic rocks. This contact can continuously be followed from north of Manning Strait through Santa Isabel and the Florida Islands, and further to the SE between

Guadalcanal and southern Malaita, to near the little Uki Island north of San Cristobal (Katz, 1980).

The tectonic contact between these two major provinces—which correspond to Hackman's (1973) Pacific Province on the one side, and the Central and Volcanic Provinces on the other—is the main boundary line beyond which, to the NE, no petroleum potential appears to exist. From the above description it is obvious that organic carbon contents are minimal in the stratigraphic column, while the limited thickness of sediments would hardly be sufficient for thermal maturation. And although there is a large number of beautifully developed anticlines both on and offshore, all the deformation took place at a very late stage only, about late Pliocene. With the rocks generally being highly indurated, dense and tight, suitable carrier beds seem to be lacking. In short, neither source nor reservoir rocks are anticipated, nor have they been seen or reported.

In the Main Solomons Province (Katz, 1980), on the other hand, the very thick sequence of mainly terrigenous sediments associated with widespread occurrences of shallow-water carbonates suggest that a genuine petroleum potential may indeed exist. The often rapid deposition, with interfingering of a variety of lithotypes, unconformities and tectonic deformation at different stages which created fold and fault structures, are all factors that are positively valued in the search for petroleum. Thus the greater part of the dotted area in Fig. 3.2—a total of some 30 000–35 000 km^2—is considered prospective.

3.2.4 New Hebrides

The depositional basins of Mio–Pliocene sediments, and thus any potential for hydrocarbons, are restricted to the central and northern parts of the New Hebrides (Fig. 2.10). From Ambrym Island to the south, the island arc as a whole is formed by a single, and simple, volcanic structure of young age (Katz, 1982; Katz, in press). Northwards, on the other hand, the arc consists largely of a double chain of islands with a deeply depressed, median basin between them. Thick Mio–Pliocene sediments are exposed on both the western and eastern islands, and also exist underneath the median basin.

This part of the New Hebrides, therefore, shows some similarities with the Solomons (Katz, 1982). The tectonic evolution, however, has been profoundly different in the two areas. In the New Hebrides the sediment thickness markedly decreases across the arc from west to east, while in the same direction the sediments become finer-grained, more pelagic and distal with respect to the main volcanic source. In addition, they include a notable component derived from a terrigenous source still further east. Paleogeographically there was, in the early to middle Miocene, an active volcanic island arc in the west, which produced lavas and volcaniclastic deposits (of both primary and secondary origin) many thousands of meters thick. Large volcanic islands were built up, around which deposition was in shallow-water but partly also terrestrial environments (carbonized wood fragments with *in situ* tree stumps; J. N. Carney, personal communication). Both carbonate reefs and volcanic-derived rocks accumulated, but the general, volcano–tectonic instability and possibly loading by lava flows caused many of these reefs to break up after short duration. Reef fragments mixed with volcanic rudites were carried by rubble avalanches into deeper water, and reworked finer materials were deposited by turbidity currents in bathyal areas between the islands. Eastwards from this high and highly active volcanic arc, a regional sedimentary basin developed which deepened to the east. Still further east it bordered a larger landmass. This was under active erosion, with some of its products being shed westwards into the deep-marine basin. Thus the regional cross-section from west to east shows an oceanic, volcanic island arc, a marginal or back-arc basin probably underlain by oceanic crust (ultramafics in Pentecost Island; Mallick and Neef, 1974), and an exposed, quasi-continental block.

Orogenic deformation, intrusion of andesitic–dioritic stocks, uplift and erosion occurred in the western island arc around the middle Miocene. This was followed by shallow-marine transgression from the north and east, with epiclastic sediments of limited thickness being deposited in narrow, longitudinal basins. These basins probably originated in fault angle depressions which formed during a process of tectonic relaxation and extension of the orogenic belt.

Eastwards the deformation rapidly diminished. Although there seem to be some gaps in the stratigraphic column, deep-marine, mainly pelagic sedimentation continued through late Miocene and into Pliocene times. But locally an intensive volcanic phase began in the latest Miocene to early Pliocene, entirely submarine at first but soon building up a new chain of volcanic islands of Plio–Pleistocene to Recent age extending long the whole New Hebrides arc. At the same time, i.e., from 3 m.y. onwards, the eastern belt has experienced intense uplift of something like 3000 m (Katz, in press), which still continues. Since Pleistocene time also the western belt has undergone renewed uplift. As a result the median basins developed which are relatively deeply depressed in the central sector (North and South Aoba basins), but only form a simple, gentle syncline further north (Fig. 2.14). The pronounced fragmentation of the whole island arc in its latest development has also affected these basins, which have additionally been reduced and separated from each other by the local penetration of young volcanics (Fig. 2.10).

The detailed analysis of the tectonic evolution, paleogeography and distribution of sedimentary facies—here given in a short summary only—provides some interesting and important insights regarding the hydrocarbon potential (Katz, 1982, in press). It suggests that the western belt is not a likely source area in the early Miocene. The very widespread and pronounced volcanic activity and the deposition of largely coarse volcaniclastic materials together with abundant lavas would have prevented or heavily diluted the formation of organic-rich sediments. A much better source potential is envisaged in the east where the quieter volcano–tectonic conditions resulted in the deposition of generally finer and well-bedded sediments. The mainly deeper-marine and pelagic environments seem to further enhance the source potential, while the local contribution of land-derived material from an easterly source may have supplied additional, carbonaceous–organic matter to the sedimentary column.

The pronounced tectonic deformation which in the middle Miocene occurred in the west, with folding and intense shearing, and a markedly greater induration of rocks including low-grade metamorphism followed by uplift and deep erosion, might have destroyed whatever accumulation of oil and gas would have existed. The younger, post-orogenic sequences only occur in local, narrow basins and are of very limited thickness. They are not a likely prospect for both the generation and accumulation of hydrocarbons.

No orogenic phase occurred in the east where sedimentation was more or less continuous, and continuously pelagic and deep-marine throughout Miocene and into Pliocene times. The more favourable conditions may have resulted in sediments of a greater source potential, which in addition had more time for thermal maturation. Heatflow from at least the latest Miocene onwards, when the present volcanic phase started, probably was greatest in the east. While early migration may partly have been directed westwards, following decreasing pressure gradients from the deep basin up-dip towards the shallower arc in the west, accumulation in that area would largely have been prevented and/or destroyed by the middle Miocene orogeny and its consequences of partly extreme distortion, high induration, and deep erosion of the rocks. Regarding possible migration routes in the younger, late Miocene to Pliocene formations, practically no conduits would have existed since in the west of the median basins these formations onlap onto deformed, earlier Miocene rocks and thus are discontinuous with other, time-equivalent formations within the western belt. On the other hand, strong uplift which began in the east about 3 m.y. ago would have caused any remaining or newly formed hydrocarbons to move into the growing folds and fault structures now developing in the east of the median basins, particularly towards Maewo and Pentecost islands (Fig. 2.10). Provided there are potential reservoir rocks available, the structured eastern margins

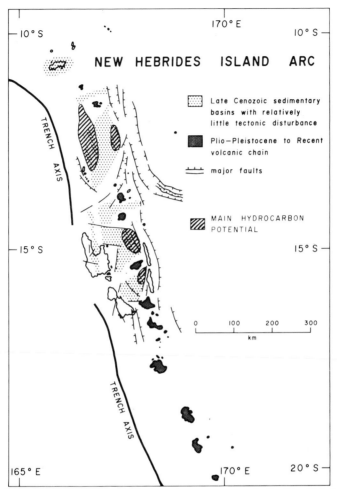

Fig. 3.5. Main areas of potential hydrocarbon occurrences in the New Hebrides.

of the North and South Aoba basins thus are the most likely loci for hydrocarbon accumulation in the central New Hebrides (Fig. 3.5) (Katz, in press).

In the northern New Hebrides, the Banks and Torres–Santa Cruz basins are underlain by early to middle Miocene rocks that across the basins are only slightly deformed. Both in the east and west they form an uplifted fault anticlinorium, while the central part of the basins is filled with an unconformable, late Miocene to Pliocene sequence (Fig. 2.14). Subsidence in this central part continued during and after the deposition of the younger sequence. If there are carrier rocks in these basins, migration and accumulation of hydrocarbons generated anywhere in the stratigraphic column could have readily occurred at more than one time. Good structural traps exist both in the east and west; stratigraphic traps associated with the late Miocene to Pliocene unconformity appear to be particularly developed in the west. With the Pliocene to Recent igneous activity along the eastern margin of the basins, and also underneath much of their more central, axial parts, a relatively high heat source may have been widely present for thermal maturation of the sediments.

In summary, probably the greatest potential is found in the Torres–Santa Cruz basin which is the largest, coherent portion of the Mio–Pliocene, median basins in the New Hebrides (Fig. 3.5). Not only are here the most favourable conditions with regard to tectonic evolution and structure, but drainage areas for any hydrocarbons generated are by far the most extensive anywhere in the New Hebrides. Given comparable thickness and lithology, i.e., source potential of sediments, the Torres–Santa Cruz basin would yield much greater volumes of hydrocarbons than any of the other basins. Last but not least, water depth is much shallower, i.e., only about 800–1700 m, compared to 2000–3000 m in the North and South Aoba basins.

3.2.5 Fiji

The most likely petroleum prospects are found in and around the big island of *Viti Levu*. However, because of repeated orogenic movements, widespread magmatic-volcanic activity and the final, regional doming-up of the island in Plio–Pleistocene time, sedimentary basins are mainly developed offshore particularly in the north (Bligh Water), west (Nadi* Bay) and east (Fig. 2.16). On land the Eocene to Miocene formations are heavily disrupted, pierced by intrusives and volcanics and strongly deformed, locally also metamorphosed; also, sedimentation was often interrupted by tectonic and volcanic disturbances. It is probable that in the tectonically quieter, more stable areas offshore, sedimentation was more continuous particularly since the Miocene, perhaps even the Eocene. Late Miocene to Pliocene sedimentary basins, which developed after the Colo Orogeny, are mainly or entirely offshore and only marginally encroach onto the land (Nadi Bay, Rewa Basin etc.; Fig. 2.16).

The first offshore petroleum licences were granted in 1969, and airborne magnetometer and marine seismic surveys were undertaken soon after (Lindner, 1972). From the very beginning of exploration, biostromal and biohermal carbonates were considered the primary targets, rather than anticlinal fold structures. In addition, it was considered that perhaps the

---

* For the spelling and pronunciation of Fijian names, see Rodda (1984).

most attractive reservoir rocks could be the rubble that was shed from either reefs or coral banks occupying elevations on the ancient sea floor. Thus important stratigraphic traps could be developed in the fore-reef slopes. One such fore-reef deposit, of middle Miocene age, is exposed on Sawa-i-Lau Island in the Yasawa Group (Fig. 2.16) [Lindner (1975); see also front cover picture in Eden and Smith (1984)]; this island is a striking topographic feature, about 700 ft (213 m) high and covering an area of about 200 acres. It is believed that many similar occurrences could be included in the sedimentary column underneath Bligh Water and elsewhere. In fact, reef-associated and other, lenticular limestone bodies are exposed in many places on Viti Levu, in formations ranging from Eocene to Miocene.

Apart from a considerable number of seismic surveys carried out by different operators during the 1970s, geochemical surveys were also undertaken offshore, in order to determine anomalies and sample active oil and gas seepages found on the sea floor. Analyses of these samples confirmed the presence of mature hydrocarbons (Stoen, 1979).

Drilling operations began in 1980. A summary of the well logs of all the seven wells drilled to date has recently been published by the Mineral Resources Department of Fiji (Eden and Smith, 1983) while in a later publication, Eden and Smith (1984) have presented a full history of exploration activities, as well as a summary of geology and hydrocarbon prospects, the licence blocks and legal and administrative arrangements of the Fiji Government. The following description has been taken mainly from these papers.

The first exploratory wells were Bligh Water No. 1 and Great Sea Reef No. 1, drilled in 1980 by Chevron Oil Co. of Fiji in the south and north of Bligh Water, respectively (Fig. 2.16) (Katz, 1981b). They reached 2742 and 2839·4 m, respectively, and bottomed in Oligocene or older rocks. Although a number of thin limestone intercalations were found in the Miocene and Oligocene, by far the greater part of the sequence consisted of volcaniclastic sediments, tuffs and agglomerates. No reefs were encountered, and the sections are generally organic-poor with no source rock character.

Near Nadi in the west of Viti Levu, the Buabua No. 1 and No. 2 wells drilled by Bennett Petroleum Corporation in December 1981 and January/February 1982, had both to be aborted because of technical difficulties after reaching 307 and 302 m, respectively. They did not drill to below the Plio–Pleistocene Meigunyah beds, which consisted of clayey silt and fine sand with some gravel beds as well as coaly layers; shell beds were very common in some horizons. Indications of gas and distillate reportedly occurred in several of the sands, and were tested in two horizons; upon perforation, a gas blow plus oil scum and water with 19 000 ppm of chloride were produced.

Later in 1982, two more wells were drilled by Bennett Petroleum Corporation in the east of Viti Levu (Katz, 1983a). The first one, Maumi No. 1, was drilled on land and reached 1591·3 m. It intersected 1003 m of clayey silt–sand with some limestone bands, of Pliocene to late Miocene age; several coal beds were encountered about 300–400 m deep. The basal portion of this section is correlated with the Suva Marl, Lami Limestone and Veisari Sandstone. Below 1003 m, the section consists of an igneous complex with spilites, volcanic breccia and glass, as well as some fossiliferous bands; from these, the age was determined as late to middle Miocene. Minor shows of methane and higher hydrocarbons were encountered through most of the section, including the basal igneous complex. The second well, Cakau Saqata No. 1, was drilled offshore to a total depth of 2272 m. Below the late Pleistocene to Recent reef limestone, 150 m thick, it drilled through 1500 m of Plio–Pleistocene, clayey silt–sandstones which particularly in the lower half are mainly volcaniclastic and/or tuffaceous. Near the base a layer of basalt was traversed. The pre-basalt sequence is of late to middle Miocene age and possibly older, and consists of tuffaceous claystone, siltstone and sandstone. In the lowermost part, a welded tuff, quartzite and basalt were encountered. Readings of methane with minor ethane and propane were obtained in the late Pliocene to early Pleistocene, with a maximum at 440 m (top Pliocene); over the same interval, minor oil fluorescence was present. Methane alone was recorded throughout the sequence.

Towards the end of 1982, another well was located in Nadi Bay in the west. This was Yakuilau Island No. 1, drilled by Worldwide Energy Corporation for OEL 9 Operators. It reached 1526 m. Since its location was very close to the two Buabua wells, the top section was virtually a repetition of what was found there. The base of the Meigunyah beds was at 318 m in Yakuilau Island No. 1, after which the well drilled over 1000 m through the late Miocene Nadi Sedimentary Group (of early Pliocene age in the very top portion of the well section; age determinations are based on planktonic foraminifera and calcareous nannofossils). The section consists of a rather monotonous, mainly grey-greenish series of volcaniclastic, clayey silt–sandstones (Fig. 2.17) which in the lower part have conglomeratic intercalations. From 1381 to 1475 m, the well log records conglomerates of middle Miocene age, whereas the very basal part of the well section again is of late Miocene to early Pliocene age. Thus a thrust or fault has been construed here (Eden and Smith, 1983). However, it is possible that the middle Miocene faunas obtained between 1381 and 1475 m are reworked and thus derived, together with the conglomerate pebbles in this same interval, from underlying Wainimala Group rocks. Coarse conglomerates with pebbles clearly originating from the Wainimala Group are known in outcrops of the Nadi Sedimentary Group, near its base some distance to the north of Nadi town.

Thus the reappearance of late Miocene to early Pliocene faunas at the base of the well section, in beds which are free of conglomerates and of a lithology that is typical of the Nadi Sedimentary Group, may simply be due to the fact that there is no admixture here of derived material, including older faunas which in the interval above would have suppressed the younger, authochtonous elements.

Thus there is no need to postulate structural complexities, and the well section could indeed represent a continuous sequence of Nadi Sedimentary Group, which at this locality would therefore be more than 1208 m thick.

There were no oil shows in Yakuilau Island No. 1 well, but some methane occurred virtually throughout the section. Peaks were 8500–9000 ppm at 1081 m, 1223 m and 1502 m, thus all in the lower part of the Nadi Sedimentary Group and probably close to the unconformity with the underlying Wainimala Group rocks. It is assumed that the methane, if not indigenous, has its source in the latter sequence below this unconformity.

Although the results of these wells are disappointing from the point of view of reservoir and source potential, which generally was poor, it is interesting to note that most of them did indeed find traces of hydrocarbons. In addition, it should be remembered that probably the more attractive prospects are in the Sigatoka and Wainimala Group rocks, i.e., in sediments which are older than the Colo Orogeny. It is in these formations of Eocene to middle Miocene age, where clean, lenticular limestone bodies are found, which were deposited in shallow water often related to reef and forereef environments; these limestones constitute the most sought-after reservoirs. Of similar age, too, are the most likely, potential source rock formations.

An appreciable thickness of sediments older than the Colo Orogeny, however, has only been drilled in two of the wells. In both these wells (Bligh Water No. 1 and Great Sea Reef No. 1) (Fig. 3.6), the respective sediments were devoid of reefal carbonates. They are dominated by a rather monotonous volcaniclastic series, and the only limestone beds encountered are thin and of deep-water origin. The problem therefore rests in the recognition and location of proper drilling targets. In this respect—and apart from the particular seismic problem of identifying reefs on seismic records—it is obvious that a better understanding of the paleogeographic conditions would be helpful. Since reef targets must be located in areas of shallow paleo-bathymetry, they are probably related to and controlled by the location and orientation of the main, volcanic island arc. A palinspastic reconstruction of Fiji, therefore, would be useful, but unfortunately there are few data available. Malahoff et al. (1982a) have shown that since the breakup of the original island arc about 10 m.y. ago, Fiji has rotated anticlockwise through at least 90° and perhaps as much as 115°. Most likely, this rotation was accompanied by internal bending, too. The age of the

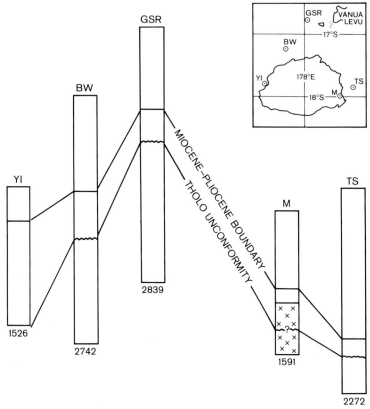

Fig. 3.6. Correlation of well sections in Fiji. Total depth in metres. YI, Yakuilau Island No. 1; BW, Bligh Water No. 1; GSR, Great Sea Reef No. 1; M, Maumi No. 1; TS, Cakau Saqata No. 1. [From Eden and Smith (1983).]

breakup, however, is precisely the age of the Colo Orogeny, the Colo Plutonic Suite being emplaced synorogenically between 11 and 8 m.y. ago (Rodda *et al.*, 1967). The age of the prospective formations, on the other hand, is of a time span immediately before this event which ranges back for another 30 m.y. Little is known of the tectonic evolution and/or basin delineations during this period.

Much better understood are the history, extent and location of basins which developed after the Colo Orogeny. Their sediment fill, which is of late Miocene to Plio–Pleistocene age, however, in most places does not seem to have sufficiently high organic carbon contents, while its burial—i.e., thermal regime—in particular, most probably has been inadequate to complete the maturation process. But there may be exceptions, and some of these younger basins may indeed be prospective, too.

Such basins mainly lie offshore, or at least in a position marginal to the updoming Fiji landmass. Where they have deeply subsided and are filled with relatively thick, terrigenous sediments containing abundant, land-derived plant material, some hydrocarbon potential may exist. As noted, e.g., in the Nadi Sedimentary Group, there may be carbonaceous inter-calations and small-scale coal seams between layers of mainly muddy silt and sand. Thus a facies may be present which, if widespread enough and under sufficient overburden, could generate hydrocarbons (probably mainly gas) in commercial quantities. And although the bulk of the sandstones in these basins probably are volcaniclastic, their derivation from volcanic and intrusive rocks of the Colo Plutonic Suite may result in a composition that locally is fairly acidic. Post-Colo sandstones which are the erosive products of dacite and tonalite are highly arkosic with clean fresh feldspar, and sometimes also quartz. Such feldspars, as seen in beds of the Nadi Sedimentary Group where often they are by far the most abundant and dominant constituents, have obviously not been subjected to prolonged weathering and chemical decomposition, and have been transported and buried fairly rapidly; as a consequence, the clay (kaolinite) content in these sediments is limited. Such sandstones may, therefore, form suitable reservoir rocks. On the other hand, there are no reef-associated, biohermal limestones in these younger basins.

One of the most promising of these basins probably is the Nadi Basin, at least in its offshore portion. Although Buchbinder and Halley (in press), in 6 outcrop samples of the Nadi Sedimentary Group near Nadi town, as well as in 7 samples of underlying Sigatoka and Wainimala Group rocks, have detected no appreciable amounts of organic carbon, this may to some extent be due to weathering, while the source rock quality may in general be better in the much thicker offshore sections than near the basin margins exposed on land. If so, the numerous clay breaks and tight, silty claystone intercalations may form good cap rocks between sandstone reservoirs, while the thick, sticky and dense clay of the overlying Meigunyah beds, 300 m thick in the Yakuilau Island No. 1 well (Eden and Smith, 1983), may provide an overall regional cover and seal of the Nadi Sedimentary Group. The availability of prospects would thus depend mainly on the existence of suitable structures. A similar young basin that formed at the edge of the emerging Fiji landmass and may have some potential is the offshore Baravi Basin in the SW of Viti Levu (Fig. 2.16) (Larue et al., 1980). Sediments are 2500–3000 m thick and deposited in a trough measuring at least some 80×30 km. The basin's proximity to an active spreading zone further west may have enhanced the hydrocarbon kitchen potential, i.e., maturation of organic matter. How-ever, the sediments are virtually undeformed and the two sedimentary units recognizable, on seismic records, are more or less conformable.

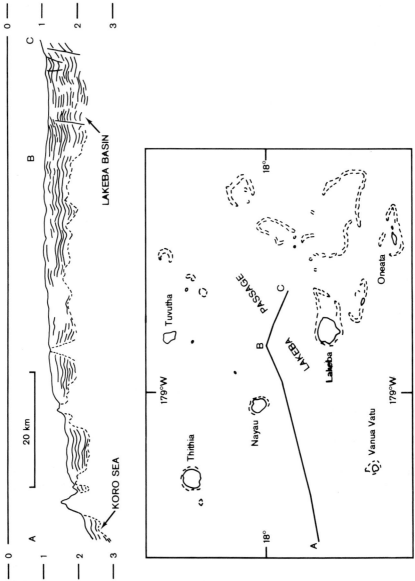

Fig. 3.7. Seismic section (tracing of Mobil Oil Co. Line 72-219) across Lakeba Passage [see Rodda (1984), regarding the spelling of Fijian names], Lau Islands, Fiji. Vertical scale in seconds reflection time.

In summary, prospects do not seem diminished by recent drilling results in Viti Levu. But they are shown to depend more closely on the paleogeography and evolution of the relevant sedimentary basins. A detailed basin analysis and synthesis of tectonic evolution should indicate areas where the most likely prospects are located.

Young basins have also formed on the *Lau Ridge*, where limestones of different ages ranging from middle Miocene to Plio–Pleistocene are exposed on the islands, where they alternate with volcanic lavas (Rodda, 1982). Most limestones are algal, or foraminiferal, dense calcilutites or marls; but true coral reef rocks also occur (Ladd and Hoffmeister, 1945). The greatest thickness exposed is 324 m. Little is known of the sedimentary sequence that exists offshore. The few seismic sections available indicate a tectonically complex structure with strong deformation and multiple intrusion of igneous bodies. Mainly along the western flank of the Lau Ridge, little deformed and generally well-bedded sediments of younger age unconformably overlie the core of the ridge, and also extend into transversal grabens which trend northeast or east across the ridge (Woodhall, in press). Such grabens are often reflected by the various boat passages between the Lau Islands and their extensive reefs, such as the Nanuku Passage, Lakeba Passage and others. The sediment sequence which fills these grabens (e.g., Lakeba Basin) (Fig. 3.7) is hardly more than 1 sec thick (reflection time) and probably is not older than middle to late Miocene, with the bulk of it being Plio–Pleistocene. Contrary to the young basins bordering Viti Levu, the sediments of which have a strong terrigenous source, these basins on the Lau Ridge probably are mainly filled with a very regular, conformable series of a single unit, which is composed of alternating hemipelagic and volcanic–tuffaceous sediments. The latter are locally derived from the multiple volcanic phases that are known to have existed on the Lau Ridge, all through the period from middle Miocene to virtually the Present (Rodda, 1982). Thus organic carbon may be a much lesser component in these sediments, than, e.g., in the Baravi or Nadi Basins in Viti Levu. Considering the source and type of the sediments, as well as their limited thickness and structure, it would seem that their hydrocarbon potential is rather low.

3.2.6 Tonga

Compared with most other island arc areas in the SW Pacific, the Tongan "arc" is a very simple feature both structurally and in its geologic evolution. Basically it forms a high platform west of the Tonga Trench, and is bordered on its western side by a young, Pleistocene to Recent belt of active volcanoes; these are associated with a number of basins and troughs (e.g.,

Tofua Trough) (Fig. 2.20) which are downfaulted from the main, Tongan platform (Fig. 2.25).

The stratigraphic sequence which underlies this platform is several kilometers thick (Katz, 1974, 1976a) and goes back to the late to middle Eocene. It is exposed on 'Eua Island (Fig. 2.23) and has been drilled in the three Kumimonu wells (Fig. 2.24). However, Oligocene rocks seem to be absent in 'Eua, where erosion on top of the late Eocene limestone has developed a considerable relief prior to the deposition of Miocene vol-caniclastics. Further west underneath Tongatapu, on the other hand, the sequence appears to be continuous including several hundred meters of marine Oligocene; but here it is entirely volcaniclastic with no limestone intercalations. Thus during Oligocene time, a structural high seems to have developed near the eastern edge of the present platform. This is supported by the gradual increase in thickness westwards, across the platform, of the entire sedimentary sequence and in particular of its younger, Mio–Pliocene portion (Fig. 2.25).

The structure of the platform is mainly controlled by a tensional regime with more or less longitudinal, normal faulting. It is typified by fault blocks of various size which are differentially tilted. In the west, close to where the platform is downfaulted towards the volcanic and trough belt, a broadly synclinal feature is often displayed (Fig. 2.25). Cross-faulting which results in a north-to-south segmentation of the platform is fairly common and of some importance, too. It is often reflected in the bathymetry which shows narrow, and relatively deep, east–west depressions across the platform; these separate most of the different island groups which thus belong to individual, structural blocks.

Since live oil seeps were first reported in 1968, much interest has been focused on the petroleum potential in Tonga, and active exploration has been carried out intermittently ever since (Tongilava and Kroenke, 1975; Katz, 1976a,b, 1978b, 1979, 1981b). The presence of a late Eocene lime-stone in 'Eua, over 100 m thick, has given rise to great hopes that a widely developed carbonate platform would underlie most of the area, with locally grown reef structures developing on top of it. In consequence, exploration has mainly been directed towards finding reef targets. The first two wells, however, which were drilled by Shell in 1971, were stratigraphic tests located on a gravity anomaly, and did not reach even the base of the Miocene (Katz, 1976a). Offshore seismic surveys by Shell and particularly Mobil Oil (Katz, 1974) had for the first time given a good regional picture of the structure and sediment thickness, but had not indicated the presence of reefs. Only on the basis of a more detailed, 48-trace vibroseis survey which in 1977 was done for Webb Tonga, Inc., on the island of Tongatapu, has it become possible to map important, seismic anomalies which were interpreted as reefs (Katz, 1978b). Three of these anomalies were drilled in the following year (Katz, 1979), but neither reefs nor a carbonate platform were encountered. However, the wells did find a considerable thickness of marine volcaniclastics of early Tertiary age, and thus confirmed the stratigraphic sequence as previously suggested (Fig. 2.24). On this basis, a detailed marine geophysical survey was carried out in 1979, with over 900 km of multichannel reflection seismic, gravity and magnetic investigations between the islands of Tongatapu and 'Eua. Covering the Tongan platform south of Tongatapu, CCOP/SOPAC ran a single-channel seismic reconnaissance survey in 1979 (Katz, 1981b; U Maung *et al.*, 1982), and in 1982 did a more comprehensive, geophysical study under a Tripartite Agreement between the governments of Australia, New Zealand and the United States (Scholl *et al.*, 1982; Scholl and Vallier, in press). These studies were continued and expanded with a further Tripartite cruise in 1984.

From all this work it has become obvious that thick sediments of early to late Tertiary age extensively underlie the Tongan platform. Also, and in spite of the negative results of previous drilling, there are good indications that reef development may indeed have occurred in some areas. While most biohermal carbonate buildups would probably be of late Eocene to Oligocene age, some may be Miocene, too. There is evidence of Miocene reef development from outcrops in the islands of the Nomuka Group north of Tongatapu (Figs. 2.23 and 2.27). In 'Eua, field work by Webb Tonga, Inc., has shown that reefs which developed on top of the late Eocene, foraminiferal platform limestone locally continued growing through Oligocene and into Miocene times, partly even into the early Pliocene (H. R. Warters, personal communication). Termination of reef growth was mainly controlled by subsidence along the flanks of the existing island–atoll complex, and/or the influx of volcaniclastic sediments. Depending on their paleogeographic location with respect to the reef and atoll, the limestones of this Oligocene to early Pliocene reef mass display a variety of local facies developments; they include high-energy coral facies, and coralgal and foraminiferal facies. Many of these limestones have excellent porosity and permeability. The total thickness of the Oligocene to early Pliocene reef complex in 'Eua is reported as over 300 m.

The sediments which cover the Tongan platform markedly decrease in thickness to the north. This is probably due to the gradual wedging-out of older sediments at the base of the sequence. Thus early Tertiary sediments seem to wedge out between Tongatapu and Nomuka; in Nomuka, Miocene sediments probably are sitting directly on basement/volcanics. Further north towards Ha'apai and Vava'u (Fig. 2.20) also the Miocene may be cutting out, so that in Vava'u the entire sediment sequence is of Plio–Pleistocene age. The sediment thickness around Vava'u is hardly more than

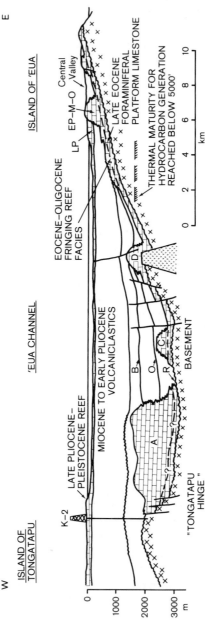

Fig. 3.8. Interpretative profile through 'Eua Channel between Tongatapu and 'Eua Islands, Kingdom of Tonga. From seismic surveys, well sections and onshore geology. K2, Webb Tonga, Inc., Kumimonu No. 2 well, drilled in 1978; B (blue), O (orange) and R (red), prominent seismic reflectors in the central 'Eua Channel area; A, C, D, some of the typical reef build-ups and primary target features; LP, late Pliocene; EP-M-O, early Pliocene–Miocene–Oligocene. [After Warters (1981).]

600 m (Mulder and Nieuwenhuizen, 1971) (Fig. 2.28). Still further north, the entire island arc structure changes its character profoundly. From the few data available [mainly from a single-channel CCOP/SOPAC reconnaissance seismic survey (Katz, 1981b) and bathymetry] it is apparent that the Tongan platform abruptly terminates immediately north of Vava'u; this termination seems to be caused by an important, transversal fault zone. To the north of it, the island arc is characterized by various high areas of apparently volcanic nature, between which there is a number of small and narrow, isolated basins filled with young sediments of only limited thickness; none reaches even 1 sec of seismic reflection time.

Accordingly, any hydrocarbon potential probably is confined to an area from Nomuka to the south. One exception may be in the Tofua Trough to the west of the Tongan platform (Fig. 2.20), which has its greatest development some distance further north where sediments are more than 1·5 sec thick, with no basement seen. The sequence includes a marked unconformity below which the sediments may be of Miocene or older age, thus representing part of the platform sequence which is downfaulted and continuing westwards underneath the Tofua Trough. If this is correct, the thick sediments in this deeply subsided trough close to a volcanic (high heat) source may have been favourably situated for hydrocarbon generation. Accumulation could have occurred in folds underneath the unconformity, and/or against the major fault zone which borders the Tofua Trough to the east (Katz, 1974).

The most prospective area of the Tongan platform as presently seen (and this is not merely a function of the greater amount and detail of exploratory work done) is in the 'Eua Channel between Tongatapu and 'Eua islands. The following information, which is based on seismic interpretation and gravity and magnetic modelling of the detailed geophysical data collected in 1979, together with geological surface data from 'Eua and subsurface well data from Tongatapu, has been provided by W. M. Barney and H. R. Warters (Denver); permission to publish is gratefully acknowledged [see also: Samuel Gary Oil Producer (1981)].

According to this information, the area of the 'Eua Channel (Fig. 3.8) was the site of a tensionally faulted shelf platform in the late Eocene, which was bounded in the north and east by an ancestral 'Eua, and in the west by a north-trending high of mainly volcanic origin ("Tongatapu Hinge"). In this western area, i.e., in the vicinities of the Kumimonu-1, -2 and -3 wells, volcaniclastic sediments accumulated, whereas platform carbonates were deposited in the shallow embayment to the east. In fact, these carbonates extended across the 'Eua Channel to the SE corner of Tongatapu, where in ditch cuttings of the Kumimonu-2 well various reefal rock types were encountered, including coral fragments, encrusting algae and particularly lime wackestone, together with lithic volcaniclastics. There is no evidence

here of pyroclasts. The algae and rare ooids in the 1800–2000 m interval
(late Eocene, Fig. 2.24) indicate very shallow water conditions of deposition. Above 1800 m (Oligocene) the rare carbonates are mainly silty,
foraminiferal lime wackestone, but it appears that carbonate deposition is
effectively "drowned" now in this area by vitric pyroclasts (volcanic glass
shards and pumice). Abundant pyroclastic material, such as crystal vitric
tuff, originating from subaerial andesitic–dacitic volcanic eruptions,
reached the depositional basin as air fall and effectively displaced or
blanketed out the shallow water carbonate deposition (Company report,
Warpet Exploration, Denver). Locally large accumulations of airborne
volcanic material were thus formed, building up the north-trending highs of
volcanic origin mentioned above.

East of the "Tongatapu Hinge" (Fig. 3.8), the shallow embayment with
carbonate deposition continued relatively undisturbed. Some faulting
caused horst blocks to move upwards, and reefs began to grow in such
positive areas of the embayment. With further tensional faulting, reefs
continued growing through the early Oligocene, now particularly also on the
subsiding margin of the "Tongatapu Hinge". In the deeply subsiding parts of
the 'Eua Channel embayment, between the upfaulted horst blocks, a silled
basin formed which probably was filled with off-reef calcareous silt and
shale, besides volcaniclastic deposits (between seismic reflectors B, O and
R; Fig. 3.8). It is possible that euxinic conditions were established in the
restricted environment of this silled basin.

This period was followed by a widespread hiatus, including local erosion
and unconformity as seen on 'Eua, and in some offshore areas as evidenced
by seismic sections. Renewed subsidence with a regional tilt to the west
occurred in the Miocene, which led to the accumulation of thick volcaniclastics and, on 'Eua, to further reef growth. There is no evidence of Miocene
reefal development under the 'Eua Channel, where the water was probably
too deep. In the wells drilled by Shell near Nuku'alofa, Tongatapu, water
depth during the Miocene was outer shelf to bathyal (Tonga Shell, 1972).

Volcaniclastic deposition continued into the early Pliocene when the
basin became shallower by either filling up or tectonic uplift. Extensive
carbonate platform deposition occurred in the late Pliocene and Pleistocene, with reef development on Tongatapu, 'Eua and locally under the
'Eua Channel. Recent, rapid subsidence of the 'Eua Channel left Plio–
Pleistocene reefs drowned on the present sea floor, while contemporaneous
reefs were uplifted on 'Eua, and also on Tongatapu where they are tilted
northwards.

From the geophysical information and regional geology, there is thus
good evidence in the 'Eua Channel to suggest that reefal mounds, patch
reefs and subordinate pinnacle reefs have formed in the late Eocene and

early Oligocene (Fig. 3.8). It is likely that porous carbonates exist in these reefs, which would constitute the principal reservoir rocks. Post-Oligocene rocks probably have no significant reservoir potential. Regarding the sealing potential of rocks surrounding the reefs, there is no direct evidence in the 'Eua Channel. However, there is ample evidence of tight impermeable sediments in the various well sections on Tongatapu. It is expected that the geochemically unstable volcaniclastics rapidly degrade to shales and volcanic clays, and thus will provide an adequate seal for reefal traps.

The silled basin with sediments deposited in a restricted and possibly euxinic environment in the bottom of the 'Eua Channel, between the various reef build-ups, is thought to constitute the source area for oil. From the seepages in both Tongatapu and 'Eua (Ohonua harbour; Tongilava and Kroenke, 1975) there is good evidence that oil is being generated in the basin, and that it is from a thermally mature source (Sandstrom and Philp, 1984). However, geochemical analyses of the well sections in Tongatapu, by Shell and Robertson Research, have found only limited quantities of organic carbon. The conclusion is that in these sections there is no potential for an effective hydrocarbon generation. On the other hand, it was shown that the little oil-prone kerogen detected would probably reach maturity below about 1500 m (5000 ft). In this respect, it is probably immaterial that today's thermal gradient in Tonga is generally reported to be low. Indeed, the bottom hole temperature in the three Kumimonu wells, measured after circulation had been stopped for 24–27 hours, indicate a thermal gradient of only 2·40–2·99°C/100 m (1·32–1·64°F/100 ft). However, Tertiary gradients and heat flow may have been, at any given place, very different from today's; considering the relative displacement, through time, of volcanic eruption centres across the arc (Katz, 1976a), this is in fact to be expected. In any case, seepages with mature oil do exist on either side of the 'Eua Channel, and their most probable source is to be looked for in the deeply buried, late Eocene to early Oligocene, restricted basin in the bottom of that channel. In this basin, obviously, the sediments must be of a different facies than those drilled in Tongatapu, and may have a much greater source potential. It is possible that an anaerobic, euxinic environment existed here, while the sediments were buried at probably sufficient depth under the Mio–Pliocene volcaniclastics to reach maturity (Fig. 3.8). Some vertical leakage, probably along fault planes, may have permitted some of the oil to reach the Plio–Pleistocene limestone at the top of the sequence, through which it migrated up-dip westward to Tongatapu and eastward to 'Eua, where the present seepages are found.

With an aim to quantify the potential for oil generation, the volume of off-reef shale facies below 5000 ft, in the late Eocene to early Oligocene 'Eua Channel basin was determined, and the ultimate generation and accumulation capacity was calculated assuming average conditions of organic carbon contents, maturity level and availability to source local accumulations. From these calculations it follows that there may be adequate oil to charge the basin. In fact, the mature source in the 'Eua Channel basin may have produced some 4–5 billion barrels of oil. Also, there is a considerable number of potential reservoir targets available, which are from 1000 to 1500 m deep and at a water depth of from 137 to 350 m [W. M. Barney and H. R. Warters, personal communication; Samuel Gary Oil Producer (1981)].

From all these facts and considerations, it appears that the prospects of the 'Eua Channel are very attractive indeed. Most likely prospects are more promising here than anywhere else in Tonga.

Further south, the Tonga platform exhibits similar structural and stratigraphic conditions (Scholl *et al.*, 1982; Scholl and Vallier, in press) and may have some potential to possibly about 23°S. Also here the total sediment thickness has been found locally to be 4–5 km and more, while both structural and stratigraphic trapping conditions (fault blocks, possible reef mounds, unconformities) have been identified. However, the water depth is considerably greater on the southern platform, i.e., around 400–600 m in most places.

3.3 The Continental Rises

3.3.1 New Caledonia–Norfolk Rise

On the big island of New Caledonia, surface indications of petroleum were already found towards the end of the last century (Paris, 1981). Near Koumac in the north, oil seepages in the serpentinized base of the overthrust peridotite massif of Tiando were confirmed in a number of shallow holes and trenches dug around the turn of the century. Nine shallow wells were drilled to a total of 1347 m in 1914–1916 and 1953–1954, but without much success. Other seepages were discovered in 1908 in Eocene limestones near Bouloupari, and in Paleocene rocks of Ouen Toro in Noumea; near the latter locality a reconnaissance well was drilled to 423 m.

Increased knowledge of the stratigraphy and tectonics in New Caledonia which was gained towards the middle of this century, including studies that more specifically were aimed at a reconnaissance of the petroleum potential (Pomeyrol, 1951), led to the implantation of two wells on the Guaro anticline near Bourail by the "Société de recherches et d'exploitation pétrolière en Nouvelle-Calédonie" (SREPNC). These wells, Guaro No. 1 and Guaro No. 2, were drilled in 1954–1955 to a total depth of 608 and 441 m, respectively. They both penetrated a rather monotonous, calcareous flysch sequence of middle to late Eocene age, which is characteristic of the central part of the Bourail basin. A number of gas and oil shows were obtained in both wells, and a production test was run in Guaro No. 2 over the interval 153–210 m. However, as results were disappointing in general, no further wells were drilled. And although additional geological and geophysical work was done in the following years—with geophysical surveys both onland and offshore—particularly in the Bourail and Noumea basins (Tissot and Noesmoen, 1958), the size of structures and general character of the sedimentary series were considered not very attractive. Since 1957, no further petroleum exploration has been carried out in New Caledonia.

Prospects in New Caledonia, if there are any, would probably be restricted to formations following the neo-cimmerian (post-Jurassic) orogeny. Late Senonian, carbonaceous shales with coal measures and associated detrital sediments, of environments that range from lagoonal-estuarine to shallow-marine, would appear as particularly interesting. The Eocene flysch basins, with thick, deep-marine turbidites and nearshore carbonate deposition along their margins, may contain a number of prospective targets. The main problem rests obviously in the finding of suitable reservoir rocks, combined with a structural setting that would allow for sufficient accumulation. While the main area for prospects seems to be on the west coast, they may be rare or non-existent on land but might be found offshore. The southern margin of the Bourail basin towards the paleo-high of the Baie de Saint Vincent is seen as one such possibility. Also, the prolongation of the Noumea basin to the SE, across the great lagoon to the south of New Caledonia, is another, extensive area which is completely unexplored but which may contain prospects. In addition, little if anything is known of Miocene sediments and their basin formation offshore. The limited indications obtained in their marginal setting where encroaching onto the present coast (e.g., at Népoui), together with the regional picture of tectonic evolution in Oligocene to Miocene times, suggest a situation that could include a considerable, Miocene potential in the offshore areas.

Further south along the Norfolk Rise, Cretaceous to Tertiary sediments are up to 3000 m thick (Dupont et al., 1975). They are generally draped over the upfaulted core of the rise, which probably consists of older, Mesozoic to Paleozoic formations. Little is known of their detailed structure or lithology, but it seems that some hydrocarbon potential cannot be ruled out. Water depth would generally be about 800–1500 m.

3.3.2 Lord Howe Rise

This major submarine feature, once part of Australia but separated from it by the opening of the Tasman Sea, has been divided in broad terms into an eastern province characterized by relatively elevated and eroded planar basement, with generally a thin sediment cover, and a western province dominated by horsts and grabens (Willcox et al., 1980, 1981). However, in the northern part where the rise is subdivided into the Lord Howe Rise proper and the Fairway Basin and Fairway Ridge (Ravenne et al., 1977b) (Fig. 1.6), sediment thicknesses of up to 3000 m have been reported (Dubois et al., 1974), and similar thicknesses may exist in local troughs of the central part. Lithologies are probably mainly pelagic, calcareous oozes, as have been drilled through most if not all the section in DSDP holes 206, 207, and 208 (Fig. 2.9). Only at site 207 on the southern Lord Howe Rise, a sandy and silt–claystone interval was found at the base of the sequence, of Maastrichtian age, which probably was deposited in a shallow-marine environment with restricted (non-oceanic) circulation (Burns et al., 1973).

In the grabens along the western part of the Lord Howe Rise, sediments up to 4000 m thick accumulated. The pre-Maastrichtian sequence may be 2000 m and more. Since the Maastrichtian unconformity has been correlated with the breakup and opening of the Tasman Sea (Willcox et al., 1980), older sediments infilling the grabens may be mainly continental, fluvio–lacustrine to deltaic and shallow-marine. Sediments of similar thickness also occur in the Lord Howe and Middleton basins between the Lord Howe Rise and the Dampier Ridge further west, and may at least partly be of similar lithology. A sedimentary wedge about 2000 m thick was deposited across the eastern margin of the Lord Howe Rise, during or before the Late Cretaceous. These sediments are mainly clastic and probably derived from the planed basement of the rise plateau to the west. They are buried below pelagic sediments about 1000 m thick on the eastern edge of the rise, and up to 3000 m in the New Caledonia Basin.

The petroleum potential of this region has been discussed in some detail by Willcox et al. (1980). According to these authors, potential source rocks may have been deposited in the pre-breakup sequence and, in places, in the shallow-marine sequence deposited immediately after breakup, as was found at the base of DSDP hole 207. The mainly continental to deltaic sediments of the western grabens may have contained a high proportion of terrestrial (humic) kerogen, which is generally gas-prone but which may also, under reducing conditions, give rise to paraffinic oils and wet gas. In addition, shallow-marine conditions in restricted basins, as may have existed in some of the grabens, could have led to the deposition of aquatic (liptinitic) kerogen, thus providing source beds with a capacity to generate oil. With up to 3000 m of potential source beds in the western grabens, buried under as much as 1000 m of additional sediments, thermal maturity of much of this material appears virtually certain. This is more so if the thermal regime was anomalously high, as is expected and normally the case in pull-apart situations. Indeed, the only heat–flow measurement on the Lord Howe Rise gave a value of about twice normal heat-flow (Grim, 1969). Regarding the Maastrichtian, shallow-marine sequence in the central part of the Lord Howe Rise, however, it is unlikely that it would have been buried sufficiently to have generated hydrocarbons. The limited overburden of a few hundred to 1000 m could hardly have provided an adequate thermal blanket. On the other hand, the sedimentary overburden on the eastern flank of the rise is thick enough (2000 m) to have matured any source material within the Maastrichtian sequence.

Potential reservoir beds may exist in numerous sandstones within the basins and grabens, and particularly in deltaic sequences on their flanks. Seismic structures interpreted as late Cretaceous to Paleocene reefs may be additional targets. Interbedded shales and overlying, pelagic oozes probably provide an adequate seal. Structural traps could have formed against the boundary faults of the grabens, but also by minor folding and faulting observed within the graben-fill. Stratigraphic traps may occur in the prograding wedge overlying the eastern margin of the Lord Howe Rise. Petroleum migrating up-dip could be trapped against the basement surface and sealed by the overlying pelagic oozes.

In summary, the deep-water areas of both the small grabens and basins in the west, including the flanks of the Middleton and Lord Howe Basins, and the eastern slope of the Lord Howe Rise which before breakup was the ancient continental slope of Australia, appear to provide long-term petroleum prospects and thus may warrant further examination. Water depth in the prospective areas is generally about 1500–2000 m.

3.4 Summary and Conclusions

A potential for hydrocarbons in the SW Pacific is restricted to the old, continental fragments of the Lord Howe and Norfolk Rises, which are characterized by a passive-margin type, tectonic environment, and to the young orogenic belts of the island arc active margins (Fig. 3.9). No potential exists in the large oceanic, so-called marginal basins. Here the very limited thickness of sediments which are mainly monotonous, pelagic oozes; the low sedimentation rates in generally well oxygenated sea bottoms which have allowed for much bioturbation; the lack of suitable reservoir rocks and trapping mechanisms; and the great water depth of generally over 4000 m are all factors which adversely affect hydrocarbon prospects.

In the Lord Howe and New Caledonia–Norfolk Rises, which have a core of Paleozoic and early Mesozoic rocks, prospects are mainly in late Cretaceous to early Tertiary sediments. In New Caledonia they are restricted to

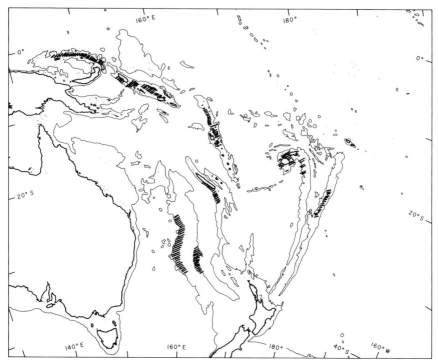

Fig. 3.9. Areas in the SW Pacific (exclusive of mainland Papua New Guinea, Australia and New Zealand) with a potential for hydrocarbon occurrences. Bathymetric contour is 2000 m. [From Katz (1984).]

the west and south where Senonian to Eocene sediments, following the neo-cimmerian (post-Jurassic) orogeny, are well developed in several big basins. Two exploratory holes were drilled on a large anticline near Bourail, in an Eocene flysch sequence where a number of gas and oil shows were obtained. However, the main potential probably lies offshore. This may include prospects in Miocene basins which formed after the late Eocene Alpine orogeny. In the Lord Howe Rise prospects are seen in the mainly continental to deltaic and shallow marine sequences in the western horst and graben province, which were deposited before the Maastrichtian continental break-up, and also in sediments immediately following this break-up. A total sediment thickness of up to 4000 m is present in these grabens. Also on the eastern side of the Lord Howe Rise, a thick sedimentary wedge was deposited during or before the Late Cretaceous, which was mainly clastic and derived from the planed basement of the rise plateau to the west. This sedimentary wedge, which represents the ancient continental slope deposits of Australia, is overlain by pelagic sediments which near the edge of the rise are 1000 m thick, increasing to 3000 m down the slope into the New Caledonia basin.

Along the active margin of the India–Pacific plate boundary, elongate basins have formed in the various island arcs. They are up to many hundred kilometres long and filled with Tertiary sediments 3–5 km thick. Because of the open-marine, volcanic archipelago setting, sediments are restricted to indigenous, mixed volcanic and organic components. Volcaniclastics occur in the form of coarse rudites to fine-grained silt-mudstone, and were deposited in large slump deposits and rubble avalanches to alternating greywacke-shale sequences and turbidites. Locally they are associated with lava flows. Organic limestones exist as carbonate platform deposits, fringing and pinnacle reefs and reef-derived breccias and calcarenites. In most areas they are subordinate, but always are extremely important as potential reservoir rocks. While no potential source rocks have been identified, environments of restricted circulation with possibly euxinic conditions may have developed in stagnant back-reef and lagoonal areas, and also in rapidly buried fore-reef environments and deep-water silled basins between rising volcanic edifices. Active volcanism may have favourably contributed to the generation and maturation of hydrocarbons, by providing for both high heat source and the conditions for anoxic environments. The presence of mature hydrocarbons, both oil and gas, has been confirmed in Tonga and Fiji.

Parts of these basins have been moderately to strongly deformed, with marginal areas being uplifted and eroded. But large areas are only little deformed and remain structurally low between the present island chains. A variety of favourable conditions for the formation of structural traps is exhibited, while stratigraphic traps or combined stratigraphic–structural traps most likely occur also, particularly in association with lenticular biohermal bodies of carbonate rocks.

The distribution and extent of prospective areas in the SW Pacific are shown in Fig. 3.9. The total area amounts to 200 000–220 000 km^2 in the island arcs from Papua New Guinea to Tonga, 30 000–40 000 km^2 in New Caledonia–Norfolk Rise and some 150 000 km^2 on the Lord Howe Rise. This is a total of 400 000 km^2. Of these prospects, very little is on land but most are offshore in water depths of a few hundred to as much as 2000–3000 m. A total of 13 wells have been drilled in the island arcs (Bougainville–Papua New Guinea, Fiji and Tonga), of which only four were offshore; four did not reach their target. Their combined drilling depth is 23 961 m; the deepest wells are in Fiji (Great Sea Reef No. 1, 2839 m) and in Tonga (Kumimonu No. 3, 2636 m). Only two wells have been drilled, for a proper target, outside the island arcs: Guaro Nos. 1 and 2 in New Caledonia, to a combined depth of 1049 m. The total drilling depth in the SW Pacific thus is 25 km.

From the above it is obvious that exploration for hydrocarbons has only just started. The potential is there, however, and there is every reason to intensify active search for these resources. With regard to their legal situation, virtually all of the prospective areas shown are in territories or 200 miles exclusive economic zones (EEZ) of one of the following countries: Papua New Guinea, Solomon Islands Republic, Vanuatu, Fiji, Kingdom of Tonga, French Territory of New Caledonia, and Australia.

Acknowledgements

This chapter, as well as the previous one on stratigraphy, could not have been written without the very active assistance from the respective government institutions in the various island countries, as well as the ORSTOM Centre and BRGM in Noumea, New Caledonia. The writer is indebted to all those who have supplied him with information and cooperated in various other ways. In particular, H. R. Warters and W. M. Barney in Denver, Colorado, provided proprietary data from recent company exploration in Tonga, without which any discussion on the Tongan oil potential would be far from complete. Their permission to freely incorporate these data is gratefully acknowledged.

4

Regional Geochemistry of Sediments from the SW Pacific

D. S. CRONAN

Applied Geochemistry Research Group, Department of Geology, Imperial College of Science and Technology, London, UK

4.1 Introduction

The literature on the geochemistry of sediments from the SW Pacific is rather limited. A number of investigations on a few samples from small individual areas have been published [see Glasby *et al.* (1979) for a review] but, until recently no general regional geochemical study on sediments encompassing substantial parts of the region had been attempted. Recognising this fact, the first Workshop on the Marine Geology and Mineral Resources of the SW Pacific, organized by CCOP/SOPAC in Suva in 1975, recommended that a compilation and evaluation of all geochemical data on sediments in the region be attempted. The present work attempts to fulfil that recommendation and both takes account of published data (Nayudu, 1971; Glasby *et al.*, 1979; Meylan *et al.*, 1982, Cronan *et al.*, 1984) and, to fill the many gaps in the sample coverage of the area, includes data on over 300 sediment samples analysed specifically for this project. A preliminary report on some of this last group of data has been published by Cronan and Thompson (1978), and that data set is considered much more fully here.

Nayudu (1971) concentrated on samples from the Antarctic Pacific, between about 120° and 180°W, with the bulk of his samples falling in the eastern part of this region. Glasby *et al.* (1979) analysed samples from the central and southern SW Pacific, together with additional samples from the vicinity of New Zealand. In the present work, additional samples from most

Sedimentation and Mineral Deposits
in the Southwestern Pacific Ocean
ISBN 0-12-195870-1

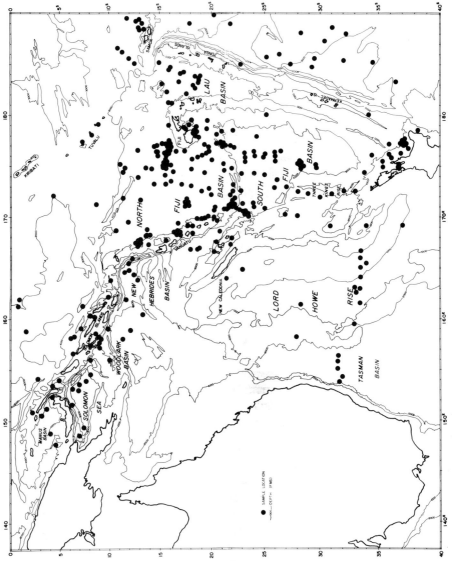

Fig. 4.1. Locations of newly analysed samples considered in this work.

of the SW Pacific collected by oceanographic and hydrographic vessels belonging to institutions in the U.S.A. and Europe, together with samples collected on CCOP/SOPAC cruises, have been analysed for Mn, Fe, Al, Ca, Ni, Cu, Pb and Zn in order to maximize the regional coverage of the area. These different data sets have been combined for the purposes of the present review, although that of Nayudu (1971) falls largely outside the area of present concern, and that of Glasby *et al.* (1979) does not include data on Ni, Cu, Pb and Zn.

The region to be considered is that bounded approximately by Australia in the west, the 160°W meridian in the east, the equator in the north and the latitude of New Zealand in the south (Fig. 4.1). Within this region is a diverse range of submarine geological settings. Included are major island arc systems with their associated trenches and marginal basins, passive basins adjacent to continental land masses, continental shelf areas, and the abyssal sea floor. Sediments from these different settings exhibit a considerable diversity in both source and nature, which profoundly affects their chemical composition.

The aims of the present study are to consider the geochemistry of sediments from the SW Pacific in an attempt to determine the regional chemical variability of the deposits and the factors determining this variability. Implications for mineral exploration are also considered.

4.2   Distribution of Samples and their Bathymetric Setting

Sample locations are plotted on Fig. 4.1. It can be seen that samples are distributed irregularly throughout the region, the major sampling being concentrated along the island arc trench system, and in the North and South Fiji Basins. In addition, a number of samples from the vicinity of New Zealand have been included. Outside of these areas, the sample distribution is patchy.

In the NW of the region, to the north of the Solomon Islands and outside the trench system, lies part of the NW Pacific Plate. Here the seafloor shows a rather undulating character with a series of basins which reach a maximum depth of about 5100 m, and also includes the Ontong Java Plateau. Samples from this region analysed in the present work range in depth from about 2300 to 4500 m. Further to the east, also on the Pacific Plate, occurs Tuvalu (Ellice Islands) and the Samoan Islands. Twelve samples from different bathymetric settings in these areas have been analysed. To the south of the Samoan Islands and east of the Tonga–Kermadec Trench, the samples analysed came from deep-water pelagic areas of relatively uniform bathy-

E

metry, although some samples were taken from near the base of the Louisville Ridge running SE from the Osbourne Seamount (175°W, 26°S) and may contain slumped material from the Ridge. Additional data from the area east of the Tonga–Kermadec Trench have been obtained from Glasby *et al.* (1979) and have been reviewed by Meylan *et al.* (1982). Between Papua New Guinea and Vanuatu (New Hebrides) occurs the Solomon Islands chain, in and adjacent to which a considerable number of samples have been taken. At the NE end of the chain, several samples have been analysed from the Manus Basin, and additional samples have been analysed from the Solomon Sea, the Woodlark Basin and the northern part of the New Hebrides Basin, all to the south of the Solomon chain. Between and around the islands themselves, the sample distribution is rather irregular. Isolated samples collected during broad-scale investigations and groups of samples collected as part of detailed CCOP/SOPAC investigations of small areas have been analysed. Amongst the latter are a number of samples taken within the New Georgia Group, some of which were from the active Kavachi submarine volcano.

Southeast of the Solomon Islands lies Vanuatu. Samples have been analysed from throughout this region, as far south as the New Hebrides Trench. As in the Solomons, isolated samples have been collected from a number of areas, including some between Vanuatu and New Caledonia, but many have also been collected in groups both nearshore and in deep water during detailed survey operations conducted by CCOP/SOPAC. A large number of samples from the New Hebrides trench region were obtained by HMS *Waterwitch* in 1896 during hydrographic survey operations, and these have also been analysed.

East of Vanuatu is the North Fiji Basin (Fiji Plateau), which, at about 3000 m average depth, is shallower than some of the other basins of the SW Pacific. The region to the NW of Fiji is the most intensively sampled in the present work and has been the subject of a separate publication (Cronan *et al.*, 1981). Elsewhere in the North Fiji Basin, the sample distribution is less dense, but small clusters of samples have been obtained from just to the south of the Fiji Islands, and from the west of the North Fiji Basin.

The South Fiji Basin has not been sampled as intensively as the North Fiji Basin. This basin is a large wedge-shaped feature which varies in depth from around 4000 m near its perimeter to about 5000 m near its centre. The samples analysed for the purposes of the present work are mostly located around a north–south trending central line, while additional data from the northwest of the basin has been obtained from Glasby *et al.* (1979).

To the west of the South Fiji Basin lie a series of basins and rises which have been only sparsely sampled for sediments. Samples from the Three Kings Rise and the Lord Howe Rise have been analysed by Glasby *et al.*

(1979) and the data are used here. Very few analyses of samples from this region have been carried out for the purposes of the present work.

To the east of the North and South Fiji Basins lie the Lau Basin and the Havre Trough. Prior to 1981, only a few samples had been collected in this region, mostly in the northern Lau Basin. These samples are described here. During May and June of 1981, a joint New Zealand Oceanographic Institute/Imperial College cruise aboard the RV *Tangaroa* obtained over a hundred additional sediment samples from this region, and preliminary data on these samples have been published by Cronan *et al.* (1982, 1984). They are not considered in detail here.

In the south of the region, data on a number of samples from the vicinity of New Zealand which were analysed by Glasby *et al.* (1979) have been included.

4.3 Analytical Methods

The method of chemical analysis of the samples analysed for the purpose of this work was atomic absorption spectrophotometry, after complete sample digestion in a mixture of hydrofluoric, nitric and perchloric acids. Eight elements, Ca, Al, Mn, Fe, Ni, Cu, Pb and Zn, were determined. The chemical data were corrected for calcium interference, and the precision of the analyses was better than ±4% at 2 SD for Ca, Al, Mn, Fe and Cu, ±5% for Zn, ±14% for Ni, and ±25% for Pb. Accuracy was checked by the analysis of several international standards. Geochemical partition analysis was additionally carried out on some of the samples, and the methods will be described in the geochemical partition section of this work.

4.4 Bulk Data Handling

In an attempt to view the overall distribution of geochemical data on SW Pacific sediments, both Cronan and Thompson (1978) and Glasby *et al.* (1979) plotted their results on histograms (Figs. 4.2 and 4.3). This approach facilitates an instant perusal of the statistical distribution of the data, and enables direct comparisons to be made with data from other sources. However, direct comparisons between the two data sets for elements in common are hampered by a number of differences between them. These included firstly that apart from Ca, Cronan and Thompson used data recalculated on a carbonate-free basis in order to illustrate variations in the non-carbonate fraction of the sediments, whereas Glasby *et al.* did not; secondly, the Glasby *et al.* data are plotted as oxides whereas Cronan and

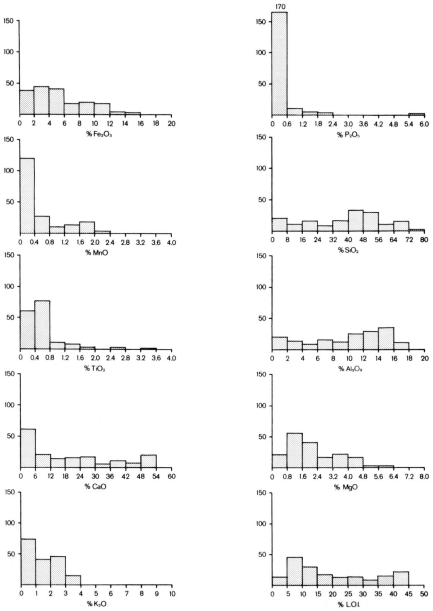

Fig. 4.2. Histograms of element oxide concentration data for SW Pacific sediments. Number of samples in each subgroup is given on ordinate. [From Glasby *et al.* (1979).]

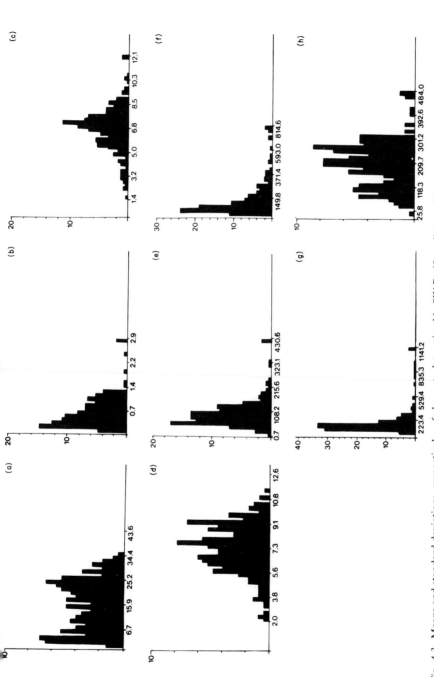

Fig. 4.3. Means and standard deviations, respectively, of elements determined in SW Pacific sediments (Ca, Mn, Fe, Al, in wt. %; remainder as ppm). (a) Ca: 15·93, 9·21; (b) Mn: 0·66, 0·75; (c) Fe: 6·73, 1·79; (d) Al: 7·33, 1·75; (e) Ni: 108·17, 107·46; (f) Pb: 149·79, 221·60; (g) Zn: 223·43, 305·93; (h) Cu: 209·70, 91·45.

Thompson plotted their data as elements; and thirdly, the horizontal intervals used on the histograms are different for the two data sets. Nevertheless, comparison of the two sets of histograms does show certain features in common, as detailed below.

The distribution of Ca is similar in both data sets, as would be expected, showing a roughly bimodal distribution with modes at either end of the distribution. Calcium distribution in marine sediments is largely a function of the distribution of calcium carbonate, which is depth-dependent. Thus, the two modes represent either samples that have been taken above the lysocline and which are rich in $CaCO_3$, and those taken below the calcium carbonate compensation depth (CCD) and which contain very little or no $CaCO_3$ (although they do contain some Ca in other mineral phases). The remaining samples between the two modes represent those taken between the lysocline and the CCD, in which variable carbonate dissolution has taken place. The slightly higher mean Ca value recorded by Cronan and Thompson than that of Glasby *et al.* is probably due to the latter containing a greater proportion of samples from below the CCD in abyssal areas to the east of the island arc system, than does the former.

The distribution of iron in the histograms illustrates the considerable range exhibited by this element within the SW Pacific. Its mean value of 6·7% found by Cronan and Thompson (Table 4.1) is somewhat higher than that found in average Pacific pelagic clays (Cronan, 1969b) and in the Glasby *et al.* samples. Manganese, too, exhibits a considerable range in both data sets. Cronan and Thompson found it to have a mean of 0·66%, which again is higher than the Pacific pelagic clay average and the Glasby *et al.* average. The highest values of Fe and Mn encountered in the region are comparable to those found in metalliferous sediments from mid-ocean ridges (Cronan, 1980). The differences in iron and manganese between the two data sets and their elevation relative to normal pelagic clays are thought to be related to a volcanic supply of these elements to some samples in the SW Pacific, as will be detailed below.

Table 4.1. Average composition of SW Pacific sediments (recalculated carbonate-free)

References[a]	Mn (%)	Fe (%)	Al (%)	Ni (ppm)	Cu (ppm)	Zn (ppm)	Pb (ppm)
1	0·66	6·73	7·33	108	209	223	149
2	0·37	6·08	9·34				
3	0·48	5·06	8·4	211	323	165	68

[a]Reference numbers: (1) data from Cronan and Thompson (1978); (2) data from Glasby *et al.* (1979); (3) average pelagic clay; data from Turekian and Wedephol (1961) and Cronan (1969b).

Aluminium and silicon in the region are close to the average values of these elements in pelagic clays, although Al is slightly depleted, on average, in the samples described by Cronan and Thompson (1978). Like the enrichment of Fe and Mn in these samples, the slight Al depletion may be a function of a volcanic supply of metals to some of the samples diluting non-volcanic Al, since it is well known that hydrothermal precipitates on the ocean floor are generally low in Al (Bostrom and Peterson, 1966). Magnesium, titanium and phosphorus in the Glasby *et al.* data set are all close to the pelagic clay average, indicating no unusual influences on their abundances, whereas K is about two-thirds of the pelagic clay average.

The minor elements considered by Cronan and Thompson (1978) are, with the exception of Pb and Zn, less than the average values for Pacific pelagic clays (Table 4.1). However, isolated high Pb values may in some samples be a result of contamination, as will be discussed below. The depletion of Ni and Cu may be the result of there being a lower proportion of Ni- and Cu-bearing minerals such as authigenic ferromanganese oxides in the SW Pacific sediments than in normal pelagic clays, as a result of its ocean margin situation and likely enhanced rate of terrigenous input.

In an attempt to deduce inter-element relationships in their samples, Glasby *et al.* (1979) subjected their data set to correlation analysis using the product moment correlation coefficient, and subsequently to cluster and factor analysis. They found significant positive correlations between Ca–LOI, Fe–Mg, Si–Al, Al–K, Si–K, Fe–Ti, Mn–Fe, Mg–Ti, Al–Fe, Mg–Mn, Mg–Al and Al–Ti, which they considered to reflect the associations of elements with carbonate phases, aluminosilicates and authigenic components of the sediments, respectively. The well known negative correlation between Ca and most other elements was also found, demonstrating that $CaCO_3$ is acting as a diluent on the phases containing the other elements (cf. Chave and Mackenzie, 1961; Cronan, 1969a). These conclusions were, in part, confirmed by factor analysis of the data which resulted in a two-factor model which accounted for 81·4% of the variance of the data (Table 4.2). There were high positive loadings of Si, Al and K on factor 1 and of Fe, Mn, Ti and Mg on factor 2. Ca was negatively loaded on both factors, more so on factor 1 than on factor 2.

On recalculating the correlation matrix for clay-rich (<10% $CaCO_3$) and for carbonate-rich (>90% $CaCO_3$) samples, Glasby *et al.* (1979) found that different patterns of major element associations emerged. In the clay-rich sediments, groupings were found between Fe–Mn–Ti–Mg–P–LOI and K–Al, and in carbonate rich sediments between Si–Al–Fe–Mn–Ti–P and Ca–K, these being both different from each other and from the associations in the total data set. These differences were considered to reflect the importance of the nature of the population subset used in determining the

Table 4.2. Varimax rotated factor loadings for total sample population of SW Pacific sediments[a]

Variable	Factor 1	Factor 2
Fe_2O_3	0·442	0·888
MnO	0·097	0·843
TiO_2	0·388	0·728
CaO	−0·923	−0·354
K_2O	0·873	0·221
P_2O_5	0·008	0·226
SiO_2	0·991	0·045
Al_2O_3	0·888	0·395
MgO	0·399	0·795
LOI	−0·952	−0·257

[a]From Glasby *et al.* (1979).

results of correlation analysis in marine sediments [cf. Cronan (1969a) and Glasby *et al.* (1974)] and, together with other observations, led the authors to conclude that the significance of the element associations found was limited, particularly in the total data set containing both carbonates and non-carbonates.

In an attempt to avoid some of the problems encountered by Glasby *et al.* (1979) in subjecting their data to correlation analysis, the data used by Cronan and Thompson has been further analysed with all elements other than Ca recalculated on a carbonate-free basis. In this way, the separate carbonate and non-carbonate populations are eliminated, and it is possible to more easily examine inter-element relationships in the non-carbonate fraction of the sediments. A varimax rotated factor loading matrix based on 261 samples is given in Table 4.3.

The factor analysis of the Cronan and Thompson data set (Table 4.3) demonstrates that five factors can account for 89·4% of the variance of the data set. Factor 1 is heavily loaded with $CaCO_3$ and can be considered to be a biogenic carbonate factor. Aluminium has a highly negative loading on this factor, which would be expected if it truly represents a biogenic factor. Factor 2 has a high loading for Fe, Mn and Cu, but a weak loading for Ni. This would tend to suggest that it is not an authigenic ferromanganese oxide factor, because Ni is enriched in such marine ferromanganese oxides. More likely it is a hydrothermal ferromanganese oxide factor, because hydrothermal ferromanganese oxides are generally depleted in Ni (Cronan, 1980) but can sometimes be enriched in Cu (Bignell *et al.*, 1976). Factor 3 has a high loading for Pb, and nothing else. This would tend to support the earlier conclusion that Pb may be acting as a contaminant in some of these

Table 4.3. Varimax rotated factor loadings for 261 sediment samples from the SW Pacific[a]

Variable	Factor 1	Factor 2	Factor 3	Factor 4	Factor 5
CaCO$_3$	0·96	0·06	0·10	0·07	−0·002
Mn	0·15	0·66	0·11	0·02	0·57
Fe	−0·20	0·90	0·07	−0·05	0·02
Ni	0·06	0·22	0·009	0·01	0·88
Al	−0·77	0·05	−0·01	−0·17	−0·41
Cu	0·31	0·75	0·12	0·02	0·31
Pb	0·14	0·15	0·97	0·01	0·03
Zn	0·16	−0·01	0·01	0·98	0·02

[a]From Cronan (1983).

sediments. A few of the samples analysed were collected with lead sounding tubes by Royal Navy hydrographic vessels towards the end of the last century, and their mode of collection has probably contaminated them with Pb. The fact that there are no other loadings on this factor suggests that no other contaminant elements have been introduced into the samples by their method of collection. Factor 4 is heavily loaded with Zn and nothing else, but whether it too has a contaminant component is not known. Interestingly, a factor analysis by Hodkinson (1985) on recently collected sediments from the SW Pacific suffering no obvious contamination has revealed that Pb and Zn act in the same manner in factor analysis as in the Cronan and Thompson data set. Evidently, therefore, possible contamination is not the only cause leading to the individualistic behaviour of these elements in the factor analysis of the Cronan and Thompson data set. Factor 5 is loaded with Mn and Ni, and negatively loaded with Al. This is probably the authigenic manganese oxide factor, since authigenic manganese oxides are characteristically rich in Ni and Cu but poor in Fe and contain negligible Al (cf. Friedrich, 1976). This factor analysis has thus demonstrated element associations with carbonate, aluminosilicate, authigenic manganese oxides and hydrothermal ferromanganese oxides. However, not only has it proved useful in elucidating the natural associations of the elements determined in the samples, but it has shown itself to be a tool for detecting possible contamination in the samples.

The computation of factor scores further clarifies the behaviour of elements in the sediments analysed. Factor scores express the contribution made to each sample by each of the factors. Most of the samples with high scores on factor 1, the supposed biogenic factor, are very enriched in calcium carbonate, containing over 80% CaCO$_3$ in some instances. Most of the samples with high scores on factor 2, the supposed hydrothermal factor, are enriched in Fe, Mn and some exhibit enrichment of Cu, and were collected in the northern part of the North Fiji Basin, in an area where submarine volcanic activity could be expected on the basis of tectonic considerations (Cronan, 1983). However, some of the North Fiji Basin samples with high scores on factor 2 also have high scores on factor 5, the supposed authigenic factor, possibly reflecting the transition from hydrothermal to authigenic ferromanganese oxides in the marine ferromanganese continuum (Cronan, 1980). All of the samples with high factor scores on factor 3 where Pb contamination is possible are old Royal Navy hydrographic samples from the British Museum collections and were collected by lead sounding tube. All are heavily enriched in lead. This would support the extraneous input of lead to the British Museum samples. All the samples with high scores on factor 4 are very enriched in Zn, but the phase in which the Zn occurs has not been identified. Most of the samples with high scores on factor 5, the supposed authigenic manganese oxide factor, are manganese enriched, or iron and manganese enriched, with considerable Ni enrichment also in some cases. With the exception of a few samples just mentioned from the North Fiji Basin which generally show Fe and Mn enrichment, all but two of these samples were taken from areas of deep water pelagic sedimentation in the South Fiji Basin or on the Pacific Plate. The two remaining samples with high scores on factor 5 came from an abyssal plain to the west of the New Hebrides.

4.5   Geochemical Partition Analysis

In a further attempt to characterize the element and element phase associations in SW Pacific sediments, a selected number of the samples analysed for the present work have been further subjected to geochemical partition analysis using the method of Chester and Hughes (1967) as modified by Cronan (1976).

The aim of geochemical partition analysis is to selectively remove by chemical techniques certain phases of the sediment for chemical analysis. In the method used here, the samples were first leached with acetic acid to take up calcium carbonate and adsorbed ions, secondly with a mixture of acetic acid and hydroxylamine HCl to dissolve amorphous iron minerals, manganese oxides and ferromanganese oxides, and thirdly with hot HCl to dissolve remaining iron oxides and to partially dissolve silicate material. Each attack was done on a fresh sample, so the proportion of the elements soluble in the different reagents was obtained by subtraction. The composition of the HCl-insoluble residue was obtained by subtracting the analysis of the HCl-soluble fraction from the bulk analysis of the whole sample.

The principal value of partition analysis in the context of the present investigation is that it can provide direct evidence over and above that given by factor analysis on the sources of metals to the sediments. For example, elements introduced into the sediments in biogenic carbonate phases will be soluble in acetic acid, those in authigenic ferromanganese oxides are soluble in hydroxylamine HCl, and those in hydrothermal iron oxides are soluble in HCl. Thus by determining the phase associations of various elements, it is possible to determine their paths of supply to the sediments. In particular, where individual samples are strongly enriched in one or more elements, partition analysis can help to determine the source of these enrichments.

4.6 Element Distributions

In this section, the regional variability of Mn, Fe, Al, Ni, Cu, Pb and Zn in the SW Pacific sediments is considered. For descriptive convenience the element concentrations have been divided into high, moderately high, moderate, moderately low and low. Numerical limits for these ranges are given in Table 4.4. All elements are described on a carbonate-free basis.

Table 4.4. Numerical limits of descriptive elements concentration ranges[a]

Description	Mn (%)	Fe (%)	Ni (ppm)	Cu (ppm)	Pb (ppm)	Zn (ppm)	Al (%)	CaCO$_3$ (%)
	0	0·81	6	22	1	38	0·45	0
Low	0·397	3·53	46	126	55	145	5·21	20
Moderately low	0·733	6·24	88	178	111	251	6·40	40
Moderate	1·06	8·96	130	230	222	357	7·59	60
Moderately high	1·40	11·7	171	281	389	569	8·79	80
High	8·44	17·1	1040	671	1390	2690	12·2	100

[a]From Cronan (1983).

4.6.1 Manganese

The distribution of "excess" manganese in the region, over that typical of average shale, should largely reflect the distribution of authigenic ferromanganese oxide minerals in the sediments, both precipitated from normal seawater or from seawater which has received hydrothermal contributions. The factor analysis demonstrates that manganese is loaded on two main factors, the supposed hydrothermal and authigenic factors. Those samples with high scores on these factors are manganese enriched. However, in the bulk of the samples in which manganese is low to moderate in concentration, it should be more equally distributed amongst a variety of phases, both detrital and authigenic.

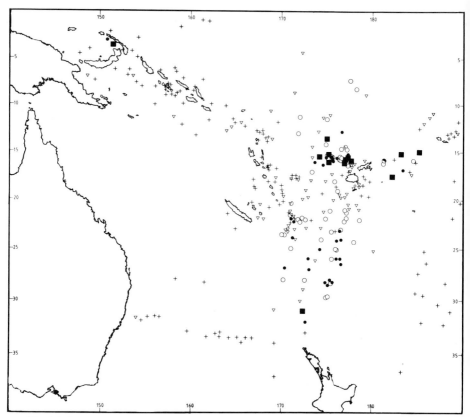

Fig. 4.4. Percent distribution of manganese in SW Pacific sediments. (+) 0–0·397; (▽) 0·397–0·733; (○) 0·733–1·06; (●) 1·06–1·40; (■) 1·40–8·44.

Manganese varies in the sediments throughout the region from less than 0·1% to over 8% (Fig. 4.4) and up to about 50% in a hydrothermal manganese crust from the Kermadec Ridge (Cronan *et al.*, 1982). Manganese values on the Pacific Plate are generally low to moderate, varying from around 0·1% up to between 0·75 and 1·0%. Values of 0·8 and 0·9% are to be found off the Ellice Islands (Tuvalu), whilst both to the north and west, and towards Samoa, manganese values drop off. On the Pacific Plate south of Samoa, Mn values are low in the Cronan and Thompson (1978) data set, but range up to near 1·4% in the samples analysed by Meylan *et al.* (1982) from the SW Pacific and Samoan Basins. These authors found Mn to increase away from the Tonga–Kermadec Ridge and Samoan areas in an easterly direction to reach maximum values south of Rarotonga in the Cook Islands and in the eastern part of the Samoan Basin. The factor scores of the Mn-enriched samples from this area in the Cronan and Thompson data set show moderate values on the authigenic factor, suggesting that the excess manganese is present in the form of authigenic ferromanganese oxides. The factor scores of the remaining samples from the Pacific Plate exhibit no marked enhancement on any particular factor.

In the region between Papua New Guinea and Vanuatu (New Hebrides), with the exception of two samples to the south of New Ireland, the manganese values in the sediments are low. These samples were taken from a wide variety of stations, both nearshore and in deep water, and include samples from close to the active Kavachi submarine volcano south of the New Georgia group in the Solomon Islands. The two samples with high manganese values, 1·05 and 2·7% Mn, respectively, were taken from near to a proposed plate boundary (Taylor, 1979) to the south of New Ireland, and may reflect hydrothermal phenomena related to volcanic activity associated with plate movement [see Cronan (1983)].

Samples from the Vanuatu chain are, for most of the group, similarly low in manganese to the Solomon Islands samples. Like the Solomons samples, they include sediments from both deep water and near-shore stations, including some close to known volcanic centres or from near-shore areas such as Port Patteson where volcanic solutions discharge into the sea via rivers (Cronan, 1983). Towards the south of the Vanuatu chain, manganese values increase to a maximum of just over 1% in the vicinity of the New Hebrides trench and the northern margin of the South Fiji Basin.

In the North Fiji Basin (Fiji Plateau) between Vanuatu and the Fiji Islands, manganese values are highly variable, ranging from less than 0·3% up to over 8%. The low values tend to be concentrated in the western half of the basin (west of 172°E) and the high values in the east. The highest values of all, between 1 and 8% Mn, are to be found in an east–west band centred around 15°S at between 173° and 177°E, to the NW of Fiji. The factor scores for the manganese-enriched samples are high on the hydrothermal factor, and, in some cases, on the authigenic factor too. This suggests both hydrothermal and authigenic supply of manganese to these sediments. The former conclusion would not be at variance with the tectonic setting of the samples which is in the vicinity of a supposed spreading centre (L. W. Kroenke, personal communications, 1981).

Many of the samples from the North Fiji Basin were included in the group of samples which were subjected to geochemical partition analysis. In the sample richest in manganese, containing 8·4% Mn, approximately 70% was hydroxylamine HCl-soluble (Table 4.5), indicating its presence predominantly in the form of ferromanganese oxide phases. In the other samples rich in managanese, containing from about 1·4 to 5% Mn, approximately 30–60% of the Mn was hydroxylamine HCl-soluble (Table

Table 4.5. Average partition of Mn in sediments from the North and South Fiji Basins (in % of total present)

	North Fiji Basin			
Partition	Station 156 (8·4% Mn)	>1·4%[a] Mn	<1·4% Mn	South Fiji Basin
Acetic acid-soluble	0·1	2·2	2·9	2·2
Hydroxylamine HCl-soluble	71	40	29	59
HCl-soluble	Trace	5·0	7·0	10·0
HCl-insoluble	29	52	60	28
Number of samples	1	5	23	15

[a]Excluding Station 156.

4.5). This is less than the 70% hydroxylamine HCl-soluble manganese often found in Mn-rich sediments from deep sea areas. This fact, coupled with the fact that little of the hydroxylamine HCl-insoluble Mn was soluble in HCl (Table 4.5), suggests that in these sediments there are substantial inputs of Mn in HCl-resistant detrital minerals. This conclusion would not conflict with the area's near continental margin situation and its being volcanically active, and thus being able to receive both terrigenous and volcanic detritus. An even small proportion of the total Mn is hydroxylamine HCl-soluble in many of the North Fiji Basin samples containing less than 1·4% Mn which, coupled with their containing 60% HCl-insoluble material (Table 4.5), suggests a substantial detrital or volcaniclastic input to them. Evidently, as Mn becomes enriched in the sediments, the excess is housed in ferromanganese oxides. The "background" Mn is largely non-authigenic.

In the South Fiji Basin, the sediments are more uniformly enriched in manganese than elsewhere in the SW Pacific region investigated, but do not exhibit the extreme enrichments found in the North Fiji Basin. Most values range between 0·7 and 1·4% Mn. Some of the more manganese-enriched samples have high scores on the supposed authigenic factor, suggesting that much of the manganese is authigenic in origin. This is confirmed by the partition analysis performed on South Fiji Basin samples, which indicates that a higher proportion of Mn is hydroxylamine HCl-soluble in these samples than in those from the North Fiji Basin (Table 4.5). The remaining Mn in these samples is present in minor amounts in the acetic acid- and HCl-soluble fractions of the sediments. Less than 30% is present in HCl-insoluble detrital minerals, a much lower proportion than in the North Fiji Basin, which would be in keeping with the greater distance of the South Fiji Basin from active volcanic areas, both terrestrial and submarine.

Prior to the joint NZOI/Imperial College work in the Lau Basin (Cronan *et al.*, 1984), relatively few samples had been investigated from this area.

However, some of those that have show considerable Mn enrichment. The additional samples analysed by Cronan *et al.* (1984) confirm these enrichments. Overall, Mn values in the Lau Basin, Havre Trough, Tonga–Kermadec Ridge region are low, averaging much less than 1% in most areas, but 0·94% in the Lau Basin. However, in the Central Lau Basin, in the vicinity of a supposed spreading centre, maximum Mn values reach 2·8%. Accumulation rates of Mn in these sediments are comparable with those in some mid-ocean ridge settings, suggesting a hydrothermal origin for the Mn enrichments.

The remaining samples from the south of the region are all low in manganese. In the case of those from off New Zealand, this would be consistent with their continental margin situation, where continentally derived detritus is abundant (Glasby *et al.*, 1979). The samples taken from the Tasman Sea contain somewhat higher values of manganese than those from the New Zealand continental margin, probably reflecting an increased proportion of authigenic ferromanganese oxides consistent with their more open ocean setting.

4.6.2 Iron

On the basis of the factor analysis, the distribution of excess iron in the region over and above that typical of average shale should largely reflect the distribution of hydrothermally precipitated iron minerals in the sediments. Many of the iron-rich sediments have high factor scores on the supposed hydrothermal factor. The partition analysis demonstrates that in the very iron rich sediment, much of the Fe is not soluble in hydroxylamine HCl but is HCl-soluble, indicating that it is not associated with ferromanganese oxides but is present in other phases such as possibly Fe silicates and oxides. However, the partition analysis also shows that much of the Fe in these sediments is also HCl-insoluble, and where iron is low overall this becomes the major proportion of the total iron. This "background" iron, like "background" Mn, must be present in detrital minerals, and these must therefore supply the majority of the Fe in the bulk of the sediments.

Iron varies considerably in concentration on the Pacific Plate (Fig. 4.5). It is lowest to the north of the Solomon Islands, and to the SE of the Kermadec Islands. Here, Fe values range up to around 3%. Highest iron values on the Pacific Plate occur to the NE of Samoa, where most samples obtained have Fe contents between 9 and 11%. This may reflect the supply of Fe-bearing volcanic detritus from the Samoa Island group. Between these samples and those low in Fe, occur sediments with moderate Fe values to the west of the Tonga–Kermadec Ridge and in the vicinity of Tuvalu.

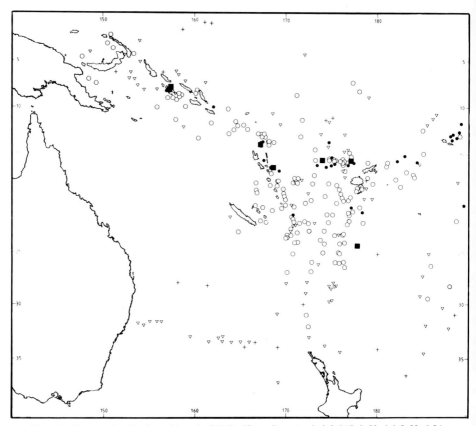

Fig. 4.5. Percent distribution of iron in SW Pacific sediments. (+) 0·815–3·53; (▽) 3·53–6·24; (○) 6·24–8·96; (●) 8·96–11·7; (■) 11·7–17·1.

The data on Fe presented by Meylan *et al.* (1982) for the Pacific Plate are similar to those described here. Highest values were found in the Samoan Basin and near Rarotonga, thought to be due to the presence of Fe-bearing volcanic minerals.

Iron tends to be moderate to low over most of the region between Papua New Guinea and Vanuatu, with the exception of a small area in the Solomon Islands. Moderate values, between 6 and 9%, occur in most samples from the vicinity of Papua New Guinea, including those which are enriched in manganese. Any hydrothermal contributions to these sediments are therefore not reflected in their iron content. In the NW end of the Solomon chain, most of the iron values in the sediments are low. However, moderate to very high values occur locally in the New Georgia Group exhibiting a maximum of 13·2% Fe. These enrichments are not associated with any known volcanic feature. Samples from the south of the New Georgia Group, including those associated with the Kavachi submarine volcano, are all moderate in their Fe content, although the volcanic-associated samples are up to 1% higher in Fe than the others. This may reflect a higher proportion of Fe-bearing volcanic detritus in these sediments. Sediments from Vanuatu are for the most part moderate in their Fe content, with most values falling between about 6 and 9% Fe. However, in two localities, Fe values considerably in excess of the regional average occur. The first of these is off Port Patteson where iron sands supplied by rivers draining volcanic areas occur on the beaches and in the offshore area. Fe values here are up to 22%. The second is off the island of Epi where an active submarine volcano occurs. Here the Fe enrichment is extreme, 27%, and is due to the presence of iron bearing hydrothermal deposits (Exon and Cronan, 1983). To the south of Vanuatu, with one exception Fe values are moderate to low.

In the North Fiji Basin, iron values vary from about 4% up to about 15%. They are generally moderate in the western part of the basin and higher in the east, with the highest value of all occurring in the area to the NW of Fiji where manganese is enriched. The factor scores for the very iron-rich samples in this area show extreme, though rather variable, values on the supposed hydrothermal and authigenic factors, suggesting both a hydrothermal and authigenic supply of Fe to the sediments there. Partition analysis confirms this conclusion (Table 4.6). Sample 156 containing 15·6% Fe contains about half of it in the hydroxylamine HCl-soluble fraction, indicating its presence in ferromanganese oxides or amorphous iron oxides, and a further 27% in the HCl-soluble fraction, indicating its probable

Table 4.6. Average partition of Fe in sediments from the North and South Fiji Basins (in % of total present)

	North Fiji Basin				South Fiji Basin	
	156[a]	>11·7%	>8·96	<8·96	33[b]	Average
Acetic acid-soluble	1·4	n.d.[c]	3·1	2·0	2·1	1·55
Hydroxylamine HCl-soluble	43	24	17·4	7·8	13·2	16·5
HCl-soluble	27	17	14·1	21·1	49·2	10·6
HCl-insoluble	28	59	65	69·1	35·5	71·3
Total	15·6	12·4	10·2	6·69	13·0	6·3
Number of samples	1	1	2	26	1	14

[a]Sample 156.
[b]Sample 33.
[c]Not determined.

presence in Fe silicates and oxides. As the total amount of Fe in the samples decreases, there is proportionally a greater decrease in the percentage of it which is hydroxylamine HCl-soluble (Table 4.6), indicating a decreasing occurrence of it in the form of authigenic ferromanganese oxides and amorphous iron oxide minerals with decreasing total iron. By contrast, the proportion of iron which is HCl-soluble varies widely from sample to sample but shows no consistent trends with decreasing iron content, indicating a rather variable contribution of iron silicate and/or iron oxides to the samples. In contrast again, the HCl-insoluble iron increases in proportion as the total iron decreases, indicating an increase in the proportion of resistant iron-bearing detrital minerals in the sediments as their total iron content falls off. In fact, the actual amount of HCl-insoluble iron in the sediments varies from only 2·6 to 6·7% with most of the values falling between 4 and 5%. Much of this iron is probably occurring in terrigenous and volcaniclastic material, and represents a relatively constant detrital input upon which the other inputs are superimposed.

Samples from the South Fiji Basin tend to have moderate Fe contents of between 6 and 9% in the northern part of the basin, decreasing towards the south. There is one high value of 13% Fe in the east in sample 33. Partition analysis of this sample shows it to contain the single largest proportion of total iron (49%) in the HCl-soluble fraction (Table 4.6). However, the bulk of the iron in the remainder of the South Fiji Basin samples is in the HCl-insoluble residue where it is probably present in resistant detrital minerals. In addition, approximately 16% of the iron in South Fiji Basin sediments is present in the hydroxylamine HCl-soluble fraction (Table 4.6), more than double the average for the majority of the North Fiji Basin samples (Table 4.6), where it is probably associated with the abundant authigenic Mn in these samples in ferromanganese oxide minerals.

Samples from the Lau Basin in the Cronan and Thompson data set tend to have moderate to moderately high iron concentrations, but not more than about 11% Fe. The data of Cronan *et al.* (1984) confirm the high iron values in the Lau Basin, where the average is 8·8% Fe. By contrast, lower Fe values occur in the Havre Trough (2·9%), and on the Tonga–Kermadec Ridge (6·15%). As in the case of the excess Mn, the excess Fe in Lau Basin sediments is thought to be hydrothermal in origin; Fe accumulation rates are close to some mid-ocean ridge values.

The samples from the south of the area tend to have low Fe concentrations, especially around New Zealand. This is probably a reflection of the generally low Fe content of the continentally derived detritus which comprises the bulk of these sediments. In the Tasman Sea where more favourable conditions for authigenic mineral formation can be expected, Fe values increase slightly.

4.6.3 Aluminium

The variation in the aluminium content of the samples studied in this work (Fig. 4.6) is rather irregular, probably reflecting the diverse influence on the deposits of terrigenous material from the island groups and the products of submarine volcanic activity.

On the Pacific Plate, the sediments vary from aluminium-poor deposits in the north of the region, through moderate values in the area of Tuvalu, to moderately high values in the vicinity of Samoa where the deposits contain up to 8% Al. The highest values of all on the Pacific Plate are to be found east of the Kermadec Ridge where fine-grained pelagic sediments rich in clay minerals occur. Meylan *et al.* (1982) also found their highest Al values east of the Kermadec Ridge and also due east of New Zealand.

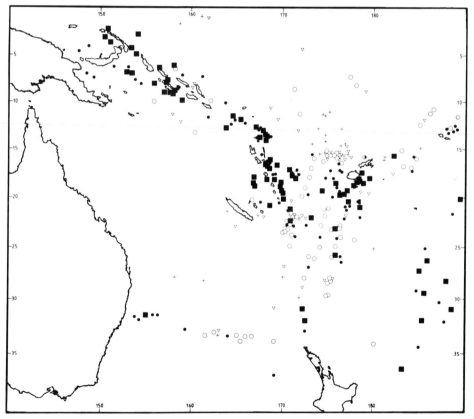

Fig. 4.6. Percent distribution of aluminium in SW Pacific sediments. (+) 0·458–5·21; (▽) 5·21–6·40; (○) 6·40–7·59; (●) 7·59–8·79; (■) 8·79–12·2.

Between Papua New Guinea and the Solomon Islands, Al varies from moderately high to high values in the vicinity of New Britain and New Ireland, increasing down the chain to the vicinity of the New Georgia Group. Further to the SE and to the south of the Solomon chain. Al values decrease to moderately low levels, increasing again to moderately high to high values at the northern end of Vanuatu.

Along the Vanuatu chain, aluminium values range from low to high, but exhibit a considerable degree of local variability probably due to the influence of the islands themselves. In the north of the group, Al values are moderately high to high with the exception of the inshore samples from Port Patteson which range in Al content from around 3% up to about 11%. Further south, most Al values are high but again there is a considerable degree of local variability, particularly in the near-shore areas. South of

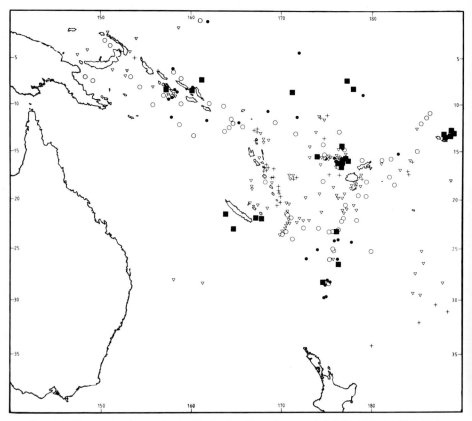

Fig. 4.7. Distribution (in ppm) of Ni in SW Pacific sediments. (+) 5·35–46·8; (▽) 46·8–88·1; (○) 88·1–130; (●) 130–171; (■) 171–1040.

Vanuatu, Al values decrease overall towards the New Hebrides Trench, but again a considerable degree of variability is evident.

Samples from the North Fiji Basin are also quite variable in their Al content, with moderately low to low values predominantly in the north of the area and moderately high to high values in the south, especially around Fiji. This is probably a reflection of the supply of terrigenous detritus from the island groups in the south. The area of intensive sampling to the NW of Fiji mostly exhibits moderate to moderately low Al values. In all the North Fiji Basin samples subjected to partition analysis, the bulk of the Al occurs in the HCl-soluble and -insoluble fractions, with the major portion generally occurring in the HCl-insoluble fraction. This confirms that the major source of Al to these sediments is detrital material.

Aluminium values in South Fiji Basin sediments tend to be lower than those just to the north. As in the case of the North Fiji Basin samples, partition analysis has shown that the bulk of the Al is held in phases such as clays and other detrital minerals.

Samples from the Lau Basin in the Cronan and Thompson data set are predominantly moderate in their Al content, with some high values in the centre of the basin where volcanic detritus occurs. Aluminium in the Lau Basin sediments analysed by Cronan et al. (1984) averages 7·4%, which is not very dissimilar to the averages recorded for the Havre Trough and Tonga–Kermadec Ridge.

Over much of the southern part of the region, Al values are again generally moderate to high, probably reflecting the input of terrigenous detritus from the land areas. However, some low values occur at relatively shallow depth on the Lord Howe Rise.

4.6.4 Nickel

On the Pacific plate, Ni values tend to be quite variable (Fig. 4.7). Low to moderately low values predominate to the east of the Tonga–Kermadec Ridge, whereas high values (around 200–300 ppm) occur near Samoa. Further east, near Rarotonga, Meylan et al. (1982) show Ni to increase to values of up to 250 ppm. Northwest towards the Ontong Java Plateau, Ni values are predominantly moderate to high with the occasional moderately low value also occurring.

Between Papua New Guinea and Vanuatu, Ni values are also quite variable, but tend to be lower in the NW of the region than in the central and SE parts. The two samples exhibiting Mn enrichment which were taken to the south of New Ireland have higher Ni values than the other samples from the same general vicinity, and the samples from the Kavachi volcano also show some enrichment in Ni relative to those surrounding. The Solomon

Sea samples tend to be moderate to moderately low in Ni, whereas those from the Woodlark Basin exhibit Ni values which are moderate to moderately high. The group of samples collected in the Kula Gulf in the New Georgia Group contain widely varying Ni contents ranging from about 50 ppm to over 250 ppm, a greater variation than occurs elsewhere in the Solomon chain.

Samples from Vanuatu are strikingly lower in their Ni content than those from the Solomon region. With one exception, all the samples analysed are low to moderately low in Ni content, with a preponderance of the low values. This might reflect compositional differences between the rocks of the two groups, since locally derived detritus is likely to supply a considerable portion of the Ni to the sediment in each region, but this has not been documented in terms of rock analyses. However, the rocks of the Solomons contain a greater proportion of ultrabasics than do those of Vanuatu, and ultrabasics tend to be Ni-enriched (Turekian and Wedephol, 1961).

A few samples were collected from the vicinity of New Caledonia, all enriched in Ni. They range from 250–760 ppm Ni. These samples do not exhibit any other metal enrichment as striking as this, and the Ni enrichment may be the result of runoff or wind-borne transport of detritus from New Caledonia, which is well known for the nickeliferous laterite that it contains.

To the south of Vanuatu, in the vicinity of the New Hebrides Trench, Ni values tend to be moderate to moderately low.

Samples from the North Fiji Basin tend to be highly variable in their Ni contents. Those from the south and the west tend to be moderate to low in Ni, as are those from the vicinity of Fiji itself, whereas those from the area NW of Fiji vary from containing low values of Ni to containing some of the highest values recorded in the SW Pacific region, in samples which are also enriched in Mn. Partition analysis has shown that, in general, more than half of the Ni in North Fiji Basin sediments is present in the HCl-insoluble fraction, with the remainder being variably distributed between the other fractions. However, in the Ni-enriched samples, the proportion of Ni present in the hydroxylamine HCl- soluble fraction increases to a maximum of almost 70% in the sample richest in Ni and Mn, indicating its presence in authigenic ferromanganese oxide phases.

In the South Fiji Basin, Ni values tend to be higher overall than in the North Fiji Basin, and reach their maximum concentrations in the central and southern regions of the basin. About one-third or more of the Ni in these samples is present in the hydroxylamine HCl-soluble fraction, indicating its presence in ferromanganese oxide minerals.

Factor scores for Ni enriched samples tend to be high on the supposed authigenic factor, further confirming its association with authigenic ferromanganese oxides.

In the Lau Basin and adjacent areas, Ni exhibits generally rather low concentrations. Its average content in Lau Basin sediments is 68 ppm, while it averages 35 ppm in the Havre Trough, and less than 30 ppm on the Tonga–Kermadec Ridge (Cronan *et al.*, 1984).

4.6.5 Copper

Copper varies quite considerably on the Pacific Plate (Fig. 4.8). The highest values, over 450 ppm, occur in the vicinity of Tuvalu and to the SW of the Gilbert Islands (over 300 ppm). Elsewhere on the Pacific Plate, Cu ranges from low values immediately east of the Tonga–Kermadec Ridge through moderately low to moderate values in the vicinity of Samoa and on the

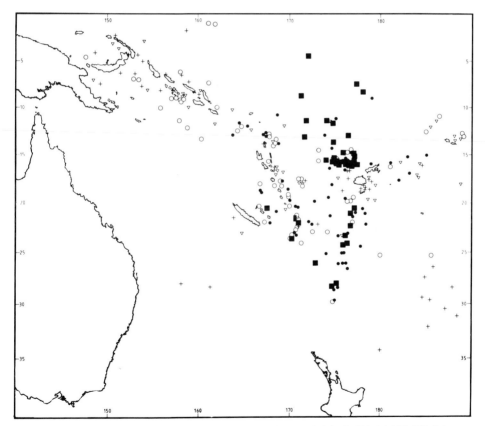

Fig. 4.8. Distribution (in ppm) of Cu in SW Pacific sediments. (+) 21·9–126; (▽) 126–178; (○) 178–230; (●) 230–281; (■) 281–671.

Ontong Java Plateau. East towards Rarotonga, Cu values increase to a maximum of over 300 ppm (Meylan *et al.*, 1982).

Between Papua New Guinea and Vanuatu, the Cu values are for the most part low to moderate. The Mn-rich samples thought to be associated with a plate boundary in the Manus Basin (Cronan, 1983) to the south of New Ireland are about double the average Cu values for the other samples in the same general area. Samples associated with the Kavachi volcano are also significantly higher in their copper content than the other Solomon Islands samples, and contain Cu contents similar to those found in the Woodlark Basin to the south. Further down the Solomon chain, towards northern Vanuatu, Cu exhibits moderately high values.

Along the Vanuatu chain, the Cu values vary considerably. They tend to be moderate to moderately high in the north, with low values off Port Patteson. Further south, Cu concentrations drop off to moderately low to low values, but increase again to the south with some high values in the vicinity of the New Hebrides Trench.

In the North Fiji Basin, the highest Cu values occur to the NW of Fiji, extending north onto the Pacific Plate. By contrast, the southern North Fiji Basin samples tend to have lower Cu values, especially on the eastern and western margins of the basin, with moderately high values near the centre where crustal extension is thought to be occurring. To the south of Fiji, most copper values are low. Many of the samples in the North Fiji Basin which are enriched in Cu have high factor scores on the supposed hydrothermal factor, and to a lesser extent on the authigenic factor, suggesting that the enriched Cu may be both hydrothermal and authigenic in origin. This suggestion is supported by the partition analysis of the Cu-enriched samples which shows that at least about half and sometimes more than three-quarters of the Cu is associated with the combined HCl and hydroxylamine HCl-soluble fractions. By contrast, in the majority of samples, somewhat less than half of the Cu is associated with the combined hydroxylamine HCl- and HCl-soluble fractions. In almost all samples other than those which are very rich in Cu, the bulk of the Cu is present in the HCl-insoluble residue, demonstrating that the principal source of Cu to the sediments in this region is Cu-bearing detrital minerals.

In the South Fiji Basin, most of the Cu values are moderate to high, although some lower values occur around its margins. Here, the high Cu samples are also high in Mn, and have high scores on the authigenic factor, suggesting the incorporation of the copper into authigenic phases, most probably ferromanganese oxides. This is supported by the partition analysis which, in the Cu enriched samples, demonstrates that about half or more of the Cu is hydroxylamine HCl-soluble. In the remaining samples, nearer one-third of the Cu is hydroxylamine HCl-soluble, with the bulk of the remainder present in the HCl-insoluble fraction where, as in the North Fiji Basin, it is largely present in detrital material.

The Lau Basin samples in the Cronan and Thompson data set contain, for the most part, moderate to moderately high Cu values. In the Cronan *et al.* (1984) data set, this is confirmed with Cu averaging over 200 ppm in the Lau Basin. In the Havre Trough it is lower, averaging 69 ppm, whereas on the Tonga–Kermadec Ridge it averages 111 ppm.

4.6.6 Lead

On the Pacific Plate, lead values vary from low to moderately low to the east of the Tonga-Kermadec Trench and in the vicinity of Samoa, to moderate and moderately high in the northern part of the region (Fig. 4.9).

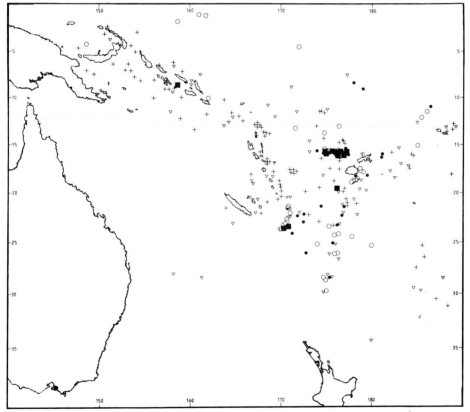

Fig. 4.9. Distribution (in ppm) of Pb in SW Pacific sediments. (+) 1–55·7; (▽) 55·7–111; (○) 111–222; (●) 222–389; (■) 389–1390.

Between Papua New Guinea and Vanuatu, most lead values are low, including those associated with volcanic activity in the New Georgia Group. One very high value of 1156 ppm Pb to the north of the New Georgia Group occurs in a sample collected with a sounding tube by HMS *Penguin* in 1894, and almost certainly results from Pb contamination from the tube.

Samples from Vanuatu are also low in lead. However, towards the south, in the vicinity of the New Hebrides Trench, lead values increase to moderate levels, and two very high values occur. These latter samples were collected with a sounding tube by HMS *Waterwitch* on a traverse of the New Hebrides Trench in 1896, and their lead values must therefore be considered suspect.

In the North Fiji Basin, samples from the south and west tend to be low in lead. One high value in this area probably results from contamination. In the area of intensive sampling to the NW of Fiji, lead values vary from low to high, and some are very high. Most of the samples were collected with lead

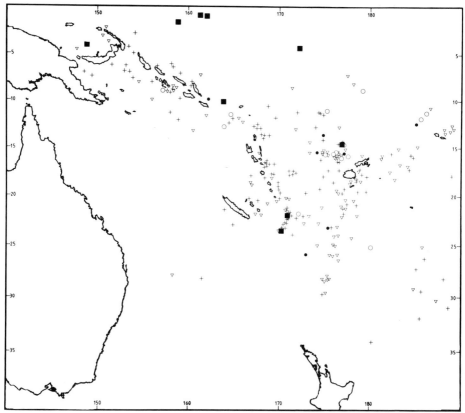

Fig. 4.10. Distribution (in ppm) of Zn in SW Pacific sediments. (+) 38·6–145; (▽) 145–251; (○) 251–357; (●) 357–569; (■) 569–2690.

sounding tubes and their lead values must also be considered suspect. Partition analysis confirms this, in that the samples with high lead content contain the bulk of their Pb in the HCl-insoluble fraction which is compatible with it being in the form of metallic lead. It is difficult to conceive what other Pb-rich HCl-insoluble phases might be present. In samples with a lower lead content, a much greater proportion is present in the other fractions of the sediments, principally the hydroxylamine HCl-soluble fraction, suggesting its association with ferromanganese oxide phases. This behaviour is opposite to the other minor metals, which tend to contain significant amounts of their totals in the hydroxylamine HCl-soluble fraction only when enriched. Interestingly, in sample 156 which is very rich in Mn and Fe, lead is present in an amount of 375 ppm, half of which is present in the hydroxylamine HCl-soluble fraction, indicating that the principal association of Pb in this sample is with ferromanganese oxides. Sample 156 was not collected with a lead sounding tube.

In the South Fiji Basin, lead values vary from low to moderately high. A few low values occur in the north, in samples not collected with lead sounding tubes, but a few moderately high values also occur in samples that were collected with sounding tubes. Towards the centre of the basin, moderate to moderately high values predominate, but most of the samples in which they occur were collected by sounding tube. Partition analysis of the South Fiji Basin samples has shown that, as in the North Fiji Basin, lead tends to be concentrated in the HCl-insoluble fraction in lead-enriched samples.

Samples from the Lau Basin tend to be moderate to low in Pb.

4.6.7 Zinc

On the Pacific Plate, to the east of the Tonga–Kermadec Ridge, Zn values are moderate to low, as they are also in the Samoa Region. To the NW, however, they increase in the region of Tuvalu, and then increase much more in the region of the Ontong–Java Plateau (Fig. 4.10).

Between Papua New Guinea and Vanuatu, most Zn values are moderately low to low, with a few higher values scattered throughout the region. There appears to be no consistent pattern in the distribution of Zn values within the area. The Mn-enriched samples from the Manus Basin are not enriched in Zn, nor are the samples associated with the Kavachi Volcano. Solomon Sea and Woodlark Basin samples differ little from samples within the Solomon chain.

Along the Vanuatu chain, the Zn values are almost uniformly moderately low to low, and are also mostly low to the south.

In the North Fiji Basin, apart from the area to the NW of Fiji, the Zn values are again moderately low to low. In the area to the NW of Fiji, where crustal extension is thought to occur, the Zn values vary from about 100 ppm up to over 2000 ppm. Partition analysis has shown that the greatest proportion of Zn in the two highest Zn samples is in the HCl-insoluble residue, indicating the presence of acid-resistant zinc-bearing phases. In the remaining samples, which are much lower in zinc, a smaller proportion is HCl-insoluble, and significant proportions are present in the other fractions of the sediment.

In the South Fiji Basin, the majority of the Zn values are also moderately low to low. Partition analysis has shown that in most cases this zinc is largely present in the HCl-insoluble residue, suggesting its presence in detrital phases.

In the Lau Basin samples analysed by Cronan and Thompson, Zn values are low to moderately low. In the Lau Basin samples analysed by Cronan *et al.* (1984), Zn averaged 165 ppm, higher than on the Tonga–Kermadec Ridge (average 113 ppm), and very much higher than in the Havre Trough (average 66 ppm). Some hydrothermal contribution of Zn in the Lau Basin is possible.

4.7   Implications for Mineral Exploration

In addition to describing the regional variability in the distribution of several elements in the SW Pacific and the controls on their concentrations, the data presented here have implications in regard to exploration for metallic minerals within the region.

The partition analyses carried out on the sediments, aided by the factor analysis, clearly demonstrate the variability of inputs to the sediments examined, and the partition analysis has been particularly useful in identifying phases in which enriched metals occur. In many cases, the "background" concentrations of the elements appear to be housed in detrital phases of one form or another, but superimposed on this are enrichments in other phases. Natural and man-made examples of this occur in the case of Mn, which is enriched in authigenic or hydrothermal manganese oxides in some samples, and in Pb, which is enriched in contaminant phases in some samples, most probably as metallic Pb. Without the partition analyses, such trends would not be so clearly discernible, nor would the contamination be so easily identifiable.

Coupled with the bulk analyses, the characterization of the nature of metal enrichments in the sediments by the partition analyses has implications in mineral exploration in that it can help to identify the source of the enrichments in individual samples. This usage in exploration for hydrothermal metalliferous sediments in the SW Pacific has been discussed in detail by Cronan (1983), and, coupled with other considerations, has led to the delineation of several areas of potential occurrence of the deposits. These include (1) the plate boundary in the Bismarck Sea, (2) the proposed spreading centres in the Woodlark and North New Hebrides Basins, (3) the submarine volcanoes off Vanuatu, (4) the proposed spreading centre in the North Fiji Basin, (5) the proposed spreading centre in the Lau Basin, and (6) the volcanically active areas of the Tonga–Kermadec Ridge.

It is evident from the data presented here that the composition of sediments from adjacent to some of the islands appears to be strongly influenced by material derived from the islands themselves. Good examples include the difference in the Ni content of sediments between the Solomon Islands and Vanuatu, and its enrichment off New Caledonia. Such differences probably reflect compositional differences between the rock types comprising the islands, and if viewed on a local scale would have implications in detecting mineralization on those islands or in their immediate offshore areas. In order to evaluate this possibility, and if successful to develop geochemical exploration methods based on it, all the sediments analysed in the present work, coupled with additional sediments from the region in CCOP/SOPAC collections, are being further analysed at Imperial College by inductively coupled plasma spectrometry for a variety of elements indicative of a wide variety of types of mineralization.

Acknowledgements

The work described in this chapter was supported by the Natural Environment Research Council. I am grateful to Richard Howarth for critical review of the manuscript.

5

Near-shore Mineral Deposits in the SW Pacific

G. P. GLASBY

New Zealand Oceanographic Institute, Department of Scientific and Industrial Research, Wellington North, New Zealand

5.1 Introduction

The Pacific Ocean is the largest feature on the earth's surface, encompassing about one-third of the area of the globe (an area greater than that of all the land above sea level) and is dotted with numerous (about 25 000) small islands (as well as a number of larger ones such as New Zealand). The essential problem of many of these islands is that they are characterized by small land areas and are resource poor (Thomas, 1967). Apart from those islands (such as New Zealand) which are continental remnants, all Pacific islands are submarine volcanoes either exposed at the surface or with raised or sea-level coral reefs. Thomas has classified Pacific islands as low islands of carbonate rock, islands of elevated reef rock, volcanic islands and islands containing ancient "continental" rocks. More detailed classifications of island geography and morphology have been given by Cumberland (1968) and Kaplin (1981). Few of these landforms are characterized by minerals of commercial value and the sea-level coral reefs provide "the most limited range of resources for human existence and are the most tenuous habitats for man in the Pacific." Exceptions to this include New Caledonia, which earns 85–95% of its export receipts from minerals (Lillie and Brothers, 1970; Paris, 1981); Bougainville, which is a major copper producer; and Fiji, which produces appreciable quantities of gold and silver (mineral production in Fiji constituted 3% of the Gross National Product, GNP, in 1977) (Colley, 1976; Colley and Greenbaum, 1980; Greenbaum, 1980); all three

Sedimentation and Mineral Deposits
in the Southwestern Pacific Ocean
ISBN 0-12-195870-1

islands are in the western Pacific. Mineral statistics for Australasia and
Oceania are given in Anonymous (1979a) and Mines Division (1979) and
selected mineral deposits described by Liddy (1972), Fisher (1973), Ridge
(1976), Colley (1978), Our Own Correspondent (1980), and Davies (1985).
The geology of a number of SW Pacific islands has been described by Kear
and Wood (1959), Phillips (1967), Wood (1967), and Bryan et al. (1972).

With the coming of independence of many of the Pacific islands over the
last decade, there has been an increasing awareness of the need to put them
on a more sound economic footing and the sea is seen as an important factor
in creating this independence (both in terms of fisheries and mineral
resources). Because of the limited resources of the islands (both in terms of
finance and skilled manpower), there has been a need to collaborate
between the island groups to explore the offshore resources. The main
vehicle for such development in the SW Pacific is CCOP/SOPAC (Committee for Co-ordination of Joint Prospecting for Mineral Resources in
South Pacific Offshore Areas) which now comprises the Solomon Islands,
Cook Islands, Tuvalu, Guam, Fiji, New Zealand, Papua New Guinea,
Tonga, Western Samoa, Kiribati and Vanuatu (with Australia, the U.S.A.,
the U.K., Japan, France, Germany and the U.S.S.R. as observers). The
organization was set up in 1971 and the Technical Secretariat is based in
Suva, Fiji. Each year, a meeting is held in one of the member states to
formulate projects and policy for the coming year and a proceedings volume
is published. This organization has done a remarkable job in increasing the
awareness of the potential of the seas amongst the island nations of the SW
Pacific (including the holding of a number of international symposia), of
carrying out exploration work for offshore minerals and putting this work on
a proper scientific basis and, perhaps most importantly for the future, in
initiating proper scientific training of local people. An example of this latter
aspect is the holding of an inshore and near-shore resources training
workshop in Suva, Fiji, in 1981 (Anonymous, 1981a). Amongst other
things, this contains useful papers on placer deposits in the islands and
environmental aspects of near-shore mining. Between 1976 and 1981,
CCOP/SOPAC organized 58 surveys to search for offshore minerals. Of
these cruises, 19 looked for precious coral, one for detrital minerals, one for
bauxite and none for aggregate. One could perhaps question this use of
shiptime in view of the importance of aggregate to many of the local
economies, and more recent cruises of CCOP/SOPAC have addressed this
problem. An indication of the lack of knowledge of the shallow-water
mineral resources of the SW Pacific islands prior to the setting up of
CCOP/SOPAC can be seen from an inspection of a bibliography of marine
geology and geophysics of the South Pacific (Kroenke and Bardsley, 1975)
which contains virtually no references to near-shore marine resources, the

only exception being Green's (1970) survey of offshore mineral resources in Fiji. In a symposium held in 1971 on metallogenic provinces and mineral deposits in the SW Pacific, the author of the chapter on mineral resources of the ocean (Burk, 1973) did not even specifically refer to the marine minerals of the SW Pacific. The major textbooks on Pacific marine geology, and on marine mineral resources from the 1960's and early 1970's (Menard, 1964; Mero, 1965; Shepard, 1973) also made virtually no mention of shallow-water mineral deposits in the SW Pacific. Perhaps most surprisingly, a major history of the Cook Islands between 1820 and 1950 makes no mention at all of the uses of the resources of the sea (Gilson, 1980). Summerhayes (1967) had, however, already reviewed the prospects for economic mineral deposits around New Zealand.

Another factor in the increasing awareness of the importance of the seas in the SW Pacific has been the International Law of the Sea Conference and the proposed declaration of the 200 mile exclusive economic zones (EEZ's) around the islands (cf. Dabb, 1981). Perhaps the principal beneficiaries of this proposal in terms of area gained relative to the original land area are the Pacific islands, particularly with the development of the concept of "Archipelagic States." For most islands, this has led to a dramatic increase in the area administered by them (Fig. 5.1). For example, the Cook Islands with total land area of 240 km^2 would have an EEZ of $1 \cdot 2$ million km^2. Certainly, the introduction of New Zealand's EEZ in 1977 (the fourth largest in the world with an area of 4 million km^2) has led to a more positive attitude to the management of marine resources there. The areas of some of these EEZ's have been listed by Gocht and Wolf (1982), and some of the problems associated with their introduction have been discussed by Prescott (1980) and Border and van Dyke (1982).

In spite of the prevailing optimism about the sea as a possible source of wealth for Pacific islands, it should always be remembered that this is necessarily constrained by the biology and geology of the region. The temperate regions of the Pacific, for example, are characterized by the low productivity, subtropical anticyclonic gyres which are essentially "oceanic deserts". Fish abundances in the deep waters outside the reefs are therefore not particularly high in mid-latitude regions of the Pacific. Even with the fish resources present, primarily tuna, these are often not fished for the benefit of the island people (G. Kent, 1980; Morgan, 1981). Similar considerations apply to the mineral resources. In the proceedings of the 1980 CCOP/ SOPAC meeting, it is stated that

major economic benefits from offshore minerals would probably come only from the development of hydrocarbons, or in the longer term, of manganese nodule accumulations. While near-shore and inshore development of minerals such as sand and gravel were of immediate local interest to some member countries, large-scale

F

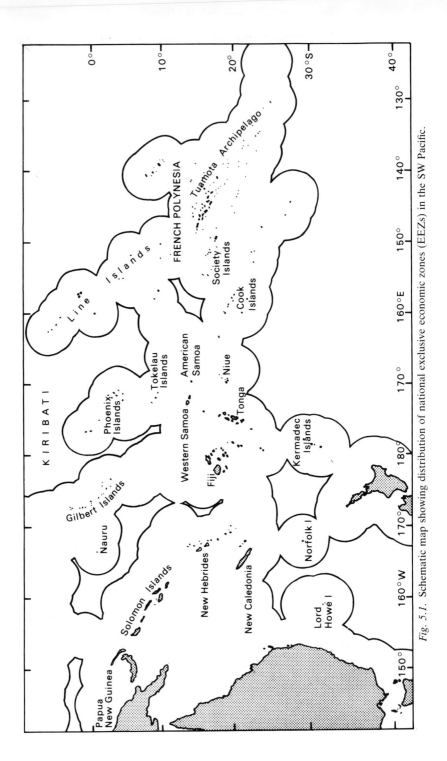

Fig. 5.1. Schematic map showing distribution of national exclusive economic zones (EEZs) in the SW Pacific.

economic benefits would not accrue from such operations. The same might be said of the working of occurrences of precious coral.

A similar view has been expressed by Richmond (1981), and, whilst any definitive statement is as yet premature, one would not usually search for hydrocarbon deposits around volcanic islands or on young ocean crust (see CCOP/SOPAC, 1980, for a review). Emery (1975), for example, has stated that "the marginal basins of the world represent a promising source of offshore oil; they are geologically interesting because most of them are areas of convergence of sea-floor crustal plates with continental plates". This, of course, is not the position in the SW Pacific where zones of convergence represent the interaction of two sets of ocean crust. The possible hydrocarbon provinces on the SW Pacific have, however, been mapped by St John (1982, 1985), [cf. Greene and Wong (1983) and Ives (1984)], and Soviet geologists have developed interesting ideas on the genesis of hydrocarbon deposits in underthrust zones (Sorokhtin and Balanyuk, 1984). The potential for marine minerals to have a decisive effect on the economies of the island nations of the SW Pacific is therefore debatable (except in the case of guano deposits as at Nauru Isand). Whilst it is probable that mining of marine deposits, particularly low-grade, near-shore deposits (such as aggregate and precious coral) will proceed and fulfil certain niches within individual island's economies, the prospects of a marine mineral bonanza for the island nations of the SW Pacific can be dismissed.

In this chapter, the problems of near-shore, shallow-water deposits such as sand and gravel, calcium carbonate, salt, heavy minerals, bauxite and precious coral will be discussed. For convenience, each mineral resource will be discussed geographically in terms of its distribution in Australia, New Zealand and the island nations of the subtropical Pacific. This division is important for near-shore, shallow-water mineral deposits because there is clearly a difference between the situation in Australia and New Zealand with broad continental shelves and the volcanic islands of the subtropical regions which have narrow (or in some cases no) continental shelves. Latitudinal variations also make a difference to mineral assemblages (and particularly reef development). In addition, there is also a difference in the level of economic development and resource requirement throughout the region. Australia is fast becoming one of the world's major mineral exporters and would tend to look at minerals principally from the point of view of export. Detailed reports of Australian mineral resources have been given in Knight (1975, 1976) and Woodcock (1980), and their distribution in the Australian EEZ presented by George (1978). Because of this, it is not proposed to review Australian mineral resources further, except where appropriate to give some general indication of Australia's production of specific minerals. New Zealand, on the other hand, is a country not well

endowed with mineral resources (particularly of metallic minerals) and
would tend to look at marine minerals (such as phosphorite or aggregate)
from the point of view of import substitution or local convenience where
land-based minerals are not available. The mineral resources of New
Zealand have been reviewed by Kear (1967), Williams (1974), and Anonymous (1975) and offshore sediment distribution by Carter (1975). The island
nations of the SW Pacific are often deficient in basic materials (such as
aggregate) and need adequate sources of these materials to satisfy their local
economies. They also tend to be tourist centres where minerals such as
precious coral could be expected to have a potential market for tourists. A
uniform picture of the SW Pacific in terms of aspirations, capabilities and
mineral resources of individual nations cannot therefore be entertained.
Potential environmental problems in marine mining must also be con-

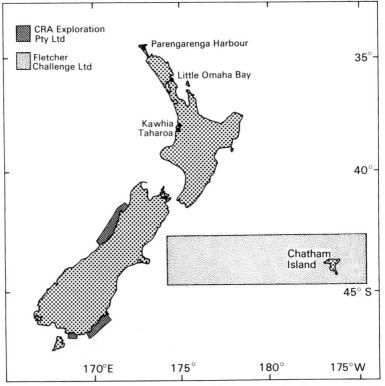

Fig. 5.2. Schematic map showing areas where offshore prospecting licences are held and
offshore mining takes place in New Zealand.

sidered, especially in the small islands of the SW Pacific where land is scarce and the concept of multiple land use (particularly for fishing and tourism) therefore essential.

By way of background, the following texts are recommended: Mero (1965); Summerhayes (1967); Archer (1973, 1974); Cruickshank (1974); Moore (1975); Noakes and Jones (1975); Glasby and Katz (1976); Alexandersson and Klevebring (1978); Halunen (1978); Burns (1979); Katz and Glasby (1979); Cronan (1980); P. Kent (1980); Bentley (1981); Carter (1981); Glasby (1982b,c), and Jouannic and Thompson (1983). As an example of offshore mineral activity within the region, a schematic map of the areas where offshore prospecting licences are held and where offshore mining takes place in New Zealand is given in Fig. 5.2. In spite of a decade of very considerable advance in the study of the offshore geology and mineral resources of the SW Pacific, it must be emphasized that much remains to be done and that our knowledge is far from complete.

5.2   Sand and Gravel

According to Alexandersson and Klevebring (1978), sand and gravel rank first in tonnage among minerals world wide, surpassing oil and coal. Both result from the natural disintegration of rocks. About 90% of all sand and gravel is used for construction or road metal. Beaches, the sea floor, lake beaches and bottoms, and river terraces and river bottoms are the major sources of sand and gravel. In areas with no natural sand and gravel, rocks are crushed to provide substitutes. Because sand and gravel are low-cost commodities with one of the lowest values on the commodity market, transport costs are a major factor in siting sand and gravel plants. The sand and gravel industry is therefore highly decentralized, each unit based on a given area. Offshore sand and gravel mining has become an increasingly important industry. In Britain, for example, offshore sand and gravel recovery, principally from the North Sea, accounted for 14% of the total U.K. production in 1978, with an annual value of £37·8 million. The U.K. offshore sand and gravel industry is now well developed and competitive; the technology for offshore sand and gravel recovery is therefore available.

In Australia, the only aggregate now mined offshore is dead coral from Moreton Bay, Queensland (Noakes and Jones, 1975). Most aggregate is supplied onshore and there has been little inducement to exploit such deposits offshore (Knight, 1976). However, Consolidated Gold Fields Australia Ltd. and ARC Marine Ltd. (1980) have recently submitted a very detailed environmental impact statement for the mining of one million tonnes of aggregate per annum off Broken Bay, Sydney.

In New Zealand, aggregate is easily the most valuable mineral recovered (excluding oil and gas) (Williams, 1974, p. 9). A report on the distribution of aggregate in New Zealand (Williams, 1974, p. 439–441, 448–455) shows that except in a few limited areas such as around Gisborne, where supplies of good quality rock are absent (the country rock being Tertiary or Cretaceous mudstones), or in Northland where deep and extensive weathering of the land has resulted in a scarcity of fresh rock, the reserves of aggregate are high (cf. Craven, 1969; Pemberton, 1979). The nature of the aggregate depends, of course, on the local geology. In Auckland it is principally basalt, in Wellington greywacke, in Taranaki andesite, in Hawke Bay limestone and in Otago schist (cf. Ellen, 1958; Kear, 1965a; Hamilton, 1967; Kear and Hunt, 1969; Skinner, 1974). Surveys of the aggregate resources in the Wellington region have been conducted by Grant-Taylor (1976), Ward and Grant (1978) and Rowe (1980). In 1979, New Zealand mined a minimum of 17·9 million tonnes of sand and gravel with a value of NZ$49·1 million (Mines Division, 1979).* Aggregate production was therefore more valuable than that of coal (NZ$45·2 million), ironsand (NZ$28·7 million) or limestone (NZ$15·1 million) and was equivalent to 32% of New Zealand's total mineral production (excluding oil or gas). To some extent, this reflects the absence of a significant metal mining industry in New Zealand. Because of the high reserves on land, there is only a limited production of aggregate off New Zealand. The areas involved are principally off Northland, Auckland and Bay of Plenty (all in the northern part of the North Island), and the locations of these operations have been mapped by Thompson (1981). The biggest of them is 1·4 km offshore near Matakana Island (off Tauranga) where 30 000 m³ of aggregate was dredged in 1980 (cf. Healy, 1977; Healy *et al.* 1977). An average of 16 500 m³ of sand was dredged annually from Little Omaha Bay (North Auckland) between 1942 and 1964 (Ward, 1977); the environmental effects of this operation are discussed later. A comprehensive report on the prospects of beach sands in Northland and Auckland has been given by Schofield and Woolhouse (1969) and a literature search conducted by Hume and Harris (1981). In addition, dredge spoil is used in certain harbours for land reclamation, although, in the case of Wellington Harbour for example, this makes up only a small fraction of the total material used (Hutchison, 1973, see also Tierney, 1977). In the harbours of the west coast of the South Island, dredging has been more extensive (Furkert, 1947). These offshore operations, however, represent less than 1% of the total aggregate production in New Zealand. Beach deposits are also mined in limited cases (Kirk, 1980; Thompson, 1981).

---

* Because mineral returns on private land need not be reported to the Department of Mines, figures for aggregate production in New Zealand must be treated as minima.

Descriptions of the principal characteristics of New Zealand beaches are given by Andrews and van der Lingen (1969), McLean and Kirk (1969), Schofield (1970), Pickrill (1977), Gibb (1978, 1979a) and Matthews (1982), of which Gibb (1979a) is the most important. Perhaps the most comprehensive treatment of late Cenozoic erosion in New Zealand (and therefore its influence on aggregate production in river beds) is given by Adams (1978). Carter (1975) describes the offshore sediments around New Zealand.

A different situation prevails with silica sands for the glass industry. For the manufacture of clear glass, the minimum silica content must exceed 99·5%. Iron should not exceed 0·035%, titanium 0·03% and chromium 0·0006% (Schofield, 1969, 1970; Williams, 1974, p. 453). These figures indicate that extremely pure silica sand is required for glass manufacture. In New Zealand, the principal glass manufacturers are N.Z. Glass Manufacturers' Company and N.Z. Window Glass Ltd of Auckland. Sand is supplied by dredging from Parengaenga Harbour in Northland. In 1979, 137 000 tonnes of sand worth NZ$780 000 were produced. This operation is the only true offshore mining presently carried out in New Zealand, although it is, of course, in a restricted harbour. Feldspar sand is also dredged from Kaipara Harbour, North Auckland, for glass making by Alex Harvey Industries as well as concrete manufacture.

In the Pacific islands, the demand for aggregate in the larger population centres has increased as a result of population increases, urbanization and westernization. Westernization, in particular, has led to the development of better roads, tourist hotels and airports as well as a change from building in traditional materials to building in concrete in urban centres, all of which require aggregate. The small size of the islands has put considerable pressure on existing resources and, in many of the major population centres, beach mining is having adverse (sometimes severe) environmental effects. This is well stated in the 1982 report on the conference on the human environment in the South Pacific held in Rarotonga. An example of this problem can be found on Tongatapu, the main island of Tonga, where beach erosion on the windward side of the island is becoming critical and other sources of aggregate are being urgently sought. One of the problems is that per capita incomes in the islands are low so that aggregate must be cheap to be used. In Tonga in 1981, for example, aggregate cost 6 Tongan dollars per tonne. This low cost precludes expensive offshore surveys or the use of sophisticated mining vessels. Nonetheless, aggregate production does not appear to be a major economic activity in these islands, as witnessed by the fact that it is scarcely mentioned in J. Carter (1981). CCOP/SOPAC is, however, involved in surveys for aggregate in Fiji, Cook Islands, Western Samoa and Tonga. This section briefly mentions some of the work being carried out (cf. CCOP/SOPAC Technical Secretariat, 1977a, b; Anonymous

1981a). It should be emphasized, however that published statistics on aggregate production in these islands appear sparse. It is hoped that this is remedied over the next few years as proper scientific surveys become available.

Fiji differs from many of the other SW Pacific islands in that aggregate is produced mainly from quarried stone by crushing. In 1979, 205 000 m³ of aggregate were produced at eight quarries (Mineral Resources Division, 1981). In addition, 50 000 tonnes of river sand were recovered in that year, and 30 687 tonnes of fine sand were also recovered from Laucala Bay, a lagoon adjacent to Suva. Total value of sand and gravel in 1977 was estimated to be about US$5 million (Anonymous, 1979a).

In Rarotonga, Cook Islands, Lewis *et al.* (1980) have described in some detail the sediment type and its distribution around the island (cf. Gauss, 1982). In their opinion, the most suitable material for construction and landfill purposes is from Avarua Channel; this material has already been lost to the environmentally sensitive beach/reef flat system and mining would not therefore have the environmental implications of mining the beach or reef flat where sediment budgets would be changed. The sand fraction of this channel material is worn coral and the gravel fraction predominantly rounded basaltic pebbles. More work needs to be done on the characteristics of this material to evaluate its usefulness.

In West Samoa, Eade (1979) surveyed Solosolo Beach on the north coast of Upolu Island about 25 km east of Apia for construction material. At that time, Western Samoa's aggregate requirements for the next 12 months were about 13 000 m³ of sand, 10 000 m³ of which were for building a complex for the South Pacific Games in 1983. The beach consisted of about 12 000 m³ of very fine black sand between high and low tide to a depth of 1 m. It was, however, considered that removal of the sand would upset the local equilibrium and expose the adjacent road to erosion. Subsequent surveys of aggregate off Samoa have been carried out by Rubin (1984a).

On Tongatapu, sand is being mined from beaches on the SE and west coasts. It has been estimated that the sand requirements of Tonga are about 8000 tonnes/yr. A survey was commenced in 1978 to study the inventory of sand in Tongatapu (Eade *et al.*, 1978). This study included reconnaissance side scan sonar, diving and sea bed sampling and has now been completed (Gauss *et al.*, 1983; Rubin, 1984b). According to Gauss (1980), Laulea Beach receded up to 10 m during a period of intensive mining between November 1979 and July 1980. Other beaches which had previously experienced mining were also showing evidence of damage with beach rock being exposed. Since demand for aggregate remains high, CCOP/SOPAC conducted a survey of possible alternatives to beach mining. The most promising find so far discovered in water depths between 6 and 30 m immediately to the south of Fafa and Velitoa Islands, about 6 km to the north of Nuku'alofa, and consists of about 4 million m³ of fine to medium, shelly carbonate sand. Laboratory tests indicate that, although not ideal, this material can be used as concrete aggregate. An alternative is the large (10 × 7·5 km) reef flats approximately 4 km west of Nuku'alofa which are exposed at low spring tide. These could be mined by excavator at low tide, but this would need careful evaluation not only with regard to upsetting the dynamics of the beach and causing local erosion but also jeopardizing a local fishery on which many local people are dependent. Several hundred people base their livelihood on this reef flats area.

In New Caledonia, Paris (1981) has noted that coral beach sands are exploited on a number of small islands off New Caledonia.

From the above statements, it is seen that aggregate supply is becoming a problem in many of the SW Pacific islands and this question needs to be examined more carefully. Fortunately, CCOP/SOPAC is undertaking that task.

5.3 Calcium Carbonate

Calcium carbonate is used commercially principally in cement manufacture. It is a very common marine mineral on shelf areas, particularly in the tropics where reef development is pronounced. Because of its ready availability on land, the bulk of calcium carbonate is mined there. Calcium carbonate is like aggregate in being a low-cost commodity. The same considerations applying to aggregate apply to calcium carbonate.

Australia is well endowed with land-based limestones deposits. Apart from coral detritus dredged from Moreton Bay, Queensland (as discussed earlier), limestone is mined exclusively on land.

In New Zealand, limestone had been used from the earliest days of European settlement for agriculture and cement (Williams, 1974, pp. 271–278). In 1979, New Zealand produced 3·5 million tonnes of limestone worth NZ\$ 15·1 million for agriculture, cement, roads and industry. Summerhayes (1967) has pointed out the high concentrations of calcium carbonate off the northern tip of New Zealand and on the Campbell Plateau (*cf.* Summerhayes, 1969a,b). According to Summerhayes, 10^{14} tonnes of calcium carbonate occur on the Campbell Plateau. Because of the ready availability of limestone within New Zealand and the high mining and transport costs offshore relative to its intrinsic value, it is unlikely that calcium carbonate will be mined off New Zealand within the foreseeable future.

One of the principal New Zealand cement companies is New Zealand

Cement Holdings Ltd which has six works (five in the South Island and one in the North Island) (cf. Commercial Editor, 1982). According to its 1980 Annual Report, the New Zealand cement industry is presently operating at a depressed level with output at only 60% of the 1975 level. This situation has been offset to some extent by exports, principally to Tahiti and Papua New Guinea but also to Tonga, Vanuatu, and the Solomon Islands. Exports increased from 3000 tonnes in 1979 to 57 000 tonnes in 1980 to 112 000 tonnes in 1981. The need for the island economies to import cement is thus demonstrated.

Carbonate sediments are recovered on a number of islands (this being the principal sediment type on many of the islands). Holmes (1980) has described the distribution of coral sand in Laucala Bay. The sand is mined by Fiji Industries Ltd principally for cement manufacture, although some is used for lime production. In 1982, 104 441 tonnes of coral sand were mined there (R. Holmes, personal communication). Much smaller amounts of limestone are produced at the Tau Limestone Quarry which produced 1306 tonnes of burnt lime in 1979 (Mineral Resources Division, 1981). Reserves at Laucala Bay are estimated to last a further 15 years. The cement satisfies all Fiji's needs and is also exported. In 1977, Fiji exported cement worth US$550 000 (J. Carter, 1981). The total value of the cement produced in that year was US$4 million (Anonymous, 1979a). Studies are presently being conducted into the environmental impact of coral mining. It appears that there may be problems in finding suitable alternatives which do not interfere with fishing, tourist amenities or navigation into Suva when the Laucala Bay deposit runs out. In spite of these impressive figures, de Bock (in Anonymous, 1981a) has questioned the quality control and standards of concrete manufacture in Fiji.

For the other islands, little statistical information on the extent of coral sand mining appears available. Nonetheless, the New Zealand Concrete Research Association has carried out tests on the suitability of coral sands from the Cook Islands, Western Samoa and Tonga for concrete manufacture and in 1982 organized a workshop in Rarotonga, Cook Islands, on making concrete from coral (Concrete Research Association, 1982). This work, however, remains of a preliminary nature.

5.4 Salt

Salt production in Australia has been discussed by Knight (1976) and Driessen (1984) and is beyond the scope of this text. Nevertheless, Australia is the most arid continent, where net annual evaporation exceeds net annual precipitation in many localities. Production of salt by the solar evaporation of seawater or lake brines therefore presents no problems, and extensive salt deposits are found within Australia. The commissioning of extensive solar evaporation plants in Western Australia in the decade prior to 1975 has led to the production of salt for export to Japan. Australia is therefore a large producer of solar salt. In 1975, 3·6 million tonnes of solar salt were produced in Western Australia.

In New Zealand, salt is produced by solar evaporation of seawater at Lake Grassmere near Blenheim in Marlborough (the northern tip of the South Island). The development of this process has been described by Reid (1976). The initial steps to establish the industry were taken in 1942 as a war-time measure, but commercial harvesting dates from 1958 when 3100 tonnes of salt were recovered. The principal requirement for the solar salt industry are high sunshine, a prevailing wind, low rainfall, proximity to the sea and a large area of flat land unsuitable for other use. Such sites are not easy to find at 42°S in a maritime climate, as an inspection of New Zealand rainfall maps would show. Lake Grassmere is one of the few possible sites and has one of the lowest rainfalls in New Zealand. In all, the solar ponds occupy 14·2 km^2. Production depends on the vagaries of climate and is therefore variable from year to year. At present, it exceeds 60 000 tonnes/yr out of a total New Zealand consumption of over 100 000 tonnes/yr. No side products are presently manufactured from the operation.

In Bali, salt is produced from seawater as a cottage industry, and a family could expect to produce about 25 kg/day (Eiseman, 1982).

5.5 Heavy Mineral Deposits

Marine heavy mineral (or placer) deposits, as their name implies, consist of heavy minerals which have been sorted from their lighter counterparts by wave action in a high-energy environment. They are found in many parts of the world as beach or drowned beach deposits. Their importance as a source of economically valuable metals is shown by the fact that two-thirds of the world's tin is produced in Southeast Asia, much of it from placer deposits, and Australia is a leading producer of titanium, zirconium and the rare earth elements from its monazite beach sands in New South Wales and southern Queensland (Alexandersson and Klevebring, 1978), although its share of the world market appears to be declining (Anonymous, 1979b, 1981b). The character of these deposits depends on a number of factors, of which the most important are:

(1) Source rock. The nature and composition of the source rock is a principal factor in controlling the nature of the heavy mineral deposit. This can vary between the granites of Southeast Asia and Cornwall which

produce the tin deposits, the granites and schists of the east coast of Australia which produce monazite, ilmenite, rutile and zircon, and the andesites of the central North Island of New Zealand which produce titanomagnetite. A suitable source rock in the hinterland is therefore an essential precursor of heavy mineral deposits. The degree of weathering of this source rock is, however, also important. In east Australia, for example, intense tropical weathering has facilitated the breakdown of the granites. On the west coast of New Zealand, on the other hand, the black sands are derived mainly from poorly consolidated andesitic lahar deposits of Late Quaternary age which outcrop in coastal cliffs and stream banks around the volcanic cones of Mt Egmont (Lewis, 1979); these too are easily erodible.

(2) A moderately wide continental shelf to trap the deposits.

(3) Mechanism of transport. River supply is often an important method of bringing heavy minerals to the sea. Depending on local conditions, heavy minerals can therefore be found either in drowned river valleys or, where longshore drift is important, dispersed along the coast from the primary input source. Generally, however, the deposits are close to the source.

(4) Wave action. Heavy mineral deposits are, almost by definition, found in high-energy environments. Sorting of these minerals is most pronounced in the swash zone. Generally, the deposits are most concentrated as beach (or dune), rather than as offshore, deposits.

(5) Late Pleistocene sea-level changes. The reason that heavy minerals are found offshore, particularly as drowned beach deposits, is due to the post-glacial sea-level changes. According to Lewis (1974), the shelf break was formed about 15 000–20 000 yr ago, when the sea was about 120 m below its present level. With each rise in sea level, a beach deposit could be formed, giving the possibility of drowned beach deposits at various present-day water depths corresponding to each stillstand. In suitable environments, these are maintained as fossil placer deposits. In New Zealand, the principal study of post-glacial sea-level changes has been given by Gibb (1979a; cf. Cullen, 1967a, 1970; Gillie, 1979).

(6) Present-day hydraulic regime. Any submarine mineral deposit is subject to the modern hydraulic regime. Lewis (1979), for example, has shown the importance of storms in the movement of sand ripples off the west coast of the North Island of New Zealand. This question of sediment movement on the New Zealand Continental shelf has been discussed in detail by Carter and Heath (1975), Carter and Herzer (1979), and Heath (1979, 1981b); these studies have shown the importance of tides and storm-driven components rather than mean circulation currents in sediment movement on the sea floor. Carter (1980) has shown that the ironsands of the North Island of New Zealand are palimpsest in that they are approaching equilibrium with the present-day hydraulic regime. In New Zealand, long-

shore drift is an important mechanism of sediment movement. According to Gibb (1978), longshore drift takes place in the nearshore zone which includes an inner surf zone extending to depths of 3–5 m and an outer wave-transport zone extending to depths of 18–65 m. For sand, 70–90% is transported in the wave-transport zone whereas for gravel 100% is transported in the surf zone. Longshore transport rates for sand vary by a factor of eight from year to year but can range up to 4.5×10^6 m³/yr.

(7) Post-glacial sedimentation is important in influencing the extent of burial of drowned beach deposits. Gibb (1977) and Carter (1980), for example, have shown that significantly larger volumes of sediment are discharged onto the shelf of the west coast of the South Island of New Zealand compared to that of the west coast of the North Island. The ironsand deposits on the west coast of the South Island have therefore been covered by modern sediments and occur in much lower concentrations in surface sediments than the deposits off the west coast of the North Island.

The above show that the genesis of offshore placer deposits is complex (cf. Hails, 1974; Barrie, 1981) and this is well illustrated by the situation off the west coast of New Zealand (Lewis, 1979; Carter, 1980). A useful review of marine placer deposits has been given by Burns (1979).

Australia is one of the world's principal producers of rutile, zircon and monazite, from the beach sand deposits of eastern Australia, which constitute part of the Eastern Australian Rutile Province. The high sand deposits extend over 400 km between the Hawkesbury River, New South Wales, and Fraser Island, Queensland, and consist of beach sands backed by sand plains. The sands are dominantly quartz but average 0·05% of heavy minerals of which rutile, zircon and ilmenite make up 90%. In 1969, rutile production was in excess of 350 000 tonnes and zircon in excess of 300 000 tonnes. This required a throughput of the sands of the order of 100 million tonnes. The distribution and origin of these beach deposits and their mining and processing have been described in detail by Blaskett and Hudson (1965), Paterson (1965), Hails (1969), Cooper *et al.* (1973), Friedrich (1974), McKellar (1975), Woodcock (1980), Morley (1981) and Towner (1984a,b). The offshore distribution of these deposits has been described by Brown (1971), Noakes and Jones (1975), Jones and Davies (1979), and Marshall (1980) and in 1980 was subject to detailed examination during cruise SO-15 of the German RV *Sonne*, results of which have been presented by von Stackelberg and Riech (1981) and von Stackelberg and Jones (1982). A sedimentation model for the region has been developed by Roy *et al.* (1980) and Roy and Thom (1981). Because these deposits are somewhat peripheral to our study of the offshore mineral resources of the SW Pacific and their offshore distribution has been considered at some length by

Australian geologists, including the discovery of gold off New South Wales and tin off Tasmania and Queensland, (cf. Anonymous, 1969; Jones and Davies, 1979), it is not proposed to discuss this topic further.

The ironsands of the west coast of the central North Island of New Zealand have been known for over a century. Indeed, sporadic attempts have been made to exploit them since 1846 (Wylie, 1937a). Chemical analyses of the ironsands were made in 1888 and showed the presence of iron and titanium oxides as the principal constituents present (Galvin, 1906). Optimism was expressed at the use of these very extensive beach and dune deposits in steel production but "the difficulty of smelting due to the mechanical conditions of the sand and the presence of titanic acid" was recognized. In fact, it was to be a further 80 years before the technology was available to mine and process these deposits (Williams, 1974; Schofield, 1975a; Pain, 1976; Watson, 1979; Carter, 1980; Graham and Watson, 1980; New Zealand Steel, undated, and references therein). This work will not be duplicated, except to say that the principal source of the coastal ironsands in the central North Island are the andesites of the Cape Egmont area. The largest deposit of ironsands is found at Taharoa with a total of 263×10^6 tonnes of titanomagnetite averaging 64% of the parent sand (Williams, 1974, p. 132). In the beaches of the west coast of the South Island and also of the North Island north of Auckland and in parts of the Coromandel Peninsula (such as Waihi Beach), ilmenite predominates (Marshall et al., 1958; Nicholson et al., 1958; Nicholson, 1967; Williams, 1974; McPherson, 1978). Grampian Mining Co. Ltd., a wholly-owned subsidiary of Fletcher–Challenge Ltd., has been investigating the use of the ilmenite-bearing beach sands near Barrytown on the west coast of the South Island for the production of TiO_2 pigment. The occurrence of rutile at Transit Beach, Fiordland, is also being investigated by Consolidated Minerals Ltd (Anonymous, 1982a). A particularly careful mineralogical study of the South Island placer deposits has been given by Hutton (1945). The offshore geology of the west coast of the South Island has been described by van der Linden and Norris (1974) and Norris (1978). The distribution of heavy mineral deposits off the east coast of the South Island (Martin, 1956; Nicholson, 1969) and the south coast of the South Island (Martin and Long, 1960; Martin, 1961) has also been recorded.

The development of an indigenous steel industry in New Zealand had to wait until 1970 when the New Zealand Steel mill at Glenbrook came into operation based on ironsands from Waikato North Head. Because of the extent of the ironsand deposits, ironsand concentrates have also been exported to Japan from Waipipi and Taharoa since 1972. The deposit is loaded on board ship as a slurry (Anonymous, 1971). A description of this early development has been given by Williams (1974), Kitt (1981) and New

Zealand Geological Survey (1981). In 1979, New Zealand Steel used 235 000 tonnes of ironsands from Waikato North Head with a value of NZ$1·7 million for local steel production and 3·2 million tonnes of ironsand with a value of NZ$27 million were mined at Waipipi and Taharoa for direct export to Japan (Mines Division, 1979). New Zealand Steel is also beginning to export steel billets to countries such as Indonesia. A substantial expansion in New Zealand Steel's capacity involving a fivefold increase in output and new technology is planned in the early 1980's (Bartlett, 1980; Bold, 1982). A NZ$481 million expansion of the New Zealand Steel plant in Auckland was announced in December 1982. It should be complete in 1986. A full account of the Taharoa ironsand project is given in *Transactions of the New Zealand Engineers*, 1974, vol. 1(2). In the case of the South Island beach ilmenite deposits, substantial effort has been directed towards processing these deposits to produce titanium dioxide (Buckenham, 1965; Nicholson *et al.*, 1966; Walker, 1967; Judd and Palmer, 1973; Palmer and Judd, 1973; Metson, 1980; Duncan and Metson, 1982; Lauder, 1983), and Fletcher Challenge Ltd has recently expressed interest in making titanium metal sponge from them but the project is only at the planning stage (Field, 1981).

With respect to the offshore deposits, two New Zealand Oceanographic Institute cruises were undertaken in 1959 and 1960 off the west coast of the North Island to survey the distribution of the ironsands and 203 stations occupied (McDougall, 1961; McDougall and Brodie, 1967). This was supplemented by further material collected in 1963 (McDougall and Brodie, 1967). A maximum concentration of the magnetic fraction of the sediments was found at a depth of about 15 fm (28 m) ranging from 3 to 36%; this was attributed to wave processes operating at the time of the late Glacial low sea level. A description of the sediments off the west coast of the North Island has been given by Lewis and Eade (1974).

In a much more comprehensive survey based on a total of 308 samples from the entire west coast of New Zealand, Carter (1980) showed that the primary source of ilmenite and magnetite in the beach sands of the South Island are acid plutonic and metamorphic rocks. Because of the higher discharges of material from the South Island rivers (Furkert, 1947; Adams, 1978; Griffiths, 1979, 1981), however, these offshore South Island ironsands are largely buried and concentrations of the heavy minerals in the surface sediments are generally low (<0·28%). The ironsand deposits off the west coast of the the North Island, on the other hand, are present in higher concentrations in the surface sediments (the grade decreasing with depth in the sediment column) and are concentrated in the inner–middle shelf area where the ironsand component of the sediment averages 3·85%. It is believed that these ironsands were deposited under littoral conditions at times of lowered sea level but have been reworked and are approaching

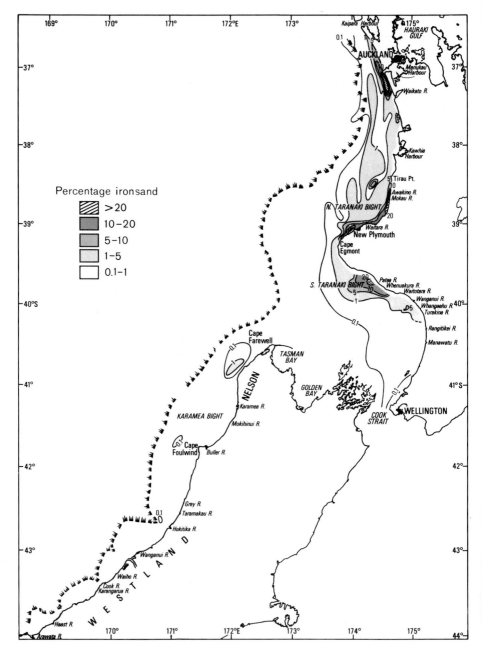

Fig. 5.3. Distribution of ironsand in the surface sediments finer than 2 mm on the west coast of New Zealand. [From Carter (1980), Fig. 3.]

equilibrium with the modern hydraulic regime; the sediments are therefore regarded as palimpsest. The distribution of the ironsand in the surface sediments of the entire west coast of New Zealand is shown in Fig. 5.3.

Gibb (1979a) has contributed significantly to our understanding of shore-line processes around New Zealand which are relevant to this study. In particular, Gibb has shown the occurrence of two stillstands in post-Glacial times, the first between 8800 and 8300 yr BP at about −22 m and the second from 7000 to 7300 yr BP at about −10 m relative to the present sea levels. Since 6500 yr BP, sea-level fluctuations have been small and seldom exceeded 1 m. Longshore drift is the predominant method of sediment transport. On the west coast of the South Island, the dominant direction of drift is northerly and is most pronounced [Gibb (1979a), Fig. 4.9]. On the west coast of the North Island, drift is dominantly southerly south of Mt Egmont and northerly north of Mt Egmont. These factors are important in accounting for the distribution of ironsand on the west coast of New Zealand.

In a particularly valuable study, Lewis (1979) has used side-scan sonar to map the ironsands offshore and shown the formation of stable ridges of black, mafic sand (ironsand) and ephemeral ribbons of light, felsic sands (river-borne material derived from the marine deposits of the Wanganui Basin and the rhyolitic tephra of the central North Island volcanoes). The migration of these bedforms is thought to be particularly influenced by periodic storms which are a characteristic of the region. A map of these bedforms is given in Fig. 5.4. A more local side-scan survey has been given by Lewis (1982). Hamill and Ballance (1985) have reported on heavy mineral sands north of Auckland.

From these various studies, it is clear that ironsands are present in considerable quantity offshore (McDougall estimated 74 million tonnes of magnetic fraction in the area he studied) and are of considerable geological interest in regard to source, relation to sea level and relation to the present-day hydraulic regime. However, they appear to have a limited economic potential in the foreseeable future because of their relatively low grade compared to that of the onshore deposits, the large reserves of the onshore deposits and the much more favourable economics of mining beach and dune compared to offshore deposits. Nonetheless, they represent an interesting deposit whose characteristics have only recently been fully documented.

Gold is also a heavy mineral, with a specific gravity of 19·3. With the presently high prices, more commercial interest is being directed towards gold. Nonetheless, because of the low concentrations of gold typically found in alluvial deposits and its heterogeneous distribution, there is a problem in the meaningful assessment of gold reserves (Nicholson *et al.*, 1966;

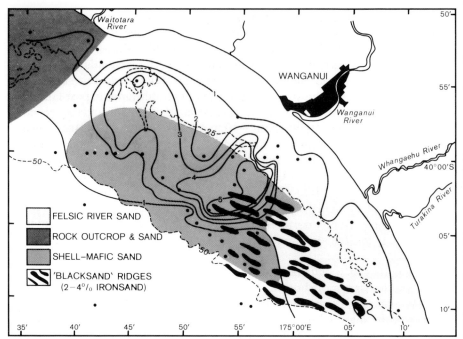

Fig. 5.4. Distribution of ironsand in sediments off Wanganui, New Zealand. [From Carter (1980), Fig. 4.]

Ingamells, 1981). In land-based deposits, for example, gold concentrations of 0·1 ppm are presently considered to be economic. In order to make a statistically valid assessment of reserves at these levels, drill hole diameters of about 1 m are required. This would make an offshore exploration programme involving drilling very expensive, except at very much the reconnaissance level where spot samples are taken. Usually, seismic measurements would be made to optimize the drilling programme by suitable selection of drill sites.

In New Zealand, alluvial gold mining was the method most frequently used for the recovery of gold during the "gold rush" days of the nineteenth century. Although the gold miner has long ceased to be a significant element in New Zealand society, a cursory inspection of the New Zealand Mining Handbook of 1906 (Galvin, 1906) shows clearly the early importance of gold to the New Zealand minerals industry and indeed to the early development of the New Zealand economy. This is well illustrated by Williams (1974), who devoted five chapters to gold in New Zealand. As shown by Williams (1974, p. 61), detrital gold was of considerable importance, making up 57% of the total gold mined in New Zealand. Principal interest was in the river beds of Westland, Marlborough and Otago in the South Island (Williams,

1974, p. 70), although some mining of the beach sands of the west coast of the South Island took place (Williams, 1974, pp. 69, 78, 138–140; McPherson, 1978). In Otago, the gold was derived from weathering of the Otago schists. The beach deposits on the west coast of the South Island are referred to as auriferous ironsand by Galvin (1906) and are predominantly ilmenite with minor magnetite, the ilmenite probably being derived from acid plutonic and metamorphic rocks in the hinterland. The heavy minerals have been concentrated by sedimentary cycles in the Pleistocene and Holocene to produce locally rich beach and dune deposits. Because the Westland shelf receives larger volumes of river-borne sediment than is the case off Taranaki, any such ilmenite deposits offshore would be rapidly buried, and this accounts for the almost complete absence of offshore ironsands here (Carter, 1980). In beach deposits, gold was discovered in the black sands south of Greymouth on the west coast of South Island in 1865 (Morley, 1981). The prevalence of the gold is, however, dependent on tides and storms showing that replenishment occurs. As far as mining goes therefore, these deposits may be regarded as transient (McPherson, 1978). In addition to gold, greenstone has been extracted from these beaches. Nonetheless, the early miners were so efficient that only limited amounts of extractable gold remain in New Zealand Quaternary Placer deposits, although with the high price of gold at present some companies have become interested in alluvial gold mining once again.

There have been some previous investigations of gold off New Zealand. The regions of potential interest are the drowned river valleys off the old gold-working areas—the west coast of the South Island, Foveaux Strait, Otago and the Coromandel Peninsula— where erosion during the glacial era may have led to accumulation of detrital gold (Katz and Glasby, 1979). Nonetheless, with its high specific gravity, gold would not be expected to be transported far offshore. Further, the possibility of reworking this material in a modern hydraulic regime or of post-depositional burial by modern sediments means that only a rich deposit would justify the exploration programme necessary to find and exploit such a resource. It seems unlikely that such an economic gold deposit will be found offshore in New Zealand. During 1976–1977, Newmont Pty Ltd and ICI formed a consortium to prospect for gold at depths up to about 75 m off the mouth of the Clutha River in Otago, one of the richest rivers in New Zealand in the gold rush days. However, they held the prospecting licence for only a few months before it was relinquished. In 1980, CRA Exploration Pty (Melbourne) was granted prospecting licences for 2 years to survey this area as well as off Southland and the west coast of the South Island (Fig. 5.2). The areas of these licences were modified somewhat in 1982. A reconnaissance programme of seafloor grab sampling and seismic profiling has been carried out by CRA, sediment samples being collected in a grid every 5 km and 1 km to a maximum depth of 150 m. The object of these surveys is to locate gold deposits which can be mined as high volume (5–10 million m^3/yr) low grade (150 mg gold/m^3) ores. In these areas, the gold is generally fine-grained (1000 μm). To mine such a deposit (if found) would be capital intensive requiring a mining ship costing of the order of $100 million. However, CRA has spent $500 000 over a period of 3 years on its gold prospecting programme and is intending similar levels of spending in future, concentrating on a deposit at a depth of about 100 m off Hokitika on the west coast of the South Island. Excellent summaries of this CRA work have been presented by Price *et al.* (1982) and Price (1983). In 1984 a vibrocoring survey of this region was undertaken. By way of background, descriptions of sediment distribution on the Otago shelf off the Clutha River (including the effect of sea-level changes and sediment dispersal off the mouth of the Clutha) have been given by Andrews (1973), Carter and Ridgway (1974) and Bardsley (1977).

A further area of interest could be off the Coromandel Peninsula in the north of New Zealand, particularly in the Thames Estuary of Waihi Beach. The Hauraki goldfield on the Coromandel Peninsula was a major gold mining area last century producing 240 tonnes of gold–silver bullion from some 40 separate mining areas between 1861 and 1952 (DSIR, 1981) and the "Thames Bonanza" has been well documented (Downey, 1935; Williams, 1974; Ridge, 1976; DSIR, 1981). Gold is thought to be present as tailings in the wharf at Thames (gold used to be mined in Thames itself) and also in the Firth of Thames as tailings from the Martha Mine. The sedimentology of the Firth of Thames has been described by Greig (1982). Gold is also known in Coromandel Harbour and in Collingwood Harbour, Nelson, both of which have auriferous catchments (gold in Collingwood Harbour is derived from the Aorere River). Capricorn Mining Co. Ltd presently has one prospecting licence and two applications for prospecting licences around the Thames mudflats in the Hauraki Gulf. CRA had an interest in the area but withdrew. From an environmental standpoint, marine disposal of tailings is an option for the proposed gold mine at Waihi in the Coromandel Peninsula (Waihi Gold Company, 1983). It should be noted that the Coromandel is perhaps the most mineralized area of New Zealand and an area of contention between mining companies and environmentalists. In addition, Western Mining has discovered significant quantities of detrital cinnabar in Whangaroa Harbour, Northland. Areas such as the Thames Mudflats and Coromandel Harbour differ from other prospects in that the prospectors are looking for high grade, low volume deposits.

For those wishing to study heavy mineral deposits off New Zealand, it is worth noting that the west and south coasts of New Zealand have been described as exposed, high energy shores (Pickrill and Mitchell, 1979; Reid, 1981). Forewarned is forearmed!

Recently CCOP/SOPAC organized cruises to look for offshore deposits in the Solomon Islands. The aim of this project was to investigate the gold content of terrigenous sediments transported from the Gold Ridge area of central Guadalcanal (an area where, as the name implies, gold is known to occur) by the Matepono River. In all, 88 samples (including 53 cores) were collected off the mouth of the Matepono River and sub-bottom reflection profiles were run. Of the 172 offshore sediment samples and four beach samples, three offshore and two beach samples contained detectable amounts of gold; the maximum gold content of the -120 mesh size fraction of the offshore sediments was 0·66 ppm. Whilst this was a very thorough study of the area (Turner et al., 1979), the results do not appear very encouraging from a commercial standpoint.

In Papua New Guinea, significant deposits of titaniferous magnetite are found in beach sands, particularly in the southeast Papuan beaches. These deposits are fully documented by Knight (1976) (cf. The Delegation of Papua New Guinea, 1981). In addition, CRA Exploration Pty (Melbourne) has recently purchased a 100% interest in three exploration tenements covering deposits of alluvial chromite located in three bays on the Marobe Coast south of Lae. Resources of 4·5 million tonnes of chromite have been established, and further evaluation, metallurgical and marketing studies have been initiated. Gold placers have also been reported (Galtier, 1984).

In Vanuatu, samples of iron-rich beach sands were collected from Port Patteson and 12 were analysed (CCOP/SOPAC Annual Report, 1981). It was established that the iron contents of the sediments were less than 15% (cf. Galtier, 1984).

In the Solomon Islands, the near-shore zone around San Jorge Island, Santa Isabel Province, was sampled for possible chromite-bearing sediments (CCOP/SOPAC Annual Report, 1982); small amounts of chromite were found (cf. Galtier, 1984).

In Fiji, Green (1970) has drawn attention to the magnetite-bearing sands on beaches near Singatoka and in areas in Nandi bay and off the coast further south (Hirst and Kennedy, 1962; Houtz and Phillips, 1963). A reserve of 1·8 million tonnes of magnetic mineral containing 94% Fe_3O_4 and 6% TiO_2 and making up 5·5% of the total material in the sand was established in the deposit near Singatoka. The possibility of this sand containing gold was discussed in view of the known gold deposits in Fiji (Whitelaw, 1967; Colley, 1976; Ridge, 1976; Colley and Greenbaum, 1980; Greenbaum, 1980) but discounted in view of the low gold contents (of the order of 0·3 ppm) in the samples analysed. The vegetation of the Singatoka dunes and the possible environmental implications of mining have been discussed by Kilpatrick and Hassal (1981). Brief summaries of the beach deposits in Fiji are given by the Mineral Resources Division, Fiji (1974) and Anonymous (1981a) and in Vanuatu by Anonymous (1981a).

As previously stated, the principal requirements for heavy mineral deposits are a suitable source rock and a high-energy environment. Many of the subtropical islands of the SW Pacific are virtually surrounded by coral reefs which dissipate energy and prevent beach formation. In some of these islands, beaches are found only on the windward side of the island. In such cases, heavy mineral beach deposits would only be expected there. Further, a suitable source of rock is required. Although andesite is a source of titanomagnetite in New Zealand and occurs in some of the Pacific islands, many islands are coral reefs. These factors taken together suggest that extensive placer deposits would not be expected off many islands in the SW Pacific.

5.6 Bauxite

Bauxite is produced by tropical weathering, and Australia is presently the world's leading producer of this mineral (Woodcock, 1980). Taylor and Hughes (1975) reported bauxite deposits in a brackish water, lake Te Nggano on Rennell Island, part of the western Solomon Islands. Bauxite deposits are also known on Waghina Island, also in the western Solomon Islands. A preliminary survey by CCOP/SOPAC off Waghina Island to search for potentially economic-grade offshore bauxite deposits, however, failed to locate any bauxite deposits in the upper 0·8 m of the sediment column (Taylor, 1977).

5.7 Precious Coral

The nature, occurrence and history of exploitation of precious coral have been described in some detail by Grigg (1974, 1976, 1977, 1981), Grigg and Bayer (1976), Eade (1978), Grigg and Eade (1981) and Wells (1981). This section is merely a brief summary of his work. Dr R. W. Grigg (University of Hawaii) is thanked for his most helpful comments on this topic.

The use of precious coral for ornamentation dates back thousands of years. Until the beginning of the nineteenth century, precious coral was known only in the Mediterranean, but today it is recovered almost exclusively in the Pacific. Unfortunately, the industry, with its reliance on tangle nets and dredges, follows the pattern of discovery, exploitation and depletion so that once large deposits of precious coral off Japan, Okinawa and Taiwan are now largely depleted.

At present, the industry is centred in Taiwan. To give some idea of scale, in 1981 some 250 Taiwanese and 100 Japanese boats were working beds in the Emperor Seamounts north of Hawaii as well as traditional grounds in the far Western Pacific (Grigg, 1981). A total of 226 000 kg of precious coral was recovered, the greatest majority from the Emperor Seamounts. The total value of coral sold in 1980 was about US$50 million (excluding the value of gold settings, etc.). It is therefore quite a sizeable industry, indeed one of the larger offshore mineral industries in the Pacific.

Most corals in the world are too soft and porous for use in jewellery. Precious corals, on the other hand, form tree-like growths, are much harder and are found in deep waters (up to 1200 m). They are coelenterates, not minerals. It is their hardness (3 on the Mohs scale, the same as ivory and pearl) which permits them to be polished into jewellery. In corals, the degree of hardness and colour are determined by the elements that make up the skeleton. Pink or red coral is about 87% calcite and 7% $MgCO_3$ with small amounts of organic matter, iron oxide and calcium sulphate. The amounts of these constituents varies widely both within and between species. Of the 26 known species of *Corallium*, only seven have a commercial value. Precious corals are classified as black, pink, gold and bamboo. Grigg and Eade (1981) have described the nature of the black and pink corals as follows:

Black corals are composed of horny protein and have annual growth rings. Most species of commercial value occur above the thermocline in warm water. In Hawaii, they are found between 25 and 100 m on hard substrata preferably limestone. Their larvae avoid bright sunlight and therefore often cluster under overhangs, in environments where currents up to 3 kts are common. The commercial species of black coral in Hawaii grow about 6 cm/yr and live for up to 100 years. They reproduce both sexually and asexually and the larvae appear to be gregarious. Their mortality rate is about 10% per annum and is approximately in balance with recruitment. They are killed largely by boring organisms, which weaken the base and cause toppling. The maximum sustained yield is about 4% per annum. Black corals are ubiquitous in the Pacific. At present, no commercial beds of black corals have been discovered in the SW Pacific but encouraging sites have been found in the Solomon Islands and to a lesser extent Vanuatu, although no really promising sites have yet been discovered. These surveys are, however, only at the reconnaissance level and further detailed study is needed. Large beds of commercial-grade black coral have, however, recently been discovered in Tonga. Black corals have also been recovered on fjord walls in New Zealand (Grange et al., 1981; Grange, 1985; Richardson, 1985); these beds presently have a protected status since 1980, although this is being challenged.

Pink corals are composed of high magnesium calcite and their colouring is related to their organic content. In the Far East, *Corallium* grows in water depths of 100–300 m. However, recent work in Midway Island has shown that in Hawaiian waters there are both "shallow" and "deep-sea" pink corals found at 400 m and 1200 m respectively on substrata of limestone or basalt. Pink corals are most abundant in areas characterized by a rise, seamount or gently-sloping terrace swept by strong currents (up to 3 knots) and free of sediment. Sediment can smother young corals and abrade the stalks of older ones causing toppling. Steep slopes where debris is moving downslope may be unsuitable for coral growth. One species of pink coral in Hawaii grows abut 1 cm/yr and lives up to 100 years. Their mortality is about 6% per year and approximately balances recruitment. Most colonies are killed by boring causing toppling. The maximum sustained yield is about 2% per year. Pink corals occur in commercial quantities only in small beds. Food supply and temperature seem to have little effect on their distribution. In the northern hemisphere they come no further south than 22°N. Gold corals are apparently confined to Alaska and Hawaii (completely different varieties). In general, the natural history of these precious corals is not well known, although they do provide a habitat for other fauna.

The manufacture of jewellery from precious coral is quite labour intensive. In Taiwan, there are about 1000 small factories. As such, a precious coral industry would make a useful supplement to the tourist industries in many of the islands of the SW Pacific. For this reason, CCOP/SOPAC began a series of surveys to locate the occurrence of precious coral in the SW Pacific and to map its abundance. Possible sites for exploration were based on existing knowledge of the bathymetry and currents of the islands and the known ecology of corals. The surveys combined bathymetric surveys of suitable sites with dredging and underwater photography. A typical example of this approach is given by Lewis *et al.* (1980). Grigg and Eade (1981) report the following success rate in recovering precious coral at various localities:

Solomon Islands	29 out 138 dredge hauls
(New Georgia Group to Makira Island)	
Santa Cruz Islands	6 out of 40 dredge hauls
Vanuatu	7 out of 74 dredge hauls
Fiji	1 out of 16 dredge hauls
Tonga	3 out of 55 dredge hauls
Samoa	6 out of 36 dredge hauls
Northern Cook Islands	3 out of 26 dredge hauls
Southern Cook Islands	none out of 2 dredge hauls
Gilbert group, Kiribati	none out of 88 dredge hauls
Phoenix group, Kiribati	none out of 7 dredge hauls

In summary, precious coral has become a moderate-sized industry in parts of the Pacific. Virtually all of the fishing activity has been concentrated in the North Pacific. The use of tangle nets and dredges plus rapid exploitation has led to rapid depletion of the bed. Only in Hawaii has a serious attempt been made to exploit these deposits in a manner which takes into account conservation principles. Exploration for precious coral in the SW Pacific has revealed no commercial beds but further exploration appears warranted. Discovery of commercial precious coral beds in the SW Pacific could be a boon to the local tourist industry if properly managed with recovery made on an ecologically sound basis by local divers. However, if large-scale outside operations become involved, the beds could be rapidly depleted with little benefit to local economies. Care should therefore be taken in developing these ecologically delicate resources.

In New Zealand, Paua shell is used in the manufacture of jewellery. The local price is NZ$800–1000/tonne, but it is hoped to export shell at NZ$2500–5000/tonne (Anonymous, 1982b).

5.8 Environmental Impact of Potential Beach and Near-shore Mining

Because marine mining is in its infancy, the potential environmental impacts are not particularly well known, and considerably more baseline data are required before definitive statements can be made (Hails, 1974; Cruickshank and Hess, 1976; Down and Stocks, 1976, 1977; Advisory Committee on Aggregates, 1976; Gayman, 1978; Bohlen et al., 1979; de Groot, 1979a,b; Gross et al., 1979; Nichols, 1979; Brinkhuis, 1980; Kirk, 1980; Alther and Wyeth, 1981; Herbich, 1981a,b; Ritchie, 1981; Zabawa et al., 1981; Howorth, 1982a; Vellinga, 1982). Nonetheless, Owen (1977) has shown that mining on the continental shelf can disrupt sediment budgets and interfere with sediment dispersal patterns, resulting in coastal erosion and the formation of navigation hazards. It can also affect offshore fisheries and this point has been stressed by Ward (1977), Morton (1981) and Anonymous (undated). Certain environments are shown to require special attention and these include fishing grounds, semi-enclosed embayments and coral reefs. The panel on Operational Safety in Marine Mining (1975) emphasized the importance of having clearly defined environmental standards set in advance of marine mining. They also state that "bays, estuaries and the inner continental shelf are among the most fragile ecosystems in the world." Because of the small land area of many of the islands in the SW Pacific, the coastal zone can be described as their most critical resource.

In fact, a coastline represents an approximate equilibrium between erosional and depositional processes (if it did not, it would either advance or retreat) (Aubrey, 1981). Removal of material can upset the material balance and disturb this equilibrum. Perhaps the classic case of this is the loss of Hallsands fishing village in South Devon in 1917 as a result of dredging shingle in the intertidal zone between 1897 and 1902, which has been particularly well documented (Hails, 1975). Another well-described case of coastal erosion due to sand mining is in Swansea Bay, Wales (Heathershaw et al., 1981). The effect of inland development of coastal erosion has been shown in California (Kuhn et al., 1980) and off the east coast of New Zealand (Gibb, 1981). In Australia, the effects of dredging about 15% of the area of

Botany Bay have been described by Jones (1981) and Jones and Caundy (1981). The importance of beach conservation in Queensland is shown by the publication of almost 50 issues of *Beach Conservation,* the newsletter of the Beach Protection Authority of Queensland. Coastal sand transport there has also been demonstrated by Jones and Stevens (1983). In New Zealand, property loss due to beach erosion has occurred at Ohiwa Harbour, Bay of Plenty (Gibb, 1977, 1979b), although not (in this case) due to offshore dredging, as well as other locations (cf. de Lacy, 1977). Erosion due to dredging has also taken place at a number of locations in the Bay of Plenty (Healy *et al.*, 1977; Harray and Healy, 1978; Dahm and Healy, 1980; Healy, 1980a,b), of which Papamoa is the most significant, and elsewhere in New Zealand (Kirk, 1977a, 1978). The dynamic nature of New Zealand's coastline has been known for many years (Holmes, 1919; Furkert, 1947), and reviews have been presented by Kirk (1977b), Kirk *et al.* (1977), McLean (1978) and Tortell and Cornforth (1982). McLean (1978), in particular, has shown that coastal erosion around New Zealand is episodic, and that changes occurred within the last century. A problem in the study of these changes is that insufficient time-frames have been recorded at sufficient locations to describe the changes well. Detailed sediment budgets in Canterbury beaches have been recorded by Kirk and Hewson (1978) and Gibb and Adams (1982).

A particularly significant example of coastal erosion as a possible result of offshore dredging has occurred at Mangatawhiri Spit in Northland, New Zealand. Erosion of the tip of the spit occurred in the early 1960's following dredging of almost 380 000 m^3 of sand from the adjacent Little Omaha Bay between 1942 and 1963 (Ward, 1977). The development of the spit as a beach resort commenced in 1971. In the winter of 1978, storms of hurricane proportion resulted in the erosion of most of the beach-front sections. As a result, the building of groynes for beach protection commenced in 1979. Because of these erosional losses, legal action was taken and the landowners sued Broadlands Ltd (the property developers) and Broadlands sued the Crown (for permitting offshore dredging too near the spit and therefore accelerating the spit erosion). These cases were settled out of court in 1980 and 1981, respectively, with a condition that the results were not to be made public. A description of these events is given by Johnston (1980). Healy (1981) has criticized the sedimentological understanding of the various consultants brought in to advise on this problem and the remedial actions offered.

Whilst the literature on Mangatawhiri Spit is somewhat contradictory (Schofield, 1967, 1975b, 1978, 1979, 1985; Riley *et al.*, 1985), it is probable that a combination of offshore dredging of sand between 1942 and 1963 and the levelling of the fore-dune by the property developers in the 1970's has contributed to the severe erosion of the spit in the vicinity of the subdivision by the storms of 1978. Other factors such as the historical sea-level rise and the possible increase in storminess may well be involved but the two man-made factors explain much of the erosion that has taken place at Mangatawhiri Spit. These ideas are not unexpected in view of Gibb's (1979a) conclusion from a nationwide survey that sediment supply has been the dominant regional factor in shoreline movement in New Zealand during the last 6500 years (i.e., since sea level stabilized) and that man's activities have significantly interferred with sediment supply at many places along the coast resulting in rapid localized erosion or accretion. To give some idea of the extent of this problem, during the past century about 56% of the open, exposed part of the New Zealand coastline has remained static, 25% eroded and 19% accreted (Gibb, 1979a). These changes show that New Zealand's coastline is dynamic and susceptible to change. Often the changes have resulted from breakwaters intercepting part, or all, of the longshore drift, particularly in the vicinity of harbours. The problems of coastal erosion due to offshore mining in New Zealand are therefore unlikely to be trivial. The large, inconclusive literature on Mangatawhiri Spit shows the problem of studying such erosion which takes place as a result of many years of activity. Environmental problems have also been encountered in mining sand from the beautiful, remote Parengarenga Harbour, particularly erosion of the adjacent Kakata Spit. These problems have been discussed by van Roon (1981) and it has been suggested that Parengarenga Harbour should become a marine reserve (Maritime Policy Branch, 1980).

By contrast, Landsea Minerals Ltd took out a prospecting licence in 1975 to assess sand and gravel reserves from a series of locations on the inner shelf off eastern Northland, New Zealand, located off rocky cliffed coasts at depths of 30–50 m. The areas were subject to very detailed study (Anonymous, 1974a,b; Bioresearches Ltd, 1974; Gibb, 1974; Ritchie and Saul, 1974, 1975; Landsea Minerals Ltd 1975; Ward, 1977; Gillie, 1979). From a detailed study of six proposed locations, Gillie (1979) argued that beach erosion would not be expected at five of these locations, principally because these areas lie off rocky areas and are not fed by beach systems. Possibly, therefore, these areas represent acceptable sites for offshore mining. However, the licence was relinquished in 1981 because the economics of the operation were not favourable.

The potential damage that can be done by unplanned offshore mining is now recognized in some countries. In Holland, dredging for sand and gravel is not permitted within 20 km of the coast and in Britain within 5 km of the coast. In New Zealand, Healy (1980a) has shown the advantages to coastal management of considering scientific factors *before* the development proceeds, and excellent reviews of this problem have been given by Gibb (1982) and Healy (1982a). In some cases, this leads to the application for development being declined. Such an example is at Mataora Beach, east Coromandel, where a scientific evaluation led to the application for a licence to mine the deposit being declined (Healy, 1982b). Gibb (1983, 1984) is developing a technique called coastal hazard mapping to facilitate the more rational use of beaches in New Zealand and the government seems to be becoming more involved in attempts to control marine erosion (Anonymous, 1984). It must be emphasized in this regard that beaches represent only a thin and fragile boundary between the land and sea. Reports by the U.S. Army Coastal Engineering Research Center (1977) and the Ministry of Transport (1980) are important texts in coastal conservation.

These factors pose problems for any rational development of offshore mining in the SW Pacific where conservation of both the coastline and the fisheries potential as well as the impact on tourism must be considered. Whilst such problems are obviously important in the developed countries of Australia and New Zealand, they are at least of sufficient size that alternative sites of offshore mining can be considered and sufficiently developed that adequate environmental safeguards can be written into any mining licence (where the political will exists). It should be pointed out, however, that this environmental awareness is late in coming. New Zealand, for example, only set up a Ministry of the Environment in 1972. Further, not all aggregate mining is up to an environmentally acceptable standard. For example, aggregate is mined in the Wellington region by open-cast mining of the greywacke cliff face at Ohiro Bay, one of the finest coastal walks in Wellington. This has caused major disfigurement of a scenic area. The country is also geologically youthful and the problems of erosion (particularly man-made) have been (and continue to be) one of the major problems in the development of the land. The problems are, however, much more acute in the small tropical island nations of the SW Pacific which have (almost by definition) limited onshore mineral resources and are usually surrounded by coral reefs which are particularly susceptible to environmental damage; they are also likely to be dependent on tourism as a major source of income for many years to come, have large coastal zones relative to their total land area and are poor. The specific environmental problems of the SW Pacific islands have recently been discussed by UNESCO (1981) and the difficulties of conservation in Micronesia described by Wilson (1976). Of particular interest here is the 1982 report of the conference on the human environment in the South Pacific held in Rarotonga which emphasized a number of environmental problems in the SW Pacific islands, one of which is beach loss and coastal erosion from aggregate mining which poses difficulties in half the countries of the region. The significant of the conference is that it attempted to formulate a common approach to the environmental problems encountered throughout the region and will undoubtedly serve as a cornerstone of future policy.

The importance of coral reefs was noted by the conference and is well stated by Johannes (1975), who points out that coral reefs are probably the most extensive shallow marine communities on earth and amongst the most biologically productive of all natural communities, marine or terrestrial. In the tropics, where man's terrestrial protein sources are often inadequate, reef fish and shellfish provide high quality protein. Reefs act as breakwaters which permit the continued existence of about 400 atolls and numerous other low lying islands; it has been estimated, for example, that the reefs dissipate 300 000 kW against the windward side of Bikini Atoll. The gross production of calcium carbonate by reef communities has also been estimated to be about 25–125 $kg/m^2/yr$. These factors, quite apart from aesthetic values, stress the importance of coral reefs to tropical islands. Nonetheless, reefs are fragile ecosystems susceptible to environmental damage, of which dredging is but one cause (cf. Brock et al., 1965, 1966; Banner, 1974; Dahl, 1977; Grigg, 1979, 1981; Chansang et al., 1981; Porter et al., 1981; Salvat, 1981; Bakus, 1983; Gomez, 1983). The problem has been starkly outlined by Rogers (1981). The convergence of population pressure, tourism and industrialization in such coastal areas is an important factor in environmental degradation. In some countries (such as the Philippines), coral seems to have been grossly over-utilized both for aggregate production and export, such that over 50% of that country's coral reefs are in progressive stages of destruction (Gomez, 1980; Gomez et al., 1981; Wells, 1982). A similar situation applies in Indonesia (Polunin, 1983). Often, this coral has been harvested illegally. In 1976, the Philippines exported 1800 tonnes of coral. These problems have been discussed at some length in Anonymous (1981a). Two principal problems resulting from the recovery of aggregate in the SW Pacific islands are noted; over-exploitation of beach resources leading to erosion and loss of beach-front properties and increasing turbidity of seawater leading to damage of the coral ecosystem (including the reef fisheries), sometimes severely. In a question and answer session at the CCOP/SOPAC meeting (Anonymous, 1981a), representatives from the Cook Islands, Fiji, Kiribati, Papau New Guinea, Solomon Islands, Tonga Vanuatu and Western Samoa all stated that aggregate recovery is a major problem in causing coastal degradation in their islands and that present regulations are inadequate to control this development properly. The problem of beach erosion due to the mining of beach sands appears to be particularly critical in the Cook Islands (Anonymous, 1983a,b). McClymont (1982) has stated that environmental impact assessment is still in its infancy and that would be the situation in many of the SW Pacific islands. This is therefore a regional problem and clearly an area where major improvements in regulatory control and practices are required. Similar coastal erosion has also caused problems in Hawaii (as at Ala Moana Beach, Waimea Beach and Kailua Bay) (cf. Campbell and Hwang, 1982; Hollett and Moberly, 1982), but here a conservation zone (extending from 400 ft to 3 miles offshore) has been created to prevent the further taking of sands and corals. Similar (or more stringent) measures would seem to be required in the SW Pacific islands. In this regard, some of the more traditional conservation measures used in Oceania might seem appropriate (Johannes, 1978).

A particularly valuable series of initial studies of coastal erosion in Kiribati and Vanuatu has recently been undertaken by Howorth (1982b–d, 1983a,b). As an example, Tarawa (Kiribati) has a long coastline per unit area of land and a high population density. Coastal erosion has been accelerated by a number of developments such as the building of causeways, extension of port facilities, building of a desalination plant, and reclamation, refuse disposal and sand/coral rock aggregate mining. These developments have implications not only for the coastline itself but also for fishing and pollution in the lagoon. Coastal protection measures presently involve the use of coral rock. The supply of these local coral boulders in south Tarawa is, however, becoming restricted, and removal of this material from the ocean reef platform may in itself encourage erosion of the ocean beaches. Although Howorth's study is preliminary, it does emphasize the importance of man-induced coastal erosion and the fragility of the shoreline equilibrium in these islands and highlights the need for scientifically based coastal surveillance and protection programmes which, because of shortage of funds and of skilled personnel, must come from outside. It also nicely outlines some of the pertinent problems involved and shows the inadequacy of some of the existing coastal protection methods, a situation which arises from a lack of understanding of the shoreline processes involved. It is therefore hoped that CCOP/SOPAC can play a role in the development of such coastal surveillance problems in the islands of the SW Pacific.

In addition to the above, Grigg (1977) has outlined the devastation to the precious coral in much of the Pacific by over-exploitation using tangle nets and dredges. It is hoped that, if commercial beds are discovered in the SW Pacific, they are subject to more scrupulous exploitation than has occurred previously.

In spite of all this, many of the SW Pacific islands are beginning to look offshore for sources of aggregate. On Tonga, for example, the use of beach deposits on the windward side of Tongatapu is approaching danger point as far as coastal erosion is concerned. Yet, suitable offshore deposits which will not facilitate coastal erosion or affect fisheries production when mined are not easy to find. It must be emphasized that many Pacific island countries are poor so that heavy capital investment in prospecting or in offshore mining equipment cannot be contemplated. (Tonga presently pays $6/tonne for its beach sand and any significant increase in cost would be a hard blow to the local economy.) A false decision could, however, significantly increase coastal erosion on these already small islands where land is at a premium and also affect fisheries production—much of which is caught inside the reef. The imposition of a 20 km or even 5 km limit for offshore mining seems a luxury that these islands can ill afford. A mistake in permitting an unwise project could, however, be environmentally very serious for some of these islands. Down and Stocks (1977) make a particularly valuable point that environmental concerns (as measured by environmental legislation) are pronounced in countries which are *affluent* and have *high public awareness* of the problems involved. Population density appears to be only a minor factor in influencing environmental legislation. The only countries which have made effective efforts to control the undesired impacts of mining are those with high per capita income. The SW Pacific islands should be particularly aware of this trap. A useful review of coastal zone management methods in various countries has been given by Mitchell (1982).

From the above, it is clear that the setting up of strict environmental guidelines for the exploitation of near-shore mineral resources is a priority for the small, tropical island nations of the SW Pacific with limited resources and delicate reef ecology if severe environmental problems are to be avoided in the not-too-distant future. Perhaps it is not too much to ask for this problem to be considered a priority aid project from the developed world.

6

Submarine Phosphatic Sediments of the SW Pacific

D. J. CULLEN

New Zealand Oceanographic Institute, Department of Scientific and Industrial Research, Wellington North, New Zealand

6.1 Introduction

Submarine phosphorites and phosphatic sediments are widespread in the SW Pacific, where they occur in a variety of environments and sedimentary associations ranging from estuarine to mid-oceanic. Specific deposits are known locally in estuaries, on the continental shelf margin and upper slope, on isolated submarine ridges, in depressions and saddles on marginal plateaux, and, in the sub-tropical and equatorial regions, on or near the tops of seamounts (Fig. 6.1). Despite the diversity of bathymetric and hydrological circumstances implicit in this distribution, an obvious common characteristic of the submarine phosphates in the region, as indeed of those elsewhere, is their association with areas of minimum sedimentation— usually isolated, scoured by current action, and, not infrequently, impinged upon by upwelling deep bottom waters.

The SW Pacific is specially favoured in having two submarine phosphatic deposits of exceptional interest, namely, some very young Quaternary phosphates off the East Australian coast, and Late Tertiary deposits on Chatham Rise, east of New Zealand, of sufficient size and quality to rate as possessing real commercial potential.

The majority of submarine phosphates in the SW Pacific are demonstrably formed by replacement of pre-Quaternary calcareous deposits, either dur-

Sedimentation and Mineral Deposits
in the Southwestern Pacific Ocean
ISBN 0-12-195870-1

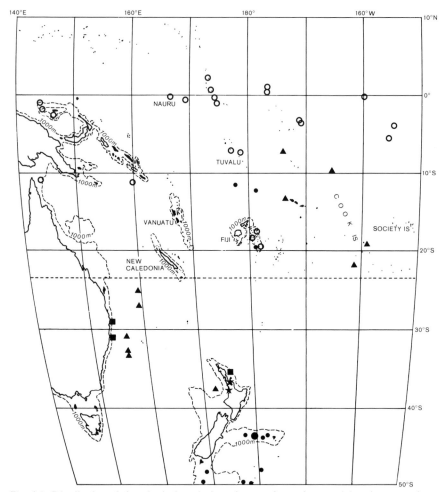

Fig. 6.1. Distribution of phosphatic deposits in various marine and terrestrial environments in the SW Pacific Ocean. (●,⬤) submarine ridges and plateaux; (▲) seamounts; (■) continental shelf margins; (○) terrestrial guano deposits; (★) estuarine deposits.

ing diagenesis or after final consolidation, and some have themselves been subsequently partly replaced, impregnated or coated by a variety of other minerals. Small, localized phosphatic concretions of Pleistocene–Holocene age, along the East Australian continental shelf margin, may have formed by precipitation from interstitial pore waters during early diagenesis, while phosphatic veneers on volcanic cobbles on Tasman Sea seamounts may have precipitated directly from sea water.

The actual phosphorus content of submarine phosphatic sediments in the SW Pacific is highly variable, ranging from a few percent to about 11·4% P

(26% P_2O_5). It has been customary to apply the term "phosphorite" only to those sediments containing in excess of 7·9% P (18% P_2O_5) (Bushinskii, 1969), but the term is now often used to describe rock or unconsolidated sediments of marine origin that contain more than 5% P_2O_5. Sediments with lower phosphate values are described as phosphatic limestones, phosphatic calcarenite, phosphatized coral, etc. Only in one area of the SW Pacific—the crest of Chatham Rise, east of New Zealand—are submarine phosphorite deposits known to attain a size and grade potentially appropriate for commercial exploitation. Because of this, the Chatham Rise phosphorites have, in recent years, been relatively intensely investigated by the New Zealand Oceanographic Institute, latterly in association with the German Federal Institute for Geosciences and Natural Resources.* Together with phosphatic sediments off the coast of northern New South Wales, they are the best known submarine phosphates in the region.

A number of criteria are suitable for the classification of submarine phosphatic sediments, such as their petrography, geochemistry, age, and sedimentary associations. For the purposes of this account, however, it is proposed to describe the phosphatic sediments of the SW Pacific in relation to their physiographic location on the sea floor, and to allow the similarities and differences exhibited by these other parameters to emerge in the course of the discussion.

6.2 Phosphatic Sediments in Estuarine Environments

6.2.1 Raglan Harbour

Phosphatic concretions have been reported by Sherwood and Nelson (1979) from a rather unusual environment—on a tidal flat in the upper reaches of the drowned river valley system of Raglan Harbour—on the west coast of North Island, New Zealand (Fig. 6.2). At this locality phosphatic concretionary bodies, which range in size from small spheres a few centimetres in diameter to lobate masses 0·55 m across, lie exposed on the surface of, or are partly embedded in, soft mud. Interspersed among them are live molluscs and dead molluscan shells.

Analysis of the concretions reveals that they consist of a cryptocrystalline matrix of disordered apatite (50% of the total), clay minerals (30%), calcite and aragonite (15%), and quartz and opaque minerals (5%). Molluscan shells enclosed within this material (Fig. 6.3) include the genera *Chione, Cyclomactra,* and *Zeacumantus,* all of which are represented in the modern local benthic fauna.

* Bundesanstalt für Geowissenschaften und Rohstoffe (BGR).

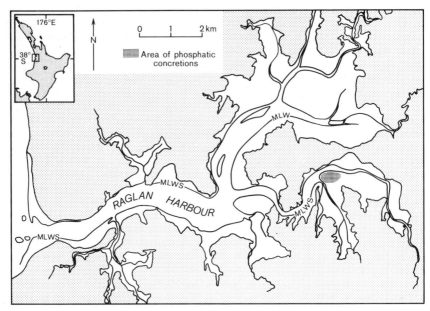

Fig. 6.2. Location of concretionary phosphatic deposits in the upper reaches of Raglan Harbour, west coast of North Island, New Zealand.

The most extensive spreads of concretions (Fig. 6.4) appear along the southern arm of the harbour inlet, peripheral to remnants of an old (<5000 yr) Holocene shore platform (Wellman, 1962), composed of an accumulation of close-spaced, loosely cemented vertical tubular burrows, and elevated some 0·3–0·4 m above the modern tidal flat surface. The exposed concretions are often heavily encrusted by barnacles and other marine organisms, and appear not to be growing at present. Concretions that occur *beneath* the platform surface, on the other hand, grade into the enclosing mud, and are thought by Sherwood and Nelson (1979) to be still forming today. They attribute the deposition of phosphate in the concretions to a prevailing highly alkaline environment created by the *post mortem* decomposition of large populations of burrowing animals. Since the burrows on the platform surface are now almost exclusively sediment-filled, this process, and any generation of phosphate in the sediment, would probably be severely restricted in the present environment, however.

There is, of course, no question of upwelling having played a part in the phosphogenesis in Raglan Harbour. In the open ocean, upwelling merely provides a mechanism for extracting phosphorus from enriched cold, bottom waters by precipitation in the warmer, shallower depth zones. In the case of Raglan Harbour, such a mechanism has been quite unnecessary. A

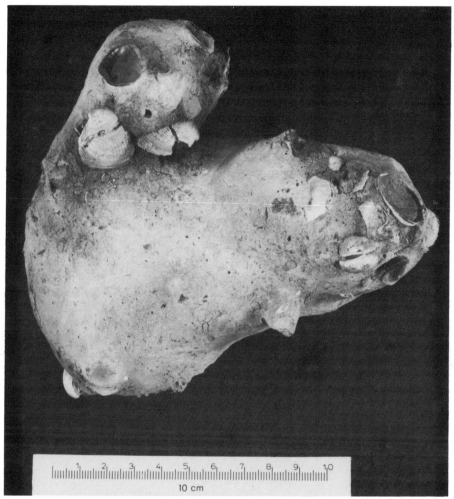

Fig. 6.3. Detail of a phosphatic concretion from the upper reaches of Raglan Harbour, showing enclosed bivalve (*Chione*) shells. (Photo courtesy of J. J. Whalan.)

plentiful supply of phosphorus seems to have been independently provided by decaying benthic organisms, and, in such a shallow coastal inlet, water temperatures would undoubtedly have been periodically raised sufficiently to encourage precipitation of any dissolved phosphates present.

As a point of passing interest, shore-line cliffs in the southern arm of the harbour and much of the hinterland are composed of lower Oligocene (Whaingaroan) sediments, similar in age to the formation that contributes so substantially to the phosphorite deposits of Chatham Rise (q.v.).

Fig. 6.4. Spreads of phosphatic concretions in Raglan Harbour. (Photo courtesy of Dr A. Sherwood.)

6.2.2. Hauraki Gulf

Phosphatic "stony aggregates" and molluscan and crustacean internal casts have been reported by Greig (1982) from the Hauraki Gulf—east of Auckland, New Zealand (Fig. 6.5). The phosphatic samples were recovered from six stations, in water depths of 26–36 m, across the northern entrance to the Firth of Thames between Ponui Island and Coromandel, and from a somewhat deeper station (approximately 50 m) off the NW tip of the Coromandel Peninsula. Petrographic and X-ray diffraction analyses of both aggregates and casts indicate that the matrix is, in each case, composed of cryptocrystalline apatite.

The phosphatic particles are not abundant, and comprise a maximum of only about 5% of the total gravel fraction of the sediments. The phosphatic stony aggregates occur mostly as rounded particles, less than 10 mm across, that contain biogenic carbonate as well as angular grains of quartz, feldspar, rhyolite and pumiceous glass. The similarity of these grains to grains in the

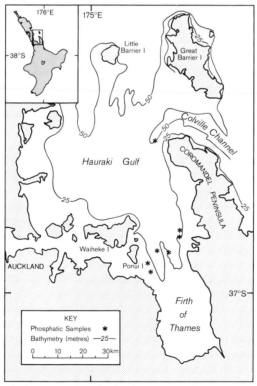

Fig. 6.5. Location of phosphatic deposits in the Hauraki Gulf and the Firth of Thames, northern North Island, New Zealand.

enclosing unconsolidated sediments has led Greig (1982) to suggest that the phosphate is autochthonous. Internal casts are usually more numerous and larger than the aggregate particles, and often measure several centimetres across. The casts are frequently complete internal moulds of recent estuarine mudflat, or subtidal to shallow-marine bivalve genera, e.g., *Chione, Nucula, Mactra, Maorimactra, Dosinia, Tawera* and *Pecten*. A small proportion (<10%) of the casts are from gastropods such as *Zeacumantus, Maoricolpus* and *Zeacolpus* which, like the bivalves are subtidal to shallow-marine genera. Phosphatized crustacean appendages, ranging from 5 to 20 mm in length, occur both loose and embedded in pale brown siltstone. Occasional fish vertebrae are also recorded.

On the basis of the associated faunal assemblage, Greig (1982) postulates that the phosphatization occurred during a low sea-level stand, possibly between 14 000 and 7250 yr BP, at which time the Firth of Thames and the southern part of the Hauraki Gulf may, in many respects, have been environmentally comparable with modern Raglan Harbour.

6.3   Phosphatic Sediments of the Continental Shelf Margin and Upper Slope

6.3.1   East Australian Coast

Two fundamentally different types of phosphatic deposit occur on the shelf margin and upper slope off the northern New South Wales coast of eastern Australia, between Port Macquarie and Evans Head (Fig. 6.6). The shelf in this region is narrow (25–40 km) and rather steeply inclined, with a smooth, convex profile.

Indurated brown, glazed, ferruginous phosphatic nodules, mostly 50–200 mm across, are relatively widely distributed where water depths lie between 200–300 m, while soft, dull grey, concretionary phosphatic masses less than 25 mm across have a restricted distribution within the 360–400 m depth zone (Table 6.1). Almost certainly, "ferruginous nodules" first described from this area by Loughnan and Craig (1962) are identical with the ferruginized *phosphatic* nodules subsequently recognized by von der Borch (1970) off Port Macquarie. Following Bowler's (1963) criteria for distinguishing between nodules and concretions, von der Borch proposed, a tripartite grouping (termed Groups A–C) of the East Australian offshore phosphate deposits on the basis of microscopic as well as megascopic features. However, his third subdivision (Group C) is no more than a conglomeratic, cemented variation of the second, i.e., the Group B ferruginized nodular phosphate. Most importantly, von der Borch recognized the

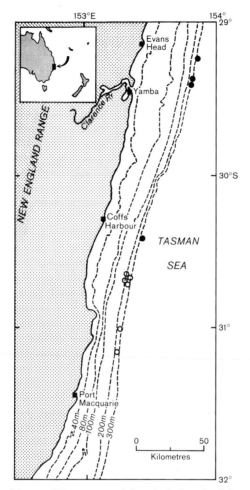

Fig. 6.6. Locations of the two different types of phosphatic deposit off the northern New South Wales coast, eastern Australia.(●) type I and (○) type II phosphorite.

essential lithological difference between the latter and the localized grey phosphatic concretions that he recovered off Evans Head, although he remained unaware of age differences, and assumed from the available paleontological and micropaleontological evidence a mid-Miocene age for both nodules and concretions.

In recent years, considerable interest has focused on the East Australian offshore phosphatic sediments, and several accounts have been published of their petrology (Marshall and Cook, 1980), geochemistry (Cook and Marshall, 1981), geochronology (Kress and Veeh, 1980; O'Brien and

Table 6.1. Characteristics of the two types of submarine phosphorite occurring off the northern New South Wales coast

	Type I Concretions	Type II Nodules
Colour	Off-white to yellow-brown and dull grey	Dark brown
Hardness	Soft, friable	Highly indurated with glazed goethitic surfaces
Size	Maximum 40 mm: mostly <25 mm	Typically 50–200 mm; sometimes cemented into slabs
Shape	Roughly spheroidal and ovoidal	Irregular, rounded to subrounded, knobbly
Depth range	360–420 m	200–300 m
Sedimentary associations	Essentially in unwinnowed sandy muds and muddy sands of the upper continental slope	A *remanié*-type deposit on an erosional "bench" at the shelf edge, associated with rounded pebbles, and phosphatized bones, fish teeth and molluscan shell
Extent	Very localized, off Evans Head, Yamba Head and Coffs Harbour	Widespread along continental shelf edge between Coffs Harbour and Port Macquarie
Age	Quaternary; some Holocene	>200 000 yr; probably mid-Miocene
Mineralogy	Silt-sized angular quartz, much biogenic carbonate, glauconite and goethite pellets	Pelletal glauconite, apatite and goethite with minor quartz, feldspar and biogenic carbonate of silt to coarse sand size; occasional pyrite framboids
	Matrix: pale brown cryptocrystalline carbonate–fluorapatite	*Matrix:* hydrated iron oxides and cryptocrystalline carbonate–fluorapatite
Geochemistry		
P_2O_5	7·5–15·3%	5·6–11·7%
Fe_2O_3	4·3–5·8%	18·4–45·0%
CaO	29·6–35·3%	10·5–26·6%
K_2O	0·6–1·05%	0·6–1·81%
Genesis	Maintenance of anoxic reducing conditions into the modern environment; possible involvement of bacterial activity	Following phosphatization in a reducing environment, subjected to strongly oxidizing conditions perhaps during a eustatic regression

Veeh, 1980) and associated sediments (Marshall, 1980). O'Brien and Veeh have re-defined a dual grouping of the deposits into Types I and II— equivalent to von der Borch's (1970) Group A and B + C, respectively—on the basis of lithologic, stratigraphic and bathymetric considerations. Recently, uranium-series datings have indicated possible Holocene ages for some of the Type I concretions. In view of this important development, a summarized account of the petrology/geochemistry, paleontology/stratigraphy, and depositional environments of the East Australian offshore nodular and concretionary phosphates would seem to be warranted.

6.3.1.1 *Petrology and geochemistry*

(a)   Type I: phosphatic concretions

Small, weakly indurated phosphatic concretions, less than 40 mm in diameter, are reported by Kress and Veeh (1980) and O'Brien and Veeh (1980)
in three samples collected from depths of 360–420 m off the northern New
South Wales coast between Evans Head and Yamba Head (Fig. 6.6).
Similar concretions have been found in a box-core sample (31°01.0'S,
153°18.7'E) collected in 1979 by the New Zealand Oceanographic Institute
in 450 m of water off Smoky Cape, some 60 km NE of Port Macquarie.
Poorly consolidated phosphatic sediment is also reported (O'Brien and
Veeh, 1980) as filling the calices of solitary corals dredged from comparable
depths off Coffs Harbour.

The smaller phosphatic concretions of this type are quite inconspicuous in
the sediments, and could very easily be overlooked or mistaken for ordinary
calcareous concretions. Indeed, their commonplace appearance provides an
object lesson in the necessity for rigorously checking the geochemistry of
concretionary masses in sediments. Perhaps if this were done more rou-
tinely, phosphatic concretions might be found to be more widespread than
hitherto suspected.

The Type I concretions range in colour from off-white to yellow-brown
and dark grey, and they have a friable, earthy consistency that tends to
become increasingly consolidated with increasing concretion size and age.
They are composed essentially of silt-sized angular quartz, relatively abundant unphosphatized foraminiferal tests and calcareous biogenic debris, and
rounded pellets of greenish glauconite and brownish goethite, with a pale
brown cryptocrystalline apatite matrix identified as carbonate–fluorapatite
by X-ray diffraction analysis. Idiomorphic apatite crystallites are occasionally revealed by the scanning electron microscope on internal surfaces of
foraminiferal tests. Von der Borch (1970) also mentions the presence of a
very thin (0·01 mm) layer of possible apatite as an exterior coating and lining
mollusc borings in the concretions, which he attributes to exposure on the
sea floor following the winnowing away of enclosing sediments. This feature
is not mentioned by later workers.

Normally, the Type I concretions contain between 7·5 and 15·3% P_2O_5,
although von der Borch (1970) reports an unusually high value of 21·2%
P_2O_5 from a sample off Yamba Head. The total iron content of these
concretions is generally low, with values ranging from 4·3 to 5·8% Fe_2O_3
(Table 6.2).

Table 6.2. Chemical composition of phosphatic concretions and nodules from the continental shelf margin and upper slope off northern New South Wales[a]

	Type I concretions				Type II nodules				Type II conglomerate:
	G7-8[b]	G7-6	G7-9[b]	G7-10[b]	1516(2)	1512	1516(3)	1516(4)	A6
SiO_2	28·30	24·61	23·50	19·40	21·91	13·15	14·24	14·96	14·58
TiO_2	0·37	—	0·30	0·28	0·22	0·19	0·18	0·23	—
Al_2O_3	4·30	3·82	3·24	2·74	2·89	3·27	2·96	3·98	3·35
Fe_2O_3	4·37	4·36	5·83	5·42	16·81	21·46	34·06	44·10	24·28
FeO	—	—	—	—	1·41	0·92	0·92	0·79	—
MnO	0·03	—	0·04	0·04	0·04	0·05	0·05	0·05	—
MgO	1·38	1·28	1·35	1·24	2·40	2·60	2·39	2·39	3·05
CaO	29·60	30·38	32·20	35·30	24·45	26·56	17·37	10·48	25·29
Na_2O	1·30	0·26	1·20	0·90	0·66	0·76	0·66	0·45	0·39
K_2O	1·05	0·60	0·97	0·93	1·81	0·60	1·30	0·96	0·82
P_2O_5	7·50	9·09	12·20	15·30	11·03	11·35	11·74	7·92	10·03
SO_3	0·69	—	0·74	0·85	—	—	—	—	—
S	—	0·14	—	—	—	—	—	—	0·29
F	1·90	2·33	1·80	1·90	—	—	—	—	3·02
LOI	19·50	24·19	17·60	17·10	—	—	—	—	17·01
H_2O^+	—	—	—	—	4·73	6·22	7·36	8·59	—
H_2O^-	—	—	—	—	2·02	2·18	2·09	1·91	—
CO_2	—	—	—	—	8·65	10·50	3·40	1·60	—
Total	100·29	101·40	100·97	100·60	99·03	99·99	98·76	98·39	102·51

[a]Data from Kress and Veeh (1980), O'Brien and Veeh (1980) and Cook and Marshall (1981). Sample locations: G7 (29°23·0′S, 153°50·0′E); 1512 (30°39·6′S, 153°19·8′E); 1516 (31°10·6′S, 153°13·9′E); A6 (30°41·0′S, 153°18·0′E).
[b]Analysis by X-ray fluorescence spectrography.

(b) Type II: ferruginized phosphatic nodules

Dark brown ferruginous phosphatic nodules that occur on the outer shelf and upper slope off northern New South Wales were first described by Loughnan and Craig (1962) and von der Borch (1970), and have since been studied in petrographic and geochemical detail by Marshall and Cook (1980) and Cook and Marshall (1981).

These Type II nodules are much more abundant and widespread than the Type I concretions, and they occur in somewhat shallower water depths, between 200 and 300 m. Moreover, they are on average appreciably larger than the concretionary phosphates, and, although small nodules (down to coarse sand size) do occur, the typical size range is from about 50 to 200 mm. They are also highly indurated, with rounded to subrounded knobbly and irregular shapes and heavily ferruginized, with a characteristic glazed goethitic surface coating. Sometimes they are cemented together by goethite to form slabs several tens of centimetres across. Indeed, from the behaviour of dredges during sampling, extensive pavement outcrops of this latter material are suspected (Marshall and Cook, 1980).

The Type II nodules are heterogeneous and contain many of the same components as the Type I concretions, with the important addition of hydrated iron oxides that have been collectively referred to as "goethite" in most published descriptions. The major components are, in fact, glauconite, apatite and goethite, with minor calcite, quartz and feldspar. Ovoidal, simple and compound pellets of glauconite and glauconitic infillings of foraminiferal tests abound in some of the less ferruginized nodules, but Marshall and Cook (1980) were unable to find any petrographic evidence that the closely associated glauconite and apatite were in any way genetically related. On the other hand, there is a gradation, with increasing ferruginization, from glauconitic to goethite-rich nodules. Goethite pellets in the latter have every appearance of being replacements of original glauconite, although it is not clear whether this substitution was achieved by *in situ* oxidation or by the introduction of allogenic iron oxides. Marshall and Cook (1980) note the presence of possible relict pyrite framboids in some goethitic pellets, and tentatively suggest authigenic pyrite as a source for some at least of the iron oxide.

Detrital quartz and feldspar grains in the nodules range in size from fine silt to coarse sand, the quartz sometimes being rimmed by glauconite, goethite or apatite. Occasionally, the quartz is partly replaced by these minerals or by calcite. Calcite is present mostly as *unphosphatized* foraminiferal tests and other biogenic detritus, but small amounts of phosphatized interstitial micrite have been observed, and, rarely, apatite replacements of skeletal carbonate. Typically, calcite and aragonite are appreciably less abundant than in the Type I concretions. As pointed out by Cook and Marshall (1981), the textural relationships of these nodules are suggestive of a complex paragenetic sequence involving several phases of phosphatization, glauconitization and ferruginization.

The matrix minerals in the Type II nodules are goethite and apatite, sometimes occurring separately and sometimes intimately associated. Quantitatively, goethite varies from an interstitial cement of minor importance to an all-pervasive constituent that obliterates the original mineralogy while leaving the sediment fabric essentially intact. The brown glazed surfaces, observed by von der Borch (1970) in nodules of this type, result from a superficial concentration of goethite of comparable high intensity. The apatite, as in Type I concretions, is a cryptocrystalline carbonate–fluorapatite that occurs as an interstitial matrix or cement, and more rarely as foraminiferal chamber infillings and ovoidal pellets. Sometimes the apatite is unadulterated, and occasionally there is evidence—in the form of idiomorphic hexagonal plates and acicular crystals of undoubted apatite

(Marshall and Cook, 1980)—suggestive of direct precipitation in intergranular voids. Frequently, however, the apatite is present in intimate intermixture with goethite, which appears to have been introduced later. The internodular cement in conglomeratic varieties, for instance, is an apatite–goethite mixture, possibly a replacement of a former calcareous, micritic cement.

Cook and Marshall (1981) report overall P_2O_5 values in the Type II nodules as being between 5·6 and 11·7%—a slightly lower range than in the Type I concretions—although von der Borch (1970, Table II) quotes values up to 20·7% P_2O_5. The matrix composition varies between nearly pure high-grade apatite with 29% P_2O_5 and only 4% FeO, and high-grade goethite containing 61% FeO and only 4% P_2O_5. As might be expected from the virtually ubiquitous presence of goethite in the nodules, total iron contents tend to be high, with a range between 18·4 and 45% Fe_2O_3 (Cook and Marshall, 1981). The enrichment of some Type II nodules in glauconite, and their impoverishment in apatite relative to the Type I concretions, are also reflected by higher K_2O values (up to 1·8%) and lower CaO values (sometimes as low as 10·5%) respectively.

6.3.1.2   *Sedimentary associations and age*

(a)   The older nodules

Von der Borch's early work (1970) tacitly assumed that the Type I concretions and Type II nodules were essentially of the same age. Planktonic and benthic foraminifera (including *Globoquadrina dehiscens, Globigerinoides bispherica* and *Orbulina*) were cited, in association with the fossilized remains of the crab *Ommatocarcinus corioensis*, as being indicative of a Miocene—possibly a mid-Miocene—age for both types of phosphatic particles. However, apart from the very obvious lithological differences between the two types, and the small but significant differences in the depth environments of each, the sedimentary associations of the concretions and the nodules are quite distinct. Whereas the Type I concretions are associated with normal continental slope sediments—foraminiferal and glauconitic sandy muds and muddy sands—Type II nodules occur on a submarine erosional "bench", in association with rounded pebbles, phosphatized bone fragments and sharks' teeth, and slightly phosphatized bivalve shells. Curiously, no mention is made in any description of these sediments of ferruginization of these associated materials. Although von der Borch (1970) has suggested that there may have been some winnowing by current action of the Type I concretions, this is incidental and non-ferruginous phosphatic concretions have been recovered *in situ* in cores over 1 m below the sediment surface. The true *remanié* deposits are undoubtedly the Type II nodules. Radiometric ages have been determined, using the uranium-series disequilibrium method for both the Type I concretions and the Type II nodules. The ages derived for the latter invariably exceed the 200 000 yr effective limit of the method, and the phosphatization of the nodules may indeed have occurred as early as the Middle Miocene as suggested by von der Borch (1970) and O'Brien and Veeh (1980). However, selective leaching of uranium isotopes from phosphorites has been reported by Kolodny and Kaplan (1970), and Kress and Veeh (1980) cite an example where a concretion with a known uranium-series age of 57 000 yr has reduced uranium in its outer layers which have an apparent age greater than 200 000 yr (G7-4, Table III). In view of this discrepancy, it may be advisable to reserve judgement on the date of phosphatization of the Type II nodules, and to realize that this could possibly be much later than Middle Miocene. Indeed, Marshall and Cook (1980) suggest that the East Australian phosphatic deposits comprise a "continuous spectrum", ranging from unabraded concretions to worn ferruginous nodules and highly ferruginized, cemented polymict nodule conglomerates.

(b) The younger concretions

Kress and Veeh (1980) and O'Brien and Veeh (1980) have no hesitation in drawing attention to the very young ages determined for many of the Type I phosphatic concretions by uranium-series disequilibrum studies (Table 6.3). Most exciting of all is the recording by O'Brien and Veeh (1980) of Holocene ages less than 5000 years for some of the Type I concretions. If these datings are substantiated, the East Australian outer shelf margin could become the third area in the world, after the Namibian and Peru–Chile continental margins, where marine phosphatic sediments are known to have formed recently, and where they may still be forming today (Burnett and Veeh, 1977; Burnett *et al.*, 1982; Bremner, 1980). Radiocarbon dating of contemporary shell and foraminiferal tests associated with the concretions in piston-core samples appears to confirm the uranium-series ages (G.W. O'Brien, personal communication). Moreover, preliminary analytical results obtained from core samples indicate that the uranium-series ages of Type I concretions have a normal stratigraphic distribution in the sediment column. Thus, when the sediment sequence is complete, concretions with very young ages are invariably found in the top 100–150 mm; those with intermediate ages occur at intermediate levels; while the oldest-dated concretions lie well down in the core sequence. At New Zealand Oceanographic Institute station P904 (29°18.0'S, 153°50.3'E) off Yamba Head, for

Table 6.3. Uranium-series analyses of phosphatic concretions and nodules from the East Australian continental margin[a]

Sample number[b]	Depth (m)	Type	U (ppm)	Th (ppm)	Activity ratio $^{234}U/^{238}U$	$^{230}Th/^{234}U$	Age (×10³ yr)	Corrected age (×10³ yr)
G7-1	384	I	84	2.5	1·10 ± 0·01	0·41 ± 0·02	57 ± 4	55 ± 4
G7-3	384	I	58	1.1	1·09 ± 0·01	0·38 ± 0·02	52 ± 3	50 ± 3
G7-4a	384	I (Centre)	121	10·0	1·10 ± 0·01	0·41 ± 0·02	57 ± 4	52 ± 4
G7-4b	384	I (Surface)	53	9.0	1·07 ± 0·01	1·18 ± 0·06	>200	>200
G7-8	384	I	116	3.7	1·12 ± 0·01	0·21 ± 0·01	≤ 25	21 ± 1
G7-9	384	I	90	2.7	1·10 ± 0·01	0·33 ± 0·02	≤ 43	40 ± 3
G7-10	384	I	132	2.2	1·08 ± 0·01	0·79 ± 0·05	≤162	162 ± 23
G7-12	384	I	156	3.8	1·10 ± 0·01	0·18 ± 0·01	≤ 21	17·5 ± 1
G7-17	384	I	120	4.6	1·02 ± 0·01	0·98 ± 0·04	>200	>200
G7-29	384	I	87	2.8	1·10 ± 0·01	0·49 ± 0·02	≤ 71	68 ± 2
G16-2	365	I	99	9.2	1·17 ± 0·03	0·062 ± 0·006	≤ 7	0
G16-3a	365	I	75	4.7	1·19 ± 0·04	0·079 ± 0·006	≤ 9	2 ± 0·5
G16-3b	365	Coral	4·87	<0·06	1·14 ± 0·02	0·15 ± 0·007	18	18 ± 0·5
S7-3	412	I	153	4.2	1·11 ± 0·01	0·45 ± 0·01	≤ 64	60 ± 2
S12-1	376	I	229	7.3	1·14 ± 0·03	0·075 ± 0·007	≤ 8·5	5 ± 1
G18-5	210	II	149	3.9	0·99 ± 0·01	0·94 ± 0·028	>200	>200
1516-A	241	II	87	4.9	0·93 ± 0·01	0·95 ± 0·036	>200	>200
B1	Unknown	II	119	7.0	1·00 ± 0·01	0·96 ± 0·05	>200	>200
A6	210	II (Conglomerate)	50	17.0	0·98 ± 0·01	0·96 ± 0·04	>200	>200

[a]From Kress and Veeh (1980) and O'Brien and Veeh (1980). Errors based on counting statistics (±1σ). Age corrected for common thorium using: $^{230}Th_{cor} = {}^{230}Th_m - [4 \times {}^{232}Th_m - (-\lambda_{230})t]$ where $^{230}Th_m$ and $^{232}Th_m$ are measured activities (dpm/g) of the respective isotopes, λ_{230} is the decay constant of ^{230}Th, and t is the time elapsed since addition of common ^{230}Th.

[b]Sample locations G7 (29°23·0′S, 153°50·0′E), A6 (30°41·0′S, 153°18·0′E), 1516(31°10·6′S,153°13·9′E), G18(SE of Coffs Harbour), G16(ESE of Coffs Harbour), S7 (off Evans Head), S12 (off Yamba), B1 (location unknown).

instance, Type I concretions at 0·35 m, 1·22 m and 1·46 m below the top of the core have been dated as 72 000, 177 000 and 240 000 yr BP, respectively. Elsewhere, concretions 0·87 m below the sediment surface have ages of about 150 000 years. These provisional figures would seem to indicate a mean sedimentation rate roughly between 5 and 10 mm/1000 yrs off the East Australian coast during the Late Quaternary. It may also be significant that these dates for phosphatic concretion formation show some coincidence with periods of low sea-level stand, relating to the last and penultimate glaciations.

6.3.1.3 Genesis

(a) Phosphatization

The conditions that are normally accepted as prerequisites for the formation of phosphatic deposits on the sea floor (other than the ready availability of fine-grained calcareous sediment or biogenic debris suitable for replacement) include an absolute minimal sedimentation rate, relatively shallow depth, and an appropriate supply of phosphorus—either directly derived from ambient sea water or provided by enriched pore waters in the surrounding fine sediments. Phosphogenesis is also normally associated with anoxic, reducing conditions on the sea floor.

(i) *Sedimentation rates.* Throughout the world, low sedimentation rates in the appropriate depth zone are mostly achieved in areas remote from supplies of terrigenous detritus, as, for instance, on isolated submarine ridges and seamounts. In this respect the East Australian offshore phosphate province is anomalous, in that it is located only 25–40 km off a coast where a number of moderately large rivers debouch from a mountainous hinterland that rises to over 1500 m. Significantly, the site of the youngest phosphatic concretions described by O'Brien and Veeh (1980) lies immediately opposite the estuary of the Clarence River, which drains a large proportion of the northern New England Range. This is in strong contrast to the locations of the only other known Holocene marine "phosphorites", off the virtually waterless desert coasts of Namibia and Peru (Baturin, 1969; Baturin et al., 1972; Burnett, 1977), where offshore deposition of land-derived sediment would be expected to be negligible. In the case of the East Australian offshore phosphate province, however, it could be that locally the sedimentation rate on the outer shelf and upper slope is drastically reduced by the exceptionally powerful East Australia Current (Wyrtki, 1966; Boland and Hamon, 1970; Hamon and Tranter, 1971), which sweeps southward along the New South Wales coast as a series of anticyclonic eddies.

(ii)   Hydrological conditions. Associated with this current in the region between latitudes 29° and 32°S (i.e., roughly between Evans Head and Port Macquarie), small-scale seasonal patches of upwelling cold water have been reported by Rochford (1972, 1975). This could be significant as dissolved phosphates in low concentrations behave abnormally in that their solubility is greater in colder waters (Clark and Turner, 1955), and areas of upwelling are likely to be locally enriched in dissolved phosphate nutrients. Such areas thus frequently become centres of high phytoplankton productivity and generally increased biological activity, and indeed a relationship between phosphate deposition and upwelling has been widely acknowledged [e.g., Baturin and Pokryshkin (1980)]. It may not be entirely fortuitous that fossil cetacean bones and fish teeth occur together with the older phosphatic nodules off the New South Wales coast. These possibly reflect the heightened incidence of mortality in locally increased faunal populations, within the biologically productive upwelling regions during Late Tertiary–Early Quaternary times, although this phenomenon is not obviously registered in the more recent phosphatic sediments.

It is of interest to note here that the Recent phosphatic sediments off Nambia and Chile–Peru lie along *west-facing* continental margins, where their formation is thought to be related to high biological productivity in upwelling zones of the cold, north-bound Benguela and Humboldt currents that emanate from the Antarctic. The East Australian situation differs in that, there, the phosphates have developed off an *east-facing* coast, in an area swept by south-flowing surface currents originating in warm tropical regions of the western Pacific. However, the coastward increase in the phosphorus levels of surface waters along the New South Wales coast, and the accompanying decrease in surface water temperature noted by Rochford (1972), are presumed to indicate upwelling and mixing with a deep, cold undercurrent.

(iii)   Bacterial involvement. The modern hydrological/biological regime along the northern New South Wales coast features only very limited upwelling and low productivity, and O'Brien *et al.* (1981) have sought to explain this fact, and the dearth of contemporary megascopic biogenic debris associated with the East Australian Late Pleistocene–Holocene phosphatic concretions, by invoking a process of bacterial assimilation of phosphorus directly from sea water. O'Brien *et al.* (1981) claim that the carbonate–fluorapatite in these more recent concretions is located within cylindrical to ovoidal bodies, supposedly recognized as bacterial cellular structures. The latter are morphologically uniform and show no internal features that could be regarded as nuclei. It is suggested that a specific type of organism, described as a facultative chemolithotropic pseudomonad, is responsible for these structures, and has been involved in the phosphogene-

sis by intracellular accumulation of phosphorus which, as a result of *post mortem* alteration, has been converted to carbonate–fluorapatite. However, Baturin and Dubinchuk (1979), after exhaustive scanning electron microscope studies of recent and ancient phosphorites from many parts of the world, have described identical ovoidal and fusiform bodies as phosphatic segregations that apparently develop inorganically at an early stage of lithification, and typify phosphatic sediments of Holocene and (to a less extent) late Pleistocene age. Baturin and Dubinchuk clearly differentiate them from nucleate structures that they unhesitatingly identify as bacterial cells and other organogenic fabrics. In view of the very considerable time-span (>240 000 yr) over which the concretions are thought to have been forming, and the very recent ages of some of them, it seems reasonable to expect that representatives of the conjectural pseudomonad bacterial population might still be living on the New South Wales continental margin today. Isolation and culture of such organisms would naturally enhance the credibility of the O'Brien et al. (1981) hypothesis.

(iv)   Eustatic effects. Kress and Veeh (1980) and O'Brien and Veeh (1980) have drawn attention to the apparent correspondence between the dates of formation of certain of the East Australian phosphatic concretions, and periods of maximum lowering of sea level during the Pleistocene glaciations. Carter (1978) had earlier proposed an association between phosphatic nodule genesis in Victoria, late in Miocene times, and the eustatic regressive phase that accompanies the formation and expansion of the Antarctic ice cap in the late Miocene. Following the hypothesis of Pevear (1966) and Riggs (1979), Kress and Veeh (1980) further suggest that, during such episodes of lowered sea level, phosphates entrapped within estuarine deposits along the northern New South Wales coast may have been released, by renewed erosion, into offshore surface waters. This, in turn, could have locally increased offshore biological activity and the flux of phosphorus to the sea floor. While such a mechanism may account for the phosphorus incorporated in some of the older nodules and concretions, it would appear not to explain the supply to the younger concretions over the last 10 000 yr. During this time sea levels have varied little.

(b)   Ferruginization

The intense ferruginization of the older, Type II phosphatic nodules off the New South Wales coast is attributed by Cook and Marshall (1981) to exposure, after initial phosphatization under reducing conditions, to a strongly oxidizing environment on the sea floor. This is indicated particularly by the high Fe_2O_3/FeO ratio of about 20 in the nodules, a whole order of magnitude greater than for normal shelf and slope sediments. Again, strongly oxidizing conditions are most likely to prevail in shallowed waters during eustatic regressive phases, although the present-day depth range of the ferruginized nodules is such as to exclude any possibility of actual subaerial exposure during the Pleistocene low sea-level stands. Various sources have been proposed for the iron in goethitic sediments. Borchert (1960, 1965) favours direct derivation from oceanic waters by a process of concentration in organic-rich sediments; Emelyanov (1971) proposes a hinterland provenance for the iron, the terrestrial climatic regime having a major influence upon its ultimate mineralogy; and oxidation of pyrite framboids and/or glauconite pellets has been suggested (Marshall and Cook, 1980; Cook and Marshall, 1981) as a source of goethite in the older nodules off Eastern Australia, although quantitatively these indigenous minerals are not adequate to account for the total amount of iron present in the nodules.

6.3.2   Northern New Zealand

6.3.2.1   *Barrier bank*

Glazed ferruginous nodules, containing up to 16% P_2O_5, have been reported by J. V. Eade (personal communication) at depths of about 350 m on Barrier Bank, some 40 km NE of Great Barrier Island, northern New Zealand (Fig. 6.7). The nodules, which are thought to be derived from a Tertiary formation of unknown age, are associated with a polymict agglomeration of rock types that includes volcanic and sedimentary varieties, in a matrix of foraminiferal sand.

Barrier Bank is regarded as a segment of continental shelf downfaulted along a submarine erosional trough. Phosphatization and ferruginization could therefore have occurred in depths shallower than are found in the present-day environment of Barrier Bank.

The brief description that is available of these ferruginous phosphatic nodules is very reminiscent of the older, Type II nodules off the East Australian coast, and probably there are many analogies in their mode of formation. Unfortunately, at the present time, our knowledge of the petrology, age, distribution, and sedimentary associations of the Barrier Bank phosphatic nodules is extremely scant.

6.4   Phosphorites on Marginal Ridges and Plateaux

This section deals with phosphatic sediments occurring at intermediate depths—mostly between about 300 and 1300 m— on submarine ridges and

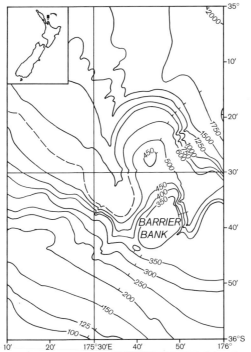

Fig. 6.7. Morphology and location of Barrier Bank, a site of nodular phosphatic deposits on the shelf edge, off northern North Island, New Zealand. Depths in metres.

plateaux composed of continental or quasi-continental crust, marginal to the main Pacific oceanic basin.

6.4.1 Chatham Rise, east of New Zealand

Hard, dense phosphorite nodules and granules, typically coated by a thin greenish-black to brownish-black glauconitic* layer, extend patchily along some 400 km of the crest of Chatham Rise between longitudes 177°E and 177°W, in water depths of 300–450 m (Fig. 6.8). The phosphorite nodule horizon, which is widely buried beneath thin, unconsolidated superficial muds and sands, ranges up to 0·70 m thick, although its average thickness is probably closer to 0·15–0·20 m. Mostly, the sea floor on the crest of Chatham Rise varies from smooth to gently undulant, but locally a sharper microrelief, somewhat reminiscent of a terrestrial "karst" topography, is associated with the phosphorite deposits (Fig. 6.9). The term "phospho-

---

* Glauconite is a potassium–iron–aluminium silicate, found in many parts of the world in sequential association with *remanié*-type phosphorites of various ages.

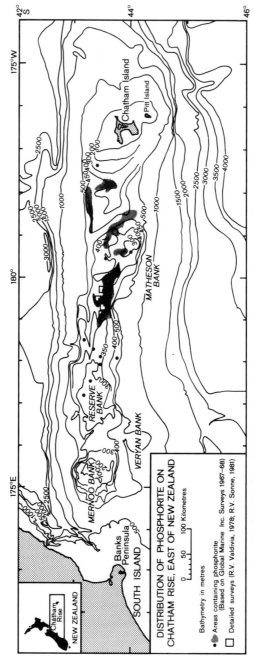

Fig. 6.8. Distribution of nodular phosphorites along the crest of Chatham Rise, east of New Zealand.

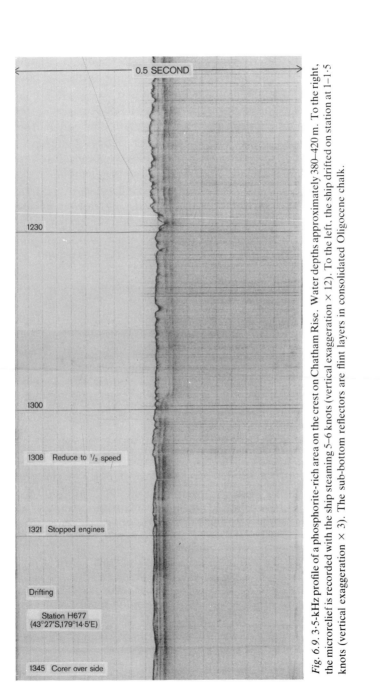

0.5 SECOND

1230

1300

1308 Reduce to ¹/₂ speed

1321 Stopped engines

Drifting

Station H677
(43°27'S,179°14·5'E)

1345 Corer over side

Fig. 6.9. 3·5-kHz profile of a phosphorite-rich area on the crest on Chatham Rise. Water depths approximately 380–420 m. To the right, the microrelief is recorded with the ship steaming 5–6 knots (vertical exaggeration × 12). To the left, the ship drifted on station at 1–1·5 knots (vertical exaggeration × 3). The sub-bottom reflectors are flint layers in consolidated Oligocene chalk.

rite" is here used advisedly, as a high proportion of the nodules and granules on the Rise do indeed have a P_2O_5 content in excess of 18%. The nodules are, in fact, phosphatic replacements of ancient limestones that, at some stage in their history, had undergone prolonged exposure on the sea floor.

The first description of nodules on Chatham Rise by Reed and Hornibrook (1952), based on samples collected by RRS *Discovery II*, recognized their phosphatic composition and a Miocene age for the parent foraminiferal limestone. Many years elapsed before the widespread occurrence of phosphatized Oligocene chalk among the phosphorite nodules was reported by Cullen (1978a, 1980). Further descriptive details and a preliminary attempt at explaining the genesis of the phosphorites in terms of sea-water temperature and chemistry, were provided by Norris (1964). Following a wide-ranging survey of the distribution of the phosphorite deposits on Chatham Rise by Global Marine Inc. of California in 1967–1968, Pasho (1976) was able to present a more comprehensive account of their distribution, geological setting, petrology and geochemistry.

Publication of this work coincided with the growing realization in New Zealand—an agricultural/pastoral country with a high consumption of phosphorus for fertilizer use—that the nation's traditional sources of rock-phosphate on Nauru and Ocean Island (in the equatorial western Pacific) and on Christmas Island (in the eastern Indian Ocean) were rapidly depleting. In view of New Zealand's lack of indigenous land-based phosphate deposits, the N.Z. Oceanographic Institute commenced, in 1975, a programme for the quantitative assessment of the offshore phosphorite resource on Chatham Rise, and for a definitive appraisal of all aspects of its geology, geochemistry and agronomic potential (Cullen, 1975, 1978a, b, 1979a, b, 1980; Cullen and Singleton, 1977). The project culminated in late 1978 with the involvement of the German Federal Institute for Geosciences and Natural Resources, and the completion of an intensive survey of part of the deposit, concentrating 687 sampling stations and several hundred kilometres of echo-sounding and profiling tracks within some 230 km^2 of sea floor (Kudrass and Cullen, 1982). In 1981, comparable collaborative surveys were again undertaken by the N.Z. Oceanographic Institute and Federal Institute, this time with the active participation of both New Zealand and German industrial representatives.

The outcome of these investigations has been the determination of provisional reserves of approximately 100 million tonnes of phosphorite in what is considered the most promising area (between 179°10′ and 179°50′E) (Cullen, 1979a), and the demonstration of the agronomic suitability of the phosphorite. Experiments by McKay *et al.* (1980) indicate that, on pasture land, Chatham Rise phosphorite—crushed but with no further processing—promotes grass growth at least as well as an equivalent amount of superphosphate fertilizer, but remains active in the soil 3–4 times longer. Thus, there is a potential minimum of 50 yr supply of phosphate for New Zealand in this one sector of Chatham Rise, and the possibilities have proved sufficiently attractive for a major New Zealand commercial group— Fletcher Challenge Ltd—to assume prospecting rights over a large expanse of the Rise. Assessments of the phosphorite deposits continue in New Zealand and the Federal Republic of Germany, and it is feasible that the Rise could become the site of the world's first "deep-sea" mining venture. Mining costs and technological developments, as described by Trondsen and Mead (1977) in their evaluation of California offshore phosphorites, are also encouraging, but local problems still to be overcome include the maintenance of adequate mining/recovery rates on a deposit known to be patchy, and the possible deleterious effects of a sticky "ooze" substrate on the operation of suction or jet-lift dredges.

6.4.1.1 Petrology and geochemistry

(a) Petrology

The phosphorites on Chatham Rise range from sand-sized particles of 1·0 mm to irregular slabs and blocks over 150 mm across, the vast majority being nodules within the 10–40 mm size range (Fig. 6.10). The phosphorite density varies, inversely with particle size, from about 2·4 to 3·0 g/cm^3. Sorting of the phosphorite with respect to size is, at best, crude, and most nodule deposits appear to be completely unsorted with random intermixtures of all available size fractions.

Nodule shapes are highly irregular. Some show evidence of rounding, some are angular and blocky with obvious fracture surfaces, and others are perforated by tubular borings up to 10 mm across. The latter are lined by the black coating layer already mentioned, and are clearly an original feature of the nodules. Mostly the burrows are filled by unconsolidated Holocene muddy sand: in a few localities, however, the filling is an indurated calcarenite micropaleontologically dated (A. Edwards, personal communication) as early Pleistocene. Clearly, both the burrows and the black coating are pre-Pleistocene. According to Pasho (1976), the nodules have been subjected to several episodes of fracturing, distinguishable by differences in angularity, and by the colour and amount of coating material on the fracture surfaces. Pasho's claim that the degree of angularity is, in general, proportional to the extent of recent fracturing is not strictly true. In many phosphorite nodules, marked angularity is undoubtedly a primary characteristic.

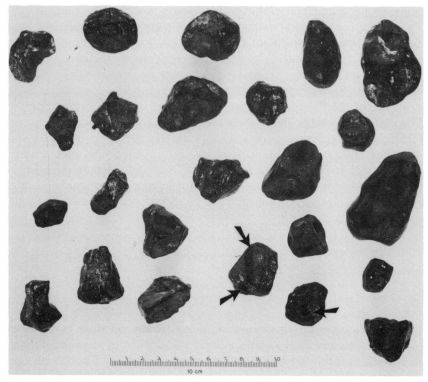

Fig. 6.10. Detail of phosphorite nodules from Chatham Rise. Note the angular to subangular shapes of many of the nodules, the remnants of burrow structures (arrows), and the characteristic dark, glossy glauconitic surface layer. (Photo courtesy of J. J. Whalan.)

Internally, the nodules are composed of a pale grey fine-grained indurated chalk or foraminiferal limestone, extensively mottled by greyish-orange to moderate yellowish-brown (10 YR 7/4–10 YR 5/4)* phosphatized patches. Not infrequently a crude concentric colour zonation, with the darkest material concentrated towards the periphery of the nodules (Fig. 6.11), indicates the pattern of replacement of individual nodules—from the outside inward. Widespread green mottling in many of the nodules extends inward also from the dark coating layer, as do greenish-black fissure fillings, both reflecting late-stage impregnations by glauconite.

Two mineral phases dominate in the phosphorites—relict calcium carbonate (calcite) left over from the incomplete replacement of the parent chalk/foraminiferal limestone, and authigenic apatite. As might be expected from the replacement origin of the phosphorites, the proportions of these

* Rock colour code advanced by the Rock-Color Chart Committee (1975).

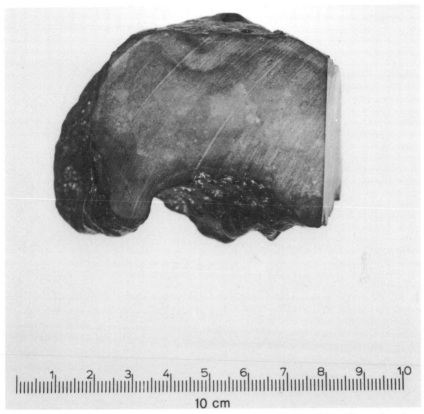

1 2 3 4 5 6 7 8 9 10

10 cm

Fig. 6.11 Section through a large phosphorite nodule from Chatham Rise. The crude colour zonation, with the pale core surrounded by a darker, more phosphatic periphery, is clearly visible, as is the near black glauconitic coating. (Photo courtesy of J. J. Whelan.)

two minerals—indicated by the respective CO_2 and P_2O_5 values (Table 6.4)—tend to vary inversely with one another.

The calcite occurs as relict patches of foraminiferal tests and interstitial cryptocrystalline matrix, which grade into areas of partial or complete replacement by a yellowish to brownish amorphous phosphate. The foraminiferal tests, which exhibit no obvious signs of pre-depositional weathering or abrasion, range in size up to about 0·6 mm when they are easily visible to the naked eye. The phosphate mineral has been identified as a carbonate–fluorapatite (francolite) with approximate composition $Ca_8 (PO_4)_4(CO_2)F_2$ plus minor Na and Mg, and is broadly comparable with marine apatites off California (Dietz *et al.*, 1942) and South Africa (Birch, 1980).

Other minerals present in the phosphorites include glauconite, authigenic opaque minerals such as pyrite and goethite, and a variety of "detrital"

Table 6.4. Chemical composition (wt%) of nodular phosphorites from Chatham Rise, east of New Zealand[a]

Compound	Station H955 (sample 1)[b]				Station H955 (sample 2)[b]				Station VA 291[c]							Bulk sample
	1	2	3	4	5	6	7	8	9	10	11	12	13	14	15	16
P_2O_5	23.8	23.0	23.0	21.3	23.7	23.0	21.9	19.5	20.10	21.12	22.31	21.31	19.18	19.67	20.56	21.97
CO_2	5.4	9.2	10.9	14.5	5.9	9.7	12.3	16.8	—	—	—	—	—	—	—	—
H_2O^+	3.5	3.2	3.1	2.4	3.6	3.2	3.0	2.5	2.71	2.57	2.40	2.27	2.08	1.88	1.96	1.20
H_2O^-	0.44	0.33	0.32	0.37	0.53	0.41	0.32	0.42	—	—	—	—	—	—	—	—
SiO_2	11.2	7.4	4.5	4.0	11.2	7.4	4.0	3.9	17.30	14.81	10.52	6.54	4.61	3.39	2.06	9.83
Al_2O_3	1.0	0.55	0.48	0.50	0.99	0.62	0.44	0.47	1.66	1.32	0.99	0.60	0.45	0.29	0.24	1.11
FeO	0.49	0.32	0.23	0.17	0.51	0.27	0.15	0.14	—	—	—	—	—	—	—	—
Fe_2O_3	5.2	4.7	3.8	2.8	5.5	5.2	5.4	3.2	7.83	7.07	5.66	4.89	3.36	2.16	1.42	5.39
TiO_2	0.03	0.02	0.01	0.04	<0.01	0.01	<0.01	0.01	—	—	—	—	—	—	—	—
MnO	0.01	0.01	0.01	0.02	0.01	0.01	0.02	0.02	—	—	—	—	—	—	—	—
CaO	41.9	44.0	46.2	47.6	40.6	44.3	45.8	47.5	35.34	37.89	41.48	45.30	47.36	48.28	50.78	42.33
MgO	1.4	1.0	0.86	0.76	1.4	1.0	0.76	0.71	2.66	2.30	1.57	0.94	0.64	0.43	0.23	1.48
K_2O	1.5	1.1	0.72	0.59	1.7	1.0	0.56	0.59	1.36	2.00	2.06	1.81	1.61	1.75	1.70	0.73
Na_2O	0.97	0.88	0.89	0.81	0.92	0.91	0.86	0.77	1.81	2.17	2.81	2.86	2.52	2.95	2.37	1.21
SO_3	1.9	1.8	1.7	1.6	1.8	1.6	1.6	1.5	—	—	—	—	—	—	—	2.88
F	3.0	2.9	2.9	2.7	3.0	3.0	2.9	2.6	—	—	—	—	—	—	—	—
Cl	<0.1	<0.1	<0.1	<0.1	<0.1	<0.1	<0.1	<0.1	—	—	—	—	—	—	—	—
LOI	—	—	—	—	—	—	—	—	7.96	7.92	9.41	13.28	17.75	19.17	19.12	12.19

[a] From Cullen (1980) and Kudrass and Cullen (1982). Particle sizes:

Columns 1, 5: particles less than 6.5 mm
Columns 2, 6: particles 6.5–13.0 mm
Columns 3, 7: particles 13.0–26.0 mm
Columns 4, 8: particles 26.0–52.0 mm
Column 9: particles 1–2 mm
Column 10: particles 2–4 mm
Column 11: particles 4–8 mm
Column 12: particles 8–16 mm
Column 13: particles 16–32 mm
Column 14: particles 32–64 mm
Column 15: particles greater than 64 mm
Column 16: average of all particles greater than 1.0 mm from 63 stations sampled by RV Valdivia

[b] Location 43°33·69'S, 179°21·54'E; depth 410 m.   [c] Location 43°29·24'S, 179°25·29'E; depth 393 m.

grains of quartz, feldspar and glass shards, with rarer epidote, zircon, sphene and hornblende (Pasho, 1976). Glauconite appears in a number of forms, the oldest being bright green, rounded clastic grains that, seemingly, were incorporated as primary components in the parent limestone. Some glauconite infillings of foraminiferal tests may have resulted from early diagenesis of the chalks and limestones; others clearly post-date the phosphatization and relate to the greenish-black glauconitic rims and fissure fillings that mark the final stages in the petrogenesis of the phosphorite nodules. The fact that the youngest glauconites on Chatham Rise are those with the darkest colour is not consistent with Pratt's (1971) suggestion, reiterated by Pasho (1976), that depth of colour in glauconites is a function of age. On the contrary, the colour of the glauconites on Chatham Rise seems to be more probably governed by geochemical factors which, while reflecting the mineralogical maturity of the glauconites, are not necessarily indicative of their absolute age.

It is significant that volcanic glass shards occur so commonly among the "detrital" grains in the phosphorites (Pasho, 1976). Although the possibility remains of purely aeolian dispersal for some of the smallest of the detrital grains, the likelihood of a volcanic source and a pyroclastic origin for some at least of these particles can no longer be ignored. Norris (1964) has described glass shards from the unconsolidated superficial sediments on Chatham Rise, and fission-track dating of these (D. Seward, personal communication) has yielded a series of Quaternary ages that equate with the known Taupo eruptions of central North Island, New Zealand. As will be explained later, the Chatham Rise phosphorites are considerably older than Quaternary, and it is evident that volcanic activity (albeit remote) has been recorded in them, and in the associated sediments, over a very long period.

(b) Geochemistry

As might be expected from their mineralogy, the nodular phosphorites are composed essentially of P_2O_5, and CaO and CO_2, with SiO_2, Fe_2O_3 and F making up much of the remainder (Table 6.4). The phosphorus contents of individual phosphorite particles—even those from a single sampling point—are quite unpredictable, and likely to range from <6·7% (<15% P_2O_5) to the reported maximum of 11·6% (26% P_2O_5). It has been estimated (Kudrass and Cullen, 1982) that the deposits in the area of commercial interest, between 179°E and 180° contain on average 9·6% phosphorus (21·5% P_2O_5). This is appreciably less than the phosphorus content of New Zealand's currently imported rock phosphates, but slightly higher than that of the processed superphosphate fertilizer so widely used throughout the country. This fact, plus a demonstrated high chemical reactivity in soils, renders the direct-application usage of crushed, unprocessed Chatham Rise phosphorite agronomically equivalent to that of superphosphate. Indeed, the lower solubility of the phosphorite means that it remains active in the soil much longer than superphosphate, and is less likely to pollute streams and lakes.

It has been observed (Pasho, 1976; Cullen, 1980) that the highest phosphorus values tend to occur in the smallest particles (especially those between 1 and 4 mm across), presumably reflecting more complete phosphatization in those size grades. Although fine-grained phosphorite seldom forms a major constituent of the Chatham Rise deposits, it would seemingly be beneficial for recovery of this grade of phosphorite to be as complete as practicable in any future mining operation. Unfortunately, the situation is not quite so simple, as the 1–4 mm sediment size-grade also carries a high proportion of biogenic carbonate debris, and there could be problems in separating this material from the fine-grade phosphorite.

Normative silica ranges up to about 11·5% but probably averages no more than 4–5%. It occurs in combination with potassium, iron, aluminium and magnesium in the glauconite coatings and impregnations, in the very small quantities of "detrital" silicate minerals, as scattered opalline radiolarian and diatom skeleta, and as occasional microcrystalline/cryptocrystalline quartz segregations that seemingly pre-date the phosphatization. Ferric iron is present, in intimate association with the phosphate, as finely disseminated limonite and goethite, which together impart the characteristic brownish and yellowish hues of the phosphorite. Ferrous iron occurs as a component of the associated glauconite. Fluorine, which consistently amounts to some 2–3% of the phosphorite, presumably resides entirely within the fluorapatite lattice.

The most noteworthy trace elements in the Chatham Rise phosphorites are uranium (Cullen, 1978b) and the rare earths (Cullen, 1980), both of which, like phosphorus, show marked quantitative variations in the different particle-size grades. Uranium values, in particular, are of concern on two counts—the possibility of commercial extraction, and the environmental effect of its application in fertilizers. The uranium distribution is extremely erratic, even within individual nodules, and ranges from a few parts per million up to a maximum of 435 ppm (Table 6.5). A provisional estimate of the average uranium content of the nodular phosphorites is 150–200 ppm, a concentration at which extraction is commercially viable. High uranium values most commonly occur in nodules 10–30 mm across, and Cullen (1978b) has also noted that phosphatized Miocene limestones on Chatham Rise frequently display higher uranium values than does the phosphatized Oligocene chalk. Although it is assumed that the uranium was introduced into the parent limestone more or less simultaneously with phosphorus, it does not necessarily follow that phosphorites with the highest phosphorus values also possess the highest uranium concentrations. On the contrary, Cullen (1978b) describes a horizontally "layered" sample of phosphatized limestone from a seafloor pavement outcrop, in which the uranium maximum values occur at an appreciably deeper (10–20 mm deeper) level than the surficial phosphorus maximum (Table 6.5). Analogous distribution patterns have since been obtained by fission-track mapping of individual phosphorite nodules (Cullen and Wall, in preparation).

Kolodny and Kaplan (1970) and Burnett and Veeh (1977) have established that 80% or more of the uranium in the Chatham Rise phosphorites is present in the tetravalent state. Hexavalent uranium, one of the criteria regarded by Burnett and Gomberg (1977), in their study of offshore Florida phosphorites, as indicative of increased oxidation in a subaerial environment, is a very minor constituent on Chatham Rise.

As in the case of uranium, rare earth element concentrations tend to be higher in the smaller phosphorite particles (Cullen, 1980) (Table 6.6). From the limited number of analyses available, there appears to be no rare earth enrichment in the Chatham Rise phosphorites, as happens in some phosphatic sediments. Indeed, cerium is noticeably depleted, and europium markedly so.

6.4.1.2 Age

The phosphorite nodules on Chatham Rise are formed of Tertiary limestones of two distinct ages. On the crest of central Chatham Rise, between longitudes 179°E and 180°, upper Eocene–lower Oligocene chalk (lithologically almost identical to the Upper Cretaceous chalk facies of northwest Europe) underlies, and is the parent rock for, large spreads of phosphorite nodules. The latter are characterized by foraminiferal assemblages that include *Chiloguembelina* and *Globigerina* cf. *angiporoides*. Further east, and along the southern flank of the Rise, nodules are formed from buff-coloured lower and lower middle Miocene foraminiferal limestones containing *Globigerinoides* cf. *trilobus*, *Globigerinoides* cf. *bisphericus* and *Globorotalia miozea*.

Clearly, a major episode of phosphogenesis post-dated the early middle Miocene deposition on Chatham Rise. Less easily decided is the extent to which the lower Oligocene chalk could have been phosphatized during a mid to late Oligocene sedimentary hiatus—no sediments of this age having been recognized on Chatham Rise, although this period marked the maximum of the early Tertiary transgression in New Zealand (Suggate, 1978). On land, at either end of Chatham Rise, early Tertiary phosphatic nodules

Table 6.5. Uranium contents of phosphorite nodules, phosphatic limestone and fossil cetacean bones from Chatham Rise, east of New Zealand[a]

Sample number	Latitude	Longitude	U_3O_8 (ppm)	Uranium (ppm)	P_2O_5 (%)	Remarks
Phosphorite nodules						
H640	43°29·3'S	179°31·9'E	236·78	200·80	—	Superficially weathered nodule
H642a[b]	43°29·3'S	179°32·6'E	142·63	120·95	—	Nodule from top of piston core
H642b[b]			164·71	139·68	—	Nodule 0·3 m below top of piston core
H667	43°33·3'S	177°45·2'W	112·15	95·11	—	Superficially weathered nodule
H675[b]	43°34·8'S	178°13·2'W	148·44	125·88	—	Nodule 0·5 m below top of piston core
H955a	43°33·7'S	179°21·5'E	11·34	9·62	20·4	Particle size 25–50 mm
H955b			249·37	211·48	22·5	Particle size 12·5–25 mm
H955c			429·31	364·07	23·0	Particle size 6·25–12·5 mm
H955d			285·51	242·12	23·8	Particle size <6·25 mm
N877a[c]	43°27·1'S	179°16·4'E	268·02	227·29	—	Particle size >32 mm
N877b[c]			512·57	434·68	—	Particle size 16–32 mm
N877c[c]			359·71	305·05	—	Particle size 6·35–16 mm
N877d[c]			50·48	42·80	—	Particle size 2·36–6·35 mm
N877e[c]			137·77	116·83	—	Particle size 1·18–2·36 mm
Phosphatized limestones						
H660	43°33·0'S	179°12·7'E	103·28	87·58	6·7	Glauconitized Oligocene chalk
H667a	43°33·3'S	177°45·2'W	59·34	50·32	—	Oligocene chalk
H667b			85·01	72·09	—	Oligocene chalk
H671a	43°30·0'S	177°58·7'W	93·31	79·13	—	Miocene limestone
H671b			421·82	357·72	—	Mid Miocene limestone
H674a	43°32·9'S	178°13·6'W	329·90	279·76	—	Lower Miocene limestone
H674b			92·43	78·39	19·0	Top, 0–10 mm ⎫
H674c			187·01	158·59	17·2	10–20 mm below top ⎪ Lower Miocene limestone block
H674d			8·89	7·54	15·7	20–30 mm below top ⎪
H674e			8·63	7·32	8·6	Base, 30–40 mm below top ⎭
Fossil cetacean bone						
H646	43°33·8'S	179°29·4'E	56·58	47·98	—	Grey-buff bone
H649	43°31·5'S	179°34·0'E	47·12	39·96	—	Tan-brown bone
H651	43°35·2'S	179°30·3'E	330·27	280·08	—	Dark brown bone
H663	43°29·4'S	179°13·6'E	435·74	369·52	—	Dark brown, coarse-textured bone

[a] From Cullen (1978b). [b] Piston-core sample. [c] Box-core sample.

Table 6.6. Trace element analyses (ppm) of some phosphorites from Chatham Rise, east of New Zealand[a]

Element	N877[b]	N879[c]			VA 291[d]						
	(1)	(2)	(3)	(4)	(5)	(6)	(7)	(8)	(9)	(10)	(11)
U	58	100	92	170	177	170	198	260	403	482	383
Th	2·7	<1·0	<1·0	<1·0	3·0	2·7	3·9	5·7	5·4	4·5	1·0
Pb	23	35	53	38	27	31	27	27	18	23	19
Hf	<2·0	<2·0	<2·0	<2·0	—	—	—	—	—	—	—
Yb	<1·0	3·5	1·6	1·6	—	—	—	—	—	—	—
Er	<0·6	3·7	1·7	1·7	—	—	—	—	—	—	—
Ho	<0·3	1·4	0·53	0·53	—	—	—	—	—	—	—
Dy	<0·7	4·8	1·7	2·1	—	—	—	—	—	—	—
Tb	<0·3	1·2	0·44	0·44	—	—	—	—	—	—	—
Gd	<1·0	9·0	3·8	3·2	—	—	—	—	—	—	—
Eu	<0·5	1·0	<0·5	<0·5	—	—	—	—	—	—	—
Sm	<1·0	8·2	3·6	3·9	—	—	—	—	—	—	—
Nd	11	18	7·2	7·0	—	—	—	—	—	—	—
Pr	3·2	4·7	1·9	1·9	—	—	—	—	—	—	—
Ce	25	26	9·2	9·0	—	—	—	—	—	—	—
La	20	38	21	15	—	—	—	—	—	—	—
Ba	200	120	230	120	207	46	68	247	68	86	86
Cs	3·2	4·3	2·1	0·93	—	—	—	—	—	—	—
I	>200	>200	>200	30	—	—	—	—	—	—	—
Nb	2·9	1·4	<1·0	<1·0	—	—	—	—	—	—	—
Zr	160	48	31	36	—	—	—	—	—	—	—
Y	35	84	48	37	88	69	51	33	20	13	13
Sr	700	>1000	900	1000	1150	1238	1358	1477	1569	1671	1674
Rb	60	>100	40	14	98	89	69	48	38	33	27
Cu	—	—	—	—	10	6·0	7·0	5·0	2·0	2·0	1·0
Ni	—	—	—	—	22	31	30	29	27	26	14
Cr	—	—	—	—	50	74	36	19	9·0	8·0	1·0
V	—	—	—	—	78	96	85	78	56	54	45
Zn	—	—	—	—	33	31	28	29	24	21	17

[a]From Cullen (1980) and Kudrass and Cullen (1982). Particle sizes:

Column 1: bulk sample
Column 2: particles 1·16–6·35 mm
Column 3: particles 6·35–16·64 mm
Column 4: particles 16·64–64·0 mm
Column 5: particles 1–2 mm
Column 6: particles 2–4 mm
Column 7: particles 4–8 mm
Column 8: particles 8–16 mm
Column 9: particles 16–32 mm
Column 10: particles 32–64 mm
Column 11: particles greater than 64 mm

[b]43°27·1'S, 179°16·4'E, 378 m.
[c]43°28·7'S, 179°16·9'E, 375 m.
[d]43°29·24'S, 179°25·29'E, 393 m.

H

are known to occur at the base of the Paleocene–Eocene Tioriori Group on Chatham Island (Hay *et al.*, 1970), and at the base of the Whaingaroan (lower Oligocene) succession in North Canterbury and Marlborough (Lensen, 1978). In both cases, however, the phosphatization in these land exposures is very localized, and seems to have *preceded* the early Oligocene. There is no evidence of a phase of phosphatization in the region later in the Oligocene.

Morphologically and chemically, phosphorite nodules on Chatham Rise formed from the Oligocene and Miocene limestones are virtually indistinguishable. Moreover, both types occur as horizontal spreads that are apparently related to a late Miocene morphology, resurrected and modified on the modern sea floor. From the evidence available, it seems likely that local exposures of both lower Oligocene and lower Miocene calcareous sediments were subjected to a single post-early Miocene episode of phosphogenesis, and that all the phosphorite nodules on the Chatham Rise sea floor formed at that time.

The termination of phosphatization on Chatham Rise was marked by the deposition of the ubiquitous glauconitic coatings on the nodules. Although these coatings have so far proved too thin to lend themselves to radiometric dating techniques, granular glauconites in unconsolidated sediments associated with the nodules [see Norris (1964)] have been concentrated, and dated by the K–Ar method as $5 \cdot 6 \pm 1$ m.y. (Cullen, 1967b). In the absence of direct dating of the actual coating material, this is provisionally accepted as the minimum age of the phosphatization.

To summarize, it is considered most probable that the Chatham Rise phosphorites formed essentially in mid to late Miocene times, although the possibility is recognized that some may have become phosphatized as early as the Oligocene (Cullen, 1980). The Miocene interval corresponds roughly to the time of phosphatization of very similar types of submarine phosphorites elsewhere in the world, e.g., off California and southern Africa.

6.4.1.3 *Sedimentary associations*

The phosphorite nodules on Chatham Rise occur mainly within an impersistent, patchy layer, of variable thickness, that rests immediately upon an irregular, bored and burrowed surface of softened Oligocene chalk or Miocene limestone. Normally, the interstitial sediment is an unconsolidated glauconitic muddy sand. In a few localities, phosphorite nodules and granules are cemented together by a calcareous matrix that has been micropaleontologically dated as early Pleistocene (A. Edwards, personal communication). The maximum thickness of the phosphorite layer so far

Fig. 6.12. Underwater photograph of phosphorite nodules exposed on the sea floor, central Chatham Rise. Distance across base of photograph approximately 1 m.

encountered is about 0·7 m, and patches range from a few metres to a few hundreds of metres across.

The main phosphorite layer forms the basal conglomerate of a sequence of completely unconsolidated muddy sands and sandy muds that enclose a scattering of fine glauconitic grains and phosphorite granules. On the crest of Chatham Rise these superficial muds and sands comprise a condensed sequence, ranging in thickness from a few centimetres—when nodules may be exposed on the sea floor (Cullen and Singleton, 1977) (Fig. 6.12)—to about a metre. On the flanks of the Rise, there is a progressive thickening downslope to attain thicknesses of several metres (Cullen, 1978a).

(a) Glacial erratics

A variety of autochthonous and allochthonous rock fragments and biogenic debris occur in intimate association with the phosphorite nodules in the main phosphorite layer. Perhaps the most widely distributed of these fragments are the glacial erratics, which appear in practically every phosphorite-bearing sample from central Chatham Rise, although rarely in significant amounts. Recognizable erratics range in size from mere granules to boulders that occasionally reach 0·5–0·6 m across, but undoubtedly some of the smaller lithic particles in the phosphorite layer were similarly deposited from icebergs. The erratics comprise a varied collection of igneous, metamorphic and indurated sedimentary rocks that include a few types characteristic of the Antarctic continent, such as the striated block of pink "Beacon Group" sandstone first described by Cullen (1962). Individually, the erratic particles rarely show evidence of their glacial origin: most are quite irregularly shaped with unworn surfaces, but occasionally a fragment displays unmistakable faceting or a striated surface. Their most diagnostic feature, however, is the manner in which heterogeneous assemblages of volcanic, granitic, gneissic, schistose, arkosic, quartzitic and pelitic rock fragments occur jumbled together within the phosphorite layer.

The abundance of glacial erratic detritus on Chatham Rise is not surprising. Today, the crest of the Rise coincides with the general trend of an east–west convergence zone between subtropical and subantarctic water masses (Garner, 1957; Heath, 1975, 1981a), and marks not only the local northern limit for iceberg penetration into the South Pacific, but also a preferred line for the melting of icebergs that have survived thus far north. There is no reason to suppose that similiar circumstances have not persisted on Chatham Rise throughout the Holocene, during the Pleistocene epoch, and perhaps in late Tertiary times. Indeed, many erratic fragments are lightly coated by greenish to greenish-black glauconite, and a late Miocene–early Pliocene age for these is quite feasible. Recent evidence places the onset of polar glaciations as far back as Miocene times (Hayes and Frakes, 1975; Carter, 1978). In this context, it must further be remembered that sea-floor spreading and crustal rotation have continued to operate throughout the late Tertiary and Quaternary, and that, at the close of the Tertiary era, Chatham Rise probably lay some 300–400 km *south* of its present position. Moreover, its axial trend would have inclined ESE–WNW rather than east–west, placing it more athwart the eastward drift of icebergs.

(b)   Phosphatized cetacean bones and fish teeth

Together these constitute another common ingredient of the phosphorite deposits on Chatham Rise, where they occur intermingled with the nodules and glacial erratic fragments. Because of the intensity of their phosphatization, and the fact that they are partly impregnated with glauconite, the bones and teeth are assumed to be at least of late Miocene age.

The cetacean bones, the majority of which seem to derive from small to medium-sized dolphin-like creatures, range in colour from pale to dark brown, and are mostly fragmental and intensely eroded. While it is sometimes possible to recognize the general anatomical provenance of the bones—from an appendage, a skull, a vertebra, etc.—nothing suitable for a species identification has yet been recovered. However, ear (tympanic bulla) bones, one of the criteria in cetacean taxonomy, seem able to resist erosion and are relatively common: the chances of eventually recovering some sufficiently well-preserved specimens must remain high. Meanwhile, a small cetacean mandible discovered among the phosphorites has been provisionally assigned to the genus *Hyperoodon*, and its evolutionary and sedimentary significance has been discussed (Fordyce and Cullen, 1979).

The fish teeth found among the phosphorite nodules are derived from a variety of shark species, and, like the cetacean bones, they are extensively phosphatized. Although less common than bone fragments, the teeth tend to be better preserved, and usually retain their outer enamel layer.

(c) Flint concretions

In areas where the phosphorite layer oversteps onto flint-bearing horizons of the upper Eocene–lower Oligocene chalk, concretionary flint masses are also widely dispersed among the phosphorite nodules. They range from a few centimetres to some tens of centimetres across. Some of the flints are broken, but the majority are completely unworn and retain their original shapes which tend towards cylindrical and fusiform. Like the associated phosphorites, erratics and cetacean bones, many of the flints have a greenish to greenish-black glauconitic coating, rendering their appearance very reminiscent of flint clasts in the basal Eocene "Bullhead Bed" that rests on the eroded surface of the Cretaceous Chalk in parts of southern Britain (Wells and Kirkaldy, 1948).

6.4.1.4 *Genesis*

Clues to the precise nature and mode of formation of the Chatham Rise phosphorites are provided by the age of the nodules and their relatively unworn shapes, by the irregular karst-like morphology of the subjacent sea floor, by the variety of associated clasts, and by the location of the phosphorites along the crest of a prominent, but isolated and almost featureless, submarine ridge. The trace element geochemistry of the nodules also provides important evidence for the interpretation of the evolution of the phosphorite deposits.

As has been shown, phosphatization occurred between the beginning of the middle Miocene and the early Pliocene, i.e., between about 15 and 5 m.y. BP. So far as is known from sedimentary and stratigraphic records from Chatham Rise, and from the Chatham Islands themselves, this was a period of stability in the region. The lack of discordance between upper Eocene and upper Miocene–Pliocene deposits on the Chatham Islands (Hay

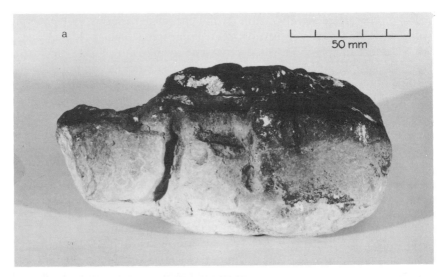

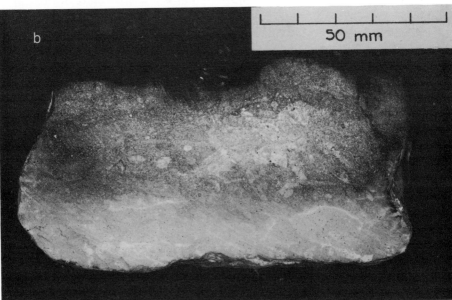

Fig. 6.13(a). Block of superficially phosphatized limestone from a pavement outcrop at Station H674 (43°32·9′S, 178°13·6′W), east Chatham Rise. (b) Section through the same block, showing the moderately phosphatic dark upper surface (20–25 mm thick, with paler core) and the basal 15–20 mm of pale phosphatic limestone. See analyses, Table 6.5. Note the bioturbation structures throughout. (Photos courtesy of J. J. Whalan.)

et al., 1970), for instance, suggests that major Tertiary diastrophism (locally termed the Kaikoura Orogeny), which produced early and middle Miocene deformation on mainland New Zealand, had little effect on central and eastern Chatham Rise. Suggate (1978) attributes the Oligocene–late Miocene sedimentary hiatus on the Chatham Islands to non-orogenic Miocene uplift, with the implication that, in shallower waters, erosion either balanced or exceeded deposition.

This, indeed, is believed to have been one of the critical factors in the formation of the Chatham Rise phosphorites, and an episode of minimal deposition accompanied by minimal erosion is envisaged as spanning middle and late Miocene times on Chatham Rise. The isolation and subdued relief of the Rise ensured that no terrigenous detritus entered the region, and that, over large areas, calcareous sediments of early Miocene and Oligocene ages remained exposed to, or in near-contact with, gently flowing bottom currents. In all probability, the latter would have consisted of deep cold water, relatively enriched in dissolved phosphates, that had been forced to ascend the lower slopes of Chatham Rise. Because of the anomalous behaviour of phosphates in solution, and the tendency for their solubility to decrease with increasing temperatures, dissolved phosphates would have precipitated out as the bottom waters passed into warmer realms on the shallow crest of the Rise. It is possible that the supply of dissolved phosphate may have been locally reinforced if, as is the case today, the Rise formed the locus of a convergence zone, with the heightened biological productivity that is usually associated with such zones. However, as has already been pointed out, Chatham Rise probably lay some 300–400 km south of its present position in the late Tertiary, so that circumstances would have been somewhat different at that time.

In addition, the concentration of some of the richest nodule deposits in broad saddles—such as that between Reserve and Matheson banks (Fig. 6.8)—may be related less to an increased winnowing effect of accelerated currents in such depressions than to the increased volume of bottom water (and hence the increased supply of phosphate) spilling over the saddles.

The overall result of these processes was the formation, over large expanses of the Chatham Rise crest, of relatively thin, phosphatic crusts on exposed calcareous sediments, which became progressively more highly phosphatized and indurated with continued exposure. Remains of such phosphatic pavements still exist locally in the eastern sector of the Rise, close to the Chatham Islands (Fig. 6.13). That these crusts were prone to intensive boring by macrobenthic organisms, before or during phosphogenesis, is evident from the abundance of relict burrows that help determine the shapes of the phosphorite nodules (Fig. 6.10) upon eventual

Fig. 6.14. Box cores (a) N834 (43°30·4′S, 179°19·6′E) and (b) N862 (43°31·4′S, 179°34·5′E) from Chatham Rise, showing phosphorite nodules resting upon the softened, burrowed surface of the Oligocene chalk. (Photos courtesy of J. J. Whalan.)

disintegration of the crusts. The dearth of contemporary molluscan shell in the sediments associated with the nodules suggests that most, if not all, of the burrowing was by soft-bodied animals such as polychaetes. There can be little doubt that the burrowing was an important factor in increasing the surface area of calcareous sediment in contact with the sea water, and hence facilitating the replacement of carbonate ions by phosphate.

The perforation of the phosphatic crust caused by burrowing also had the effect of weakening it mechanically. There is evidence—from the crude concentricity of apatite distributions in many nodules—that, even before the cessation of phosphatization, the crust had started to break up over large areas with the creation of incipient, irregularly shaped nodules. Pasho (1976) has suggested that fragmentation resulted from uplift during the Kaikoura Orogeny and subsequent *subaerial* erosion. Such an interpretation is, however, inconsistent with what is known of the oxidation state of the traces of uranium present in the Chatham Rise phosphorites. As demonstrated by Burnett and Gomberg (1977), hexavalent uranium dominates in subaerially weathered material, whereas in the Chatham Rise phosphorites, 80% of the uranium is tetravalent (Kolodny and Kaplan, 1970; Burnett and Veeh, 1977). Subaerial weathering of the Chatham Rise phosphorites in the past is thus extremely unlikely.

The mechanism of nodule formation proposed here involves carbonate dissolution that continued into Pliocene and Quaternary times, and was responsible for the removal of large volumes of exposed and thinly buried Oligocene and Miocene limestones. The resultant collapse and progressive slumping of the phosphatic crust material, on a sea floor that rapidly acquired the characteristics of a "karstic" surface, produced spreads of phosphorite nodules almost *in situ*. The effects of what is believed to be a dissolution process may be studied in box-core samples from Chatham Rise. In many of these cores, phosphorite nodules rest upon a surface of white Oligocene chalk that has become softened to the consistency of processed cream cheese, and intensely bored, to a depth of about 0·25 m (Fig. 6.14). Originally this material was regarded as redeposited chalk (Cullen, 1978a) but careful micropaleontological examination has failed to discover any contaminant post-Oligocene foraminiferal species. Below, the chalk is invariably of normal induration, and locally contains flint layers that intersect the sea-floor surface. It is difficult to escape the conclusion that some solution process, perhaps aided by burrowing and the secretion of organic acids by the infauna, has affected the chalk surface, and may have done so as far back as late Tertiary times.

It is important to emphasize here that lateral transportation of the nodules has been minimal, and, in all probability, restricted to local "creep" down the slopes of depressions in the karstic sea-floor surface, partly under the influence of bioturbational activity. The coefficient of rounding of the nodules is generally exceptionally low, and such rounding as is observed may often be attributed to solution prior to, or in the early stages of, phosphatization. Essentially, the movement of the nodules throughout late Tertiary and Quaternary times has been vertically downward with continuing dissolution of the subjacent limestone horizons. That abrasion has been minimal is

Table 6.7. Phosphatic sediments on Campbell Plateau, S and SE of New Zealand: location, age and main chemical characteristics[a]

Sample number	Latitude	Longitude	Location	Depth (m)	Age	Coating	%P_2O_5	%F
D134	48°16·0'S,	168°43·5'E	Snares Depression	668	Oligocene–lower Miocene	Glauconite	23·1	3·1
D147	49°31·0'S,	167°25·0'E	Snares Depression	574	Upper Eocene–Oligocene	None	20·6	2·9
F90	49°30·5'S,	167°40·0'E	Snares Depression	601	—	Glauconite	19·4	2·6
F122	48°06·0'S,	179°57·0'W	Bounty Platform	252	Miocene–Pliocene	None	10·0	1·3
F127	49°22·0'S,	176°16·0'E	Pukaki Saddle	1280	Upper Miocene–Pleistocene	Mn oxides	26·2	3·9
F129	49°24·0'S,	177°59·0'E	Pukaki Saddle	978	—	Mn oxides	—	—

[a]From Summerhayes (1969b).

evident from the near-perfect preservation of the thin glauconite coatings, so characteristic of the Chatham Rise phosphorite nodules.

To summarize, two critical processes were involved in the genesis of the phosphorite nodules on Chatham Rise—the phosphatization and fragmentation of a surficial crust, or "hard ground", that formed along the crest of the Rise. The phosphatization was related to a late Tertiary episode of geological stability and sedimentary stagnation, when both deposition and erosion were at a minimum. As a consequence, carbonate sediments remained in prolonged contact with upwelling phosphate-bearing bottom waters, possibly enriched through biological activity in a high-productivity convergence zone. Fragmentation of the resultant phosphatic crust was caused by collapse and subsidence, attendant upon dissolution of the underlying carbonate sediments, and aided by burrowing and bioturbation. Geochemical evidence suggests that this was an exclusively submarine process. Subsequent current action has merely been sufficient to remove the finest of sedimentary particles, and the nodular phosphorite, with its accompanying cetacean bones and fish teeth, glacial erratics, and residual concretionary flints, has accumulated as an intensely winnowed, condensed sequence of the type commonly described as a *"remainé"* or "lag" deposit. An exclusively submarine origin for such a deposit may, however, be somewhat unusual.

6.4.2   Campbell Plateau, southeast of New Zealand

Phosphatized Cenozoic foraminiferal limestones have been recorded by Summerhayes (1969b) in three distinct localities and associations on Campbell Plateau, where they occur either in local depressions in broad saddles scoured by current action, or on the margins of a wide, level platform, characterized by negligible deposition. Their distribution and geology are comparatively poorly known (Table 6.7).

6.4.2.1   *Glauconite-coated phosphorites*

In the Snares Depression, NE of the Auckland Islands and east of the Snares Islands (Fig. 6.15), nodular phosphatic limestones of two ages are recognized, the older being micropaleontologically dated as upper Eocene–Oligocene (presumably lower Oligocene), and the younger as Oligocene (presumably upper Oligocene)–lower Miocene. This closely parallels the situation on Chatham Rise, and the similarity is further enhanced by the fact that the nodules in the Snares Depression are coated by dark green

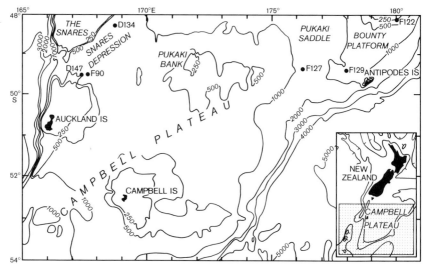

Fig. 6.15. Distribution of submarine phosphorites at N.Z. Oceanographic Institute stations D134, D147, F90, F122, F127 and F129, on Campbell Plateau, S and SE of New Zealand.

glauconite, and associated with unconsolidated glauconitic oozes. The phosphate (19·4–23·1% P_2O_5) and fluorine (2·6–3·1% F) contents of nodules in the Snares Depression also closely resemble those on Chatham Rise. Indeed, the only obvious difference lies in the depths at which the two deposits occur, those in the Snares Depression being consistently some 200 m deeper, between 574 and 668 m.

6.4.2.2 *Manganese-coated phosphorites*

Nodular phosphorites in the Pukaki Saddle occur in even greater depths, between 978 and 1280 m, and they are further distinguished by possessing some of the highest phosphate values—up to 26·2% P_2O_5—recorded among submarine phosphorites in the SW Pacific. The most notable feature of the Pukaki Saddle phosphorites, however, is that instead of being coated with glauconite, the irregular surfaces of the nodules are encrusted by up to 20 mm of black to brown manganese and iron hydrous oxides. Micropaleontological dating of the limestone cores of these nodules as upper Miocene–Pleistocene is somewhat ambiguous, and it is not at all clear whether phosphatization in this area was contemporaneous with the late Miocene phosphogenesis on Chatham Rise and in the Snares Depression, or whether it occurred significantly later. The age of the superimposed manganiferous coating helps little in this respect. Following Bonatti and Nayudu (1965),

Summerhayes (1969b) has related the manganese deposition to late Tertiary–Quaternary submarine volcanism, associated with the nearby Antipodes Islands eruptive centre [see Cullen (1969)]. Summerhayes bases this interpretation on the localism of manganese coatings to the vicinity of the Antipodes Islands, and on their exceptionally high content of supposedly volcanogenic cobalt. According to Arrhenius and Bonatti (1965), high cobalt values (Mn/Co ratios less than 300) may be attributed to rapid precipitation from seawater saturated with volcanic emanations. Nodule coatings in the Pukaki Saddle possess Mn/Co ratios as low as 31·3 (Summerhayes, 1969b). More recently, such interpretations have been refuted for the Campbell Plateau deposits by Glasby and Summerhayes (1975), who now regard slow diagenetic accretion in a "markedly oxidizing" environment as being responsible for the manganese crusts. According to these authors also, glauconite tends to be deposited on phosphatic nodules, preferentially to manganese oxides, when the environment or suitable microenvironments are rendered mildly reducing due to a plentiful supply of decomposing organic matter.

6.4.2.3 *Uncoated phosphatic limestones*

The third group of phosphatic limestones occurs on the SE margin of the Bounty Platform, in water depths of about 250 m, the shallowest recorded phosphates in the region. As yet, little is known of this occurrence except that phosphate values are low (10% P_2O_5), the limestones are *not* coated either by glauconite or manganese oxides, and the micropaleontological age is given rather indeterminately as Miocene–Pliocene.

Clearly, there is considerable scope for further research on the distribution, stratigraphy, petrology and geochemistry of the phosphorites and phosphatic limestones on Campbell Plateau.

6.4.3 Lau Ridge, southern Fiji

Rounded nodules of hard phosphatized foraminiferal limestone, containing 10–15% P_2O_5, have been briefly described by Eade (1980) from a single locality, west of Vatoa in the southern Lau Group of Fiji, between latitudes 19°30' and 20°S (Fig. 6.16). The nodules occur in association with volcaniclastic rocks in depths of 555–775 m, on an isolated knoll on the gently sloping western flank of the Lau Ridge. Many are coated by a layer, up to 7 mm thick, of manganese and iron oxides that contains relatively high amounts of the trace elements Cu (0·041%), Pb (0·250%), Zn (0·070%), Co

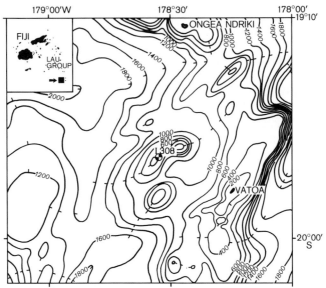

Fig. 6.16. Location of the submarine phosphorite occurrence at Station I308, WNW of Vatoa in the southern Lau Islands, Fiji.

(0·480%), Ni (0·530%), V (0·175%), and Mo (0·046%). Although nothing is known at present of the age of the limestone forming these nodules, or of the date of phosphatization, further work is planned in this area in the near future.

Meanwhile, it is clear that, in their general isolation from major sources of terrigenous sediment, and their location on the (presumably) current-swept crest of a submarine ridge, these phosphatic limestones conform to the general pattern established for the phosphorites on Chatham Rise and Campbell Plateau. One difference that may be significant, however, is the local occurrence on islands of the Lau Group of onshore guano-derived phosphate deposits (White and Warin, 1964). In this respect, the Lau Ridge submarine phosphates display an affinity with phosphatic sediments that cap seamounts and submerged ridges further north (to be described in the following paragraphs). In that they are composed essentially of foraminiferal (as distinct from coral) limestones, however, the Lau Ridge submarine phosphates are provisionally considered as bridging the gap between the "tropical" type of submarine phosphorite and those on Chatham Rise and Campbell Plateau.

The possibility of a genetic relationship between the onshore and offshore phosphatic sediments on the Lau Ridge clearly requires further investigation, and could modify this view.

6.5 Phosphatic Sediments on Seamounts and Oceanic Ridges

6.5.1 The tropical SW Pacific

North of the tropic of Capricorn (23°27'S), deposits of phosphatized coral debris locally occur on the tops of isolated seamounts, guyots and submarine ridges in the SW Pacific (Fig. 6.17). As yet, little is known of the distribution of this type of phosphatic sediment, except that its incidence seemingly becomes progressively more frequent as the equator is approached. In this respect, it parallels the distribution of onshore guano-type deposits (Fig. 6.1) on remote coral islands and atolls in the equatorial SW Pacific (White and Warin, 1964; Cook, 1974), although the possibility of any interrelationship is still a matter for debate.

As Cook (1974) has pointed out, one factor that restricts the extent of phosphatic sediment on the modern central Pacific sea floor is water depth. In these low latitudes, coral growth in the euphotic zone—i.e., in waters less than about 50 m deep—is so prolific that any incipient phosphatization is rendered virtually undetectable. It has been determined, for instance, that areas submerged at the time of the early Holocene eustatic rise in sea level will be covered by essentially non-phosphatic calcareous deposits up to 10 m thick. Thus the most likely places for submarine phosphatic sediments to be encountered in the tropical SW Pacific are on seamounts, guyots and ridges that have remained at depths exceeding 50 m throughout Holocene, Pleistocene and late Tertiary times.

Phosphatized foraminiferal limestone and coral have been recovered (Cullen and Burnett, 1986) from depths between 1000 and 2000 m in the vicinities of the Tokelau and Northern Cook Islands (Fig. 6.18). For instance, pale-coloured, indurated limestone, containing some 26% P_2O_5, comprises the cores of ferromanganese concretions at a depth of 1350 m on Kalolo Seamount (7°17·1'S, 173°50·4'W), 200 km NW of Atafu in the Tokelau Group. The phosphatized core material on this seamount is shown, by micropaleontological dating (M. Clarke, personal communication), to be late Pliocene (2·8–3·0 m.y.) in age. Similar material, also enclosed within a ferromanganese crust up to 45 mm thick and containing about 21% P_2O_5, occurs at a depth of 1780 m on Albert Henry Seamount (9°46·8'S, 165°31·1'W), approximately 130 km north of Pukapuka in the Northern Cook Islands. Large-scale isostatic subsidence of these volcanic piles seems the only logical explanation for the presence of coralline material at such depths, although independent evidence for such an event has not been scrutinized.

Further south, in the vicinities of Western Samoa and Rarotonga, phos-

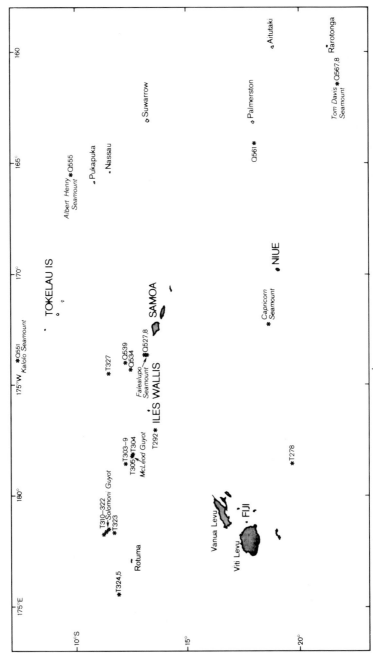

Fig. 6.17. Distribution of N.Z. Oceanographic Institute sampling stations in the tropical SW Pacific that have yielded phosphatic deposits on seamounts and guyots.

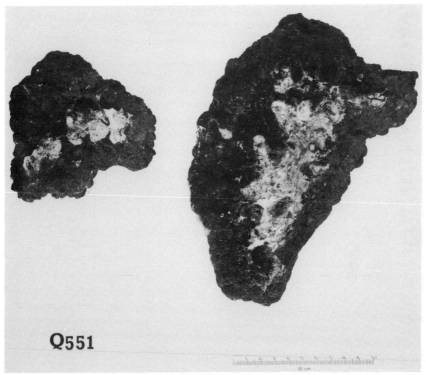

Q551

Fig. 6.18. Manganese-coated limestone from a depth of 1350 m on Kalolo Seamount, approximately 200 km NW of Atafu, Tokelau Islands. (Photo courtesy of J. J. Whalan.)

phate assumes a more diffuse form in ferromanganese concretions, and it appears that replacement of original phosphatized calcareous material by manganese and iron oxides has been more intense. Irregularly shaped, small dark concretions in a gritty calcareous sand, at a depth of 1280 m on Falealupo Seamount (13°15·3'S, 173°39·2'W) west of Western Samoa, contain small amounts of phosphate. Micronodules occurring at depths of 1000–1250 m on Tom Davis Seamount (21°44·6'S, 161°29·0'W), some 200 km WSW of Rarotonga, contain 3·2% P_2O_5, possibly in a fine-grained brownish substance (?replaced coral) disseminated throughout the nodules. Similar material has been recovered from depths of 1650–1800 m on Eclipse Seamount (19°09·6'S, 159°19.1'W), near Aitutaki. By way of contrast, Miocene limestone exposed at a depth of 950 m on Capricorn Seamount (18°41·5'S, 172°17·2'W), at a comparable latitude on the eastern margin of the Tonga Trench, contains only 0·25% P_2O_5 and has a manganiferous coating a mere fraction of a millimetre thick. The association of low P_2O_5 values with minimal manganese encrustation in this instance, when compared with the relatively high proportions of both phosphate and manganese found on other South Pacific seamounts, is suggestive of a positive relationship between the processes of phosphatization and manganese deposition.

A different phosphate association has recently been reported (Cullen and Burnett, 1986) from two guyots—Solomoni and McLeod guyots—interspersed among the line of shallow banks along the northern margin of the Fiji Plateau (Fig. 6.17). As on the isolated seamounts, phosphatized coral dredged from the smooth, pavement-like guyot surfaces is manganese-coated. However, the crystal form of the apatite in these guyot occurrences, as shown by the scanning electron microscope, is distinct from that normally found on oceanic seamounts. Instead of the relatively large, stubby, tabular, hexagonal apatite crystals that typify marine phosphorites, the apatites from Solomoni and McLeod guyots are small, elongate, acicular forms. In this respect, the latter more closely resemble the apatites of insular phosphorites, as on Nauru and Ocean Island. The presence of the mineral dolomite—unknown on the isolated oceanic seamounts—in association with the apatite on Solomoni and McLeod guyots is another point of comparison with the insular type of phosphorite. Pending further surveys, it is tentatively suggested by Cullen and Burnett (1986) that Solomoni and McLeod guyots, and by analogy perhaps the adjacent shallow banks, may represent submerged former atolls fringing northern Fiji Plateau. In the case of Solomoni and McLeod guyots the submergence was presumably, in the main, tectonically controlled.

A similar association of phosphatic Plio–Pleistocene limestone with a dolomitic facies has been described by Anglada *et al.* (1975), in water depths of 500–600 m, 20–25 km south of the Ile des Pins, New Caledonia. In this instance, the phosphatization is attributed to the action of upwelling currents.

6.5.2 The sub-tropical SW Pacific

Phosphatic mineralization, in the form of cryptocrystalline apatite with minor amounts of dahllite, has been reported by Davies and Marshall (1972) and Slater and Goodwin (1973) on some of the "Tasmantid" guyots in the western Tasman Sea, and on Gifford Guyot and Capel Bank that rise from the northern part of Lord Howe Rise (Fig. 6.19). In these areas, phosphorite occurs in two forms—as phosphatized and ferruginized non-reefal limestone, and as phosphatic veneers on cobbles of volcanic rock. On the summit platforms and upper slopes of Derwent–Hunter (30°50'S, 156°15'E), Barcoo (32°40'S, 156°20'E) and Taupo (33°15'S, 156°15'E) guyots, which

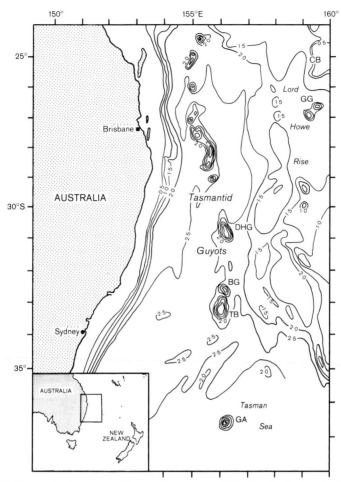

Fig. 6.19. Guyots in the western Tasman Sea, upon which phosphatic deposits have been located. BG, Barcoo Guyot; CB, Capel Bank; DHG, Derwent–Hunter Guyot; GA, Gascoyne Guyot; GG, Gifford Guyot; TB, Taupo Bank. Depths in kilometres. Based on data published by Slater and Goodwin (1973), reproduced here by courtesy of Elsevier Scientific Publishing Company.

extend in a north–south chain between latitudes 30 and 34°S, grey porous limestones occur as slabs and large fragments at depths between 300 and 375 m. These limestones, which are composed of shallow-water biogenic debris and microfossils that include algal, coral, foraminiferal, bryozoan, echinoderm and molluscan skeletal material, have been locally phosphatized and ferruginized. Unfortunately, no chemical analyses are available, and their actual phosphate content is unknown. On Gifford Guyot and

Capel Bank, fragments of a somewhat different hard, massive, yellow to
light brown foraminiferal limestone in depths between 65 and 245 m are
partly phosphatized and encrusted by manganese oxides up to 10 mm thick.
Again, no figures are available of their phosphate content. Phosphatic
deposits, overlain by up to 50 m of non-phosphatic biogenic calcareous
sediment, have also been reported by Davies and Marshall (1972) on
Gascoyne Guyot, the southernmost of the Tasmantid seamount chain.
Micropaleontological dating of the limestones on both the Tasmantid and
the Lord Howe Rise guyots provides a maximum age of late Pliocene/early
Pleistocene, which, interestingly enough, corresponds roughly to the age
obtained for the phosphatic cores of manganiferous concretions on Kalolo
Seamount in the equatorial South Pacific.

   Little information is provided on the second type of "phosphorite" on the
Tasmantid guyots, except that apatite occurs as thin, discontinuous veneers
on many basalt cobbles associated with the phosphatic limestones on
Derwent–Hunter, Barcoo and Taupo guyots.

   Yellow-brown phosphatic material, composed of fluorapatite, hydroxy-
apatite and calcite, has been reported by Glasby (1972a) as occurring in
hard, blue-black manganiferous encrustations up to 50 mm thick, obtained
from a depth of 1065 m on the crest of Aotea Seamount (37°34'S, 172°05'E).
This is an isolated volcanic pile that rises within the New Caledonia Basin,
some 160 km west of northernmost New Zealand. The phosphatic material
is, in fact, a relict foraminiferal limestone of late Oligocene–late Miocene
(most probably early Miocene) age, that has been extensively replaced by
interpolated layers and cusp-like growths of manganese oxides. Presumably
this phosphate represents merely the modified surface rind of a subjacent
solid limestone that forms a capping or partial capping to the seamount.

6.5.3 Genesis of phosphatic sediments on seamounts

From these descriptions it is evident that phosphatic deposits are widely
distributed on seamounts in the SW Pacific, and that one feature which
obviously varies is the nature of the parent carbonate material. This
diversity is seemingly related, in part at least, to water temperature and
hence to latitude. In low latitudes, for instance, the phosphate frequently
consists of altered coral or coral-reef debris; in higher, cooler latitudes the
parent carbonate comprises pelagic foraminiferal ooze or, if the original
depths were shallow, molluscan, echinoderm, bryozoan and algal skeletal
fragments.

   Cook (1974) has provided a useful summary of the overall hydrological,
biochemical and bioproductivity situation in the SW Pacific, and has sought

to explain—and predict—phosphate distributions in terms of these parameters. As has been recognized elsewhere, Cook regards as important factors in seamount phosphogenesis nutrient enrichment in seawater, high biological productivity, regional oceanic upwelling associated with the equatorial divergence, and localized dynamic upwelling due to the topographic obstruction of major current systems.

As already mentioned, Cook (1974) also focuses attention on the question of a possible genetic relationship between the terrestrial guano deposits, which abound on many islands and atolls in the equatorial Pacific, and submarine phosphates on seamounts in the same region. In view of the demonstrable rapid rate of subsidence (in the order of 0·5 mm/yr) of certain seamounts, it seems eminently possible that some of the phosphates that cap seamounts in the tropical and equatorial SW Pacific could have formed through inundation of subaerial guano deposits. Furthermore, it may be that the dissolution of terrestrial guano deposits (either in the subaerial or the submarine environment) has led to local enrichment of oceanic waters in phosphates, and that this, in turn, may have promoted local phosphatic replacement of calcareous accumulations on seamounts. Be this as it may, it must still be remembered that elsewhere in the SW Pacific, e.g., in the Tasman Sea, there appears to be no question of the involvement of guano in phosphate deposition on seamounts. In these instances, a process is envisaged that merely combined minimal deposition and erosion—and hence maximum exposure of accumulated calcareous sediment—with localized dynamic upwelling of phosphate-rich cold bottom waters. There is even evidence of *direct* precipitation of phosphate on the surfaces of volcanic pebbles (Slater and Goodwin, 1973).

Clearly, in the present state of knowledge, the origins of phosphate deposits on seamounts in the SW Pacific need to be individually assessed.

Acknowledgements

The author is indebted to Mr David A. Grieg for permission to quote from his unpublished M.Sc. thesis on Hauraki Gulf sediments, and to Dr Alan Sherwood for provision of samples and photographs of Raglan Harbour material. The remaining photographs were provided by Mr J. J. Whalan and his associates at the Science Information Division of New Zealand DSIR, Messrs G. Harrison and C. Verburg, and the diagrams were prepared by Mr R. Nowicki, also of Science Information Division.

7

Manganese Nodules in the SW Pacific

G. P. GLASBY,[1] N. F. EXON[2] and M. A. MEYLAN[3]

[1]*New Zealand Oceanographic Institute, Department of Scientific and Industrial Research, Wellington North, New Zealand,* [2]*Bureau of Mineral Resources, Geology and Geophysics, Canberra City, Australia and* [3]*Department of Geology, University of Southern Mississippi, Hattiesburg, USA*

7.1 History of Investigations

Although the most extensive hauls of manganese nodules recovered during the HMS *Challenger* Expedition of 1872–1876 were made at stations 281 (22°21'S, 150°17'W) and 285 (32°36'S, 137°43'W) in the Pacific, SE of Tahiti (Murray and Renard, 1891), it is only in the last decade that the distribution, morphology, geochemistry and genesis of manganese nodules in the South Pacific region have become reasonably well known. Examination of the nodule compositional maps of Mero (1965) and Cronan and Tooms (1969) for the entire Pacific, for example, shows an almost complete absence of data for this SW Pacific region. Indeed, only with the publication of Skornyakova and Andrushchenko's paper in 1970 did generalized maps of the nodule distribution and composition in the Pacific become available, and this was not published in English until 1974. As subsequent work was to show (Meylan *et al.*, 1978), the maps prepared by these authors for the SW Pacific were not strictly accurate. An exception to this situation was in the southern (or Antarctic) sector of the South Pacific where the work of the USNS *Eltanin* between 1965 and 1972 led to a much better understanding of the distribution of manganese deposits there (cf. Goodell *et al.*, 1971; Watkins and Kennett, 1971, 1972, 1977; Payne and Conolly, 1972; Meylan

Sedimentation and Mineral Deposits
in the Southwestern Pacific Ocean
ISBN 0-12-195870-1

and Goddell, 1976). The reason for this overall lack of interest in the SW Pacific was twofold. Firstly, the intensification of the research effort into manganese nodules did not begin until the early 1970s with the appreciation that manganese nodules may indeed be an economic resource. Secondly, the SW Pacific is remote from the principal industrialized nations (United States, Europe, Japan, Soviet Union) capable of carrying out such research, particularly when equally interesting projects could be conducted nearer to home. In addition, New Zealand did not begin serious studies of its surrounding deep-sea environment until the early 1970s and Australia has had a tradition of being more involved with continental than with offshore geology.

The expansion of research effort on the manganese nodules of the SW Pacific is therefore a direct spin-off of the expansion in oceanography in general and manganese nodule studies in particular in the 1970s. In addition, there was a growing interest during this decade by the island nations of the South Pacific in obtaining a better understanding of the nature of the mineral resources in the seas around them. The setting up of CCOP/SOPAC in 1971 is now beginning to pay dividends in terms of deep-sea mineral research with a number of small cruises to investigate nodule distribution being mounted in the late 1970s. As mentioned previously (Glasby, this volume), the subtropical islands of the SW Pacific are generally resource poor so that it is logical for these nations to look to the sea in their search for mineral resources and as a continuation of their maritime tradition. This was well illustrated during a 1974 cruise of RV *Tangaroa* (Meylan *et al.*, 1975; Bäcker *et al.*, 1976) when considerable interest was generated in the Cook Islands over the nodules discovered during that cruise SW of Rarotonga. Indeed, a proposal for nodule mining was made by a shipping firm based in New Zealand *before* the cruise returned to New Zealand and the prospect of nodules as an economic resource constituted a major issue at the following Cook Islands General Election (Anonymous, 1974). All this was in spite of the fact that the nodules ultimately turned out to be uneconomic (Bäcker *et al.*, 1976; Commonwealth Secretariat, 1976).

Despite this, there has been a steady increase in our knowledge of manganese nodule distribution and geochemistry in the SW Pacific during the 1970s. In 1972, one of the authors prepared a review of the limited amount of information then available (Glasby, 1972). This was followed by the production of a series of charts showing the distribution and contents of Mn, Fe, Ni, Cu and Co in nodules from the South Pacific (Glasby and Lawrence, 1974a–f) and a review of the significance of these data (Glasby, 1976a). A chart showing the relationship of the nodule distribution to sediment type was prepared later (Glasby and Lawrence, 1980). In addition, an assessment of the surface densities of manganese nodules in the southern sector of the South Pacific based on *Eltanin* data was undertaken (Glasby, 1976b). Of particular significance was the CCOP/SOPAC-IOC Workshop on Geology, Mineral Resources and Geophysics of the South Pacific held in Suva in 1975 in which a major effort was put into manganese nodules with nine papers being devoted to that subject (Intergovernmental Oceanographic Commission, 1975; Glasby and Katz, 1976). Subsequent workshops were held in Noumea and Suva (Intergovernment Oceanographic Commission, 1980, 1983).

The need to assess the marine resources in the surrounding oceans also led the New Zealand Oceanographic Institute to undertake two cruises to the SW Pacific Basin in 1974 (Meylan *et al.*, 1975; Bäcker *et al.*, 1976) and to the Samoan Basin in 1976 (Meylan *et al.*, 1978) using RV *Tangaroa*. The results of these cruises have been summarized by Meylan (1978), Glasby *et al.* (1981) and Glasby (1983). In 1976, CCOP/SOPAC mounted a cruise to study the distribution and composition of nodules in the South Penrhyn Basin (Landmesser *et al.*, 1976; Glasby, 1978). Subsequent CCOP/SOPAC cruises, together with assessments of all existing data, led to papers covering the Samoan and South Penrhyn Basin (Exon, 1981), the region of Kiribati (McDougall and Eade, 1981; Exon, 1982a), and the Samoan region (Exon, 1982b). Exon (1983) also reviewed all data covered in these earlier papers for an area extending from 6°N to 25°S (cf. Cronan, 1984). In addition, Aplin (1983) and Aplin and Cronan (1985a,b) have recently completed a detailed study on a suite of encrustations from the Line Islands Archipelago between 15°N and 15°S and on a suite of nodules and associated sediments from 0° to 15°S between 145° and 175°W based mainly on CCOP/SOPAC material. A complete list of CCOP/SOPAC cruises to look for nodules in the SW Pacific is given in Table 7.1.

Some sampling of nodules in the region was undertaken by ORSTOM (Noumea) (Monzier, 1975, 1976; Monzier and Missegue, 1977; Récy *et al.*, 1977b), and CNEXO (France) carried out detailed sampling of nodules and sediments in the Tiki Basin (Renard, 1976; Hoffert *et al.*, 1979; Pautot *et al.*, 1979; Defossez *et al.*, 1980; Hoffert, 1980). These data from the Tuamotu Archipelago have been evaluated from an economic standpoint by Frazer and Fisk (1980). Nodule distribution in the equatorial South Pacific and Aitutaki Passage was studied by the German vessel RV *Sonne* in 1978 (Friedrich *et al.*, 1981, 1983; Andrews *et al.*, 1983; Glasby *et al.*, 1983a,b) and the importance of the Aitutaki Passage as a route for the northward migration of Antarctic Bottom Water (AABW) was emphasized by Pautot and Melguen (1976, 1979). Pautot and Melguen (1979) also suggested that the highest abundance of nodules is found between the lysocline and CCD (carbonate compensation depth), although this idea has been disputed by Glasby (1980). In 1980, the Geological Survey of Japan carried out a

Table 7.1. CCOP/SOPAC cruises in the SW Pacific which have sought manganese nodules

Year	Location	Nodules recovered	CCOP/SOPAC cruise report	Principal scientists
1976	South Penrhyn Basin	Yes	4	Landmesser, C. W., and Kroenke, L.W.
1977	Northern Samoan Basin	No	10	Halunen, A. J.
1977	Northernmost SW Pacific Basin	Yes	11	Eade, J. V.
1977	Slope of Tonga Trench	No	12	Eade, J. V.
1979	Tonga Platform	No	17	Eade, J. V.
1978	South Penrhyn Basin	Yes	18	Eade, J. V.
1979	Western Central Pacific Basin	Yes	26	McDougall, J. C.
1979	Samoan Basin	No	35	Gauss, G. A.
1980	Southern Line Islands	No	36	Halunen, A. J.
1980	Western Central Pacific Basin	Yes	37	Gauss, G. A.
1980	Fiji Islands	No	39	Eade, J. V.
1980	North and South Penrhyn Basins	Yes	41	Gauss, G. A., and Moreton, D. L. E.
1980	South Penrhyn Basin	Yes	42	Lewis, K. B.
1980	NE Pacific Basin (Line Islands)	Yes	43	Lewis, K. B.
1981	San Cristobal Trench (Solomons)	No	49	Gayman, W.
1981	Melanesian Basin (Gilberts)	Yes	56	Exon, N. F.
1981	Western Central Pacific Basin	Yes	58	Tiffin, D.

transect from the Wake Island (at 17°N) to Tahiti (at 15°S) which crossed the Manihiki Plateau and the Penrhyn Basin in the SW Pacific (Mizuno and Nakao, 1982; Usui, 1983). The survey involved two parallel transect lines about 200 km apart and a total of 27 stations were occupied south of the equator. In 1980, RV *Sonne* also undertook a transect Tahiti–East Pacific Rise–Wellington to examine the distribution and geochemistry of manganese nodules in a previously unstudied area of the SW Pacific Basin (W. L. Plüger, personal communication). Occurrences of nodules in the Tasman Sea have also been reported by Noakes and Jones (1975), Glasby (1976b), Exon (1979), Exon *et al.* (1980) and Jones (1980). Manganese crusts of hydrothermal origin were discovered on the western flanks of the Tonga–Kermadec Ridge during a joint cruise of Imperial College, London, and the

N.Z. Oceanographic Institute in 1981 aboard RV *Tangaroa* (Cronan *et al.*, 1982). More specific projects dealing with topics such as rare earth geochemistry (Glasby *et al.*, 1978; Rankin and Glasby, 1979), Mössbauer spectra (Johnston and Glasby, 1978) of SW Pacific nodules or SW Pacific micronodules (Immel and Osmond, 1976) or petrology of basalts associated with nodules (Roonwal, 1983) have also been carried out. These have been reviewed by Glasby (1981). Photomicrographs of a few South Pacific nodules have been presented by Sorem and Fewkes (1979) and a limited number of SW Pacific nodules have been dated radiometrically (Krishnaswami *et al.*, 1982; Sharma and Somayajulu, 1982). The structure of micronodules has been documented by Lallier–Verges and Clinard (1983). In addition to marine manganese nodules, manganese mineralization related to plate tectonic processes occurs in parts of the SW Pacific. This has been well documented by Burns (1976).

The above programmes have brought together much new information and permitted for the first time a meaningful regional assessment of the principal factors controlling nodule genesis in the SW Pacific (Glasby, 1976a; Meylan, 1978; Skornyakova, 1979; Glasby *et al.*, 1981, 1982; Exon, 1983; Friedrich *et al.*, 1983) which can be usefully compared with data for the Pacific as a whole (cf. Calvert and Price, 1977; Piper and Williamson, 1977; Arrhenius *et al.*, 1979; Skornyakova, 1979). In addition, several articles have comprehensively reviewed recent developments and literature on manganese nodules in the SW Pacific (Exon, 1981, 1982a, 1983; Glasby, 1981, 1982a). The economic potential of South Pacific nodules has also been dealt with by a number of authors (e.g., Horn *et al.*, 1973; Archer, 1976, 1979; Frazer, 1977, 1980; Healing *et al.*, 1979; Frazer and Fisk, 1980; Exon, 1981, 1982a,b, 1983; Friedrich *et al.*, 1981; Glasby, 1982a; McKelvey *et al.*, 1983). In particular, Cronan (1981) and Glasby (1982a) have pointed out that the most prospective regions in the SW Pacific lie in the ocean basins within 20°S of the equator beneath the zone of equatorial high productivity (cf. Frazer and Fisk, 1980). Exon (1983) has shown that areas with both high grades and abundances are found only along the equator from 5°S northwards. The search for prospective nodules has recently been refined by Cronan (1984). According to Frazer (1980) and Friedrich *et al.* (1981), the most promising area for economic-grade nodules in the South Pacific is the Peru Basin. Nonetheless, it must be stressed that the data base for this statement remains sparse. According to M. B. Fisk (personal communication), the South Pacific has only 11 nodule occurrences recorded for every million square kilometres.

Rather than duplicate the work described in the above publications, this chapter will summarize some of the factors influencing manganese nodule distribution, morphology, mineralogy, geochemistry and genesis in the SW Pacific and present preliminary results of some recent new work on nodules in this region. The reader is referred to Glasby (1972, 1976a, 1981, 1982a) and Exon (1981, 1982a,b, 1983) for detailed literature surveys of previous studies of SW Pacific nodules.

7.2   Controls on Nodule Distribution

Perhaps the most important factor controlling the distribution of manganese nodules on a regional scale in the SW Pacific is the bathymetry. It is well known that the principal factor controlling nodule abundance is sedimentation rate, nodules being found in highest abundance where sedimentation rates are less than a few millimetres per 10^3 yr (cf. Frazer and Fisk, 1981). Nodules therefore form in greatest abundance in deep-ocean basins far from land where the supply of terrigenous and volcaniclastic sediments is low and the region lies beneath the CCD.

If one looks at a bathymetric map of the SW Pacific, one sees immediately the importance of the Pacific–Indian plate boundary in controlling nodule distribution. To the west of this (i.e., the Tonga–Kermadec Trench) lies a series of relatively shallow basins (typical depths being somewhat in excess of 4 km) where sedimentation rates are higher due to the fact that the basins lie close to the CCD or are in relative proximity to a source of terrigenous sedimentation (cf. Berger *et al.*, 1976). Manganese nodules therefore tend not to be abundant in areas such as the Lau Basin, South Fiji Basin, Fiji Plateau, Coral Sea and Tasman Sea, except in the Southern Tasman Sea where high bottom current velocities are locally important (Payne and Conolly, 1972). This fact has been known for a considerable time (Murray, 1902, 1906). Some of these basins (e.g., Lau Basin, South Fiji Basin) are also geologically very youthful, which again precludes nodule formation. On the other hand, nodules form in considerable abundance in the basins of the Pacific Plate, particularly in the SW Pacific, Samoan and South Penrhyn Basins, since these basins are generally deeper (and situated on older crust) and further from major landmasses so that the combined effects of terrigenous, volcaniclastic and calcareous sedimentation are low. Within the SW Pacific and Samoan Basins, Meylan (1978) has shown that nodule distribution has a well-defined western boundary which coincides with a demarcation between different sedimentary regimes (cf. Meylan *et al.*, 1982). Nodules tend to occur principally on medium to dark brown silty clays. In the western sector of the basins, the input of terrigenous sediments from New Zealand or of volcaniclastic material from the volcanoes of the Tonga–Kermadec Ridge raises sedimentation rates above the threshold for high-density nodule formation. Nodules are found in abundance beneath the Antarctic Circumpolar Current where sedimentation rates are again low due to the influence of the strong bottom currents (Goodell *et al.*, 1971; Glasby, 1976b). Very close to the equator the high plankton productivity means that sedimentation rates are high and nodules rare there (Exon, 1983; Friedrich *et al.*, 1983). Sedimentation rate is therefore a principal factor in controlling nodule distribution in the SW Pacific. This contention is supported by inspection of the Pacific sediment isopach map of Ludwig and Houtz (1979) which shows clearly that nodules occur in abundance almost exclusively where the sediment thickness is less than 100 m. Chester and Aston (1976) have shown that, on average, sedimentation rates in the South Pacific are the lowest in the world ocean. This is particularly true in remote, deep areas such as parts of the SW Pacific Basin and again favours manganese nodule formation. Of course, nodules and crusts do occur in plateaux and seamount environments in this region such as on the Manihiki Plateau (Heezen *et al.*, 1966; Bezrukov, 1973; Exon, 1981), Campbell Plateau (Glasby and Summerhayes, 1975), Aotea Seamount (Glasby, 1972), off New Caledonia (Monzier, 1976) and on the seamounts of the Tuamotu Archipelago (Glasby, 1976a; Frazer and Fisk, 1980). These nodules and crusts are often, though not always, cobalt-rich (Glasby, 1976a; Arrhenius *et al.*, 1979; Frazer and Fisk, 1981; Halbach *et al.*, 1982) and their mode of origin is somewhat different from that of normal deep-sea nodules. The Co-rich crusts have recently attracted interest as a possible commercial source of cobalt.

A second important factor in controlling nodule distribution in the SW Pacific is the distribution of bottom currents which result in erosion of the underlying sediment and an effective lowering of the sedimentation rate, thereby creating conditions favourable for nodule growth (cf. Gordon, 1975; Glasby and Read, 1976). In the SW Pacific, Pautot and Melguen (1979) have reviewed the migration path of the Antarctic Bottom Water (AABW) which acts as a western boundary current east of New Zealand. On reaching the Samoan Basin, the water mass bifurcates and passes through either the Samoan Passage (Johnson, 1974; Hollister *et al.*, 1974; Lonsdale and Spiess, 1977) or the Aitutaki Passage (Pautot and Melguen, 1976, 1979) and thence northward. A very substantial account of the movement of this water at the exit of the Samoan Passage is given by Lonsdale (1981). According to Lonsdale, the region beneath the AABW in the SW Pacific is starved of terrigenous sediments. Although the influence of bottom water movements on nodule growth is now widely accepted (Glasby and Read, 1976; Exon, 1981, 1983; Glasby *et al.*, 1982), it is equally true to say that the precise migration path of the AABW in the SW Pacific has not yet been fully assessed and further research on this topic is required.

In a recent article, Glasby *et al.* (1982) have suggested that, because of the low abundance of benthic fauna on the sea floor in the mid-latitude regions, bioturbation is not sufficient to maintain manganese nodules at the sediment surface there. As a result, the intensification of the AABW in the SW Pacific some 3·5 m.y. BP becomes the baseline for initiating the low-sedimentation environment necessary for large-scale nodule growth in the SW Pacific. Nodules from the SW Pacific are therefore thought to be younger and smaller on average than those from the equatorial North Pacific.

A further factor in nodule distribution is the availability of potential nucleating agents. This subject has been discussed at length by Meylan (1978) who showed that, in the *Tangaroa* survey area, the nodule nuclei are typically glassy volcanic material in various stages of alteration. Only a few percent of the nodules have a different type of nucleus, such as a shark's tooth. Mn-stained pumice was collected throughout the study area, but most pumice, even that in the coarse fraction of sediments associated with nodules, lacks any appreciable manganese encrustation. Similarly, crystal-line basalt rock fragments were also common in the sediment coarse fraction but were not seen as primary nodule nuclei. Post-depositional replacement of the nodule nuclei by ferromanganese oxides is, however, common in this region.

Other factors which are necessary for the growth of manganese nodules, such as well-oxidized bottom waters (Glasby and Read, 1976; Meylan, 1978), are usually satisfied in a low-sedimentation, deep-sea environment such as found in the SW Pacific and Samoan Basins (cf. Warren, 1971, 1973).

Meylan (1978) has also suggested that nodules from the SW Pacific represent a kinechronous lithologic facies (i.e., they may have formed initially at a sediment surface that has been buried by sediments on which the nodules are now found). This idea is now widely accepted (von Stackelberg, 1979; Piper and Fowler, 1980; Glasby *et al.*, 1982).

7.3   Nodule Deposits of Various Basins

The nodule deposits of seven of the South Pacific basins are now relatively well understood and are briefly described below. All these basins lie on the Pacific Plate where sedimentation rates are generally low and nodules widespread. Most of the data in these short summaries are extracted from Exon (1981, 1982a). The geographical locations of the seven basins is given in Fig. 7.1. Summary information on these basins including water depth, nodule abundance and nodule grades is given in Table 7.2.

7.3.1   Melanesian Basin

The Melanesian Basin extends NW from north of the Fiji Plateau, and is bounded by the Gilberts–Tuvalu island chain to the NE and the Ontong–

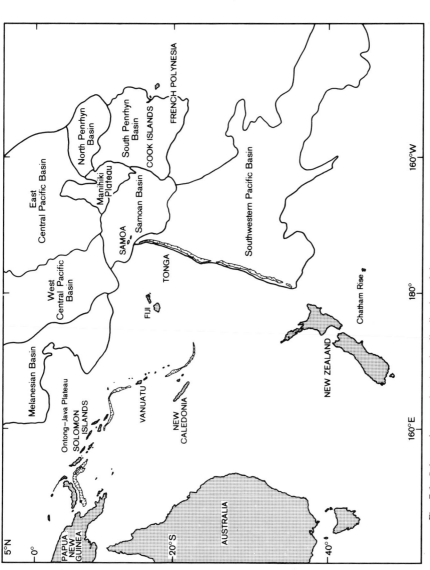

Fig. 7.1. Schematic map showing the distribution of the seven basins considered in this chapter.

Table 7.2. Summary information on nodules from Pacific Plate Basins[a]

Basin	Water depth (m)	Stations analysed	Average nodule metal grades (weight %)							Average nodule density (kg/m²)	Nodule shape, size, surface texture
			Mn	Fe	Mn:Fe	Ni	Cu	Co	Ni + Cu + Co		
Melanesian	4500–5300	4	16·4	16·4	1·07	0·46	0·43	0·27	1·16	Unknown	Unknown
West Central Pacific	5000–6000	14	14·7	14·6	1·08	0·37	0·32	0·20	0·89	Unknown	Unknown
East Central Pacific	5000–6000	57	20·3	10·7	1·89	0·80	0·74	0·21	1·73	7·83	Variable size and shape, smooth and rough textures
North Penrhyn	5000–5500	18	18·6	11·5	1·62	0·60	0·47	0·25	1·32	ca. 1	Small–medium size, varied shape
South Penrhyn	4800–5700	48	15·2	15·4	0·99	0·41	0·22	0·36	0·99	ca. 15	Small–medium size, varied shape
Samoan	5000–5600	9	16·3	14·6	1·13	0·34	0·20	0·28	0·67	Moderate	Small, spherical and polynucleate, microbotryoidal texture
Southwestern Pacific[b]	4500–5500	18	16·2	18·6	0·87	0·43	0·24	0·40	1·07	Moderate	Small–medium size, dominantly spheroidal shape
Equatorial NE Pacific (Si ooze)[c]	5000–5300		22·4	8·1	2·76	1·16	1·02	0·25	2·43	Variable	Medium–large size, spherical and discoidal shapes dominate

[a]Largely after Exon (1981, 1982a).
[b]Data limited to that of Exon (1981).
[c]Data from Horn et al. (1972, 1973).

Java Plateau to the SW. Only the Ellice Subbasin, west of Tuvalu, is as deep as 5000–5300 m. Shallower areas are floored with calcareous ooze. Below 5000 m, the normal sediment is red clay.

Manganese nodule data are sparse in the basin but nodules are present in 30% of stations. Almost nothing is known about nodule abundances. The average grade (Ni plus Cu plus Co) from only four analyses is 1·16% and the maximum grade is 1·47%.

The bulk of the basin, lying in less than 5000 m of water and floored with calcareous ooze, is most unlikely to contain ore-grade nodules. However, Exon (1982a) has suggested that the deeper Ellice Subbasin, floored with clay and lying beneath the southern edge of the equatorial zone of high plankton productivity, is worthy of further investigation.

7.3.2   West Central Pacific Basin

This basin extends NNW from north of Samoa to east of Makin. It is bounded by the Gilberts–Tuvalu island chain to the west and the Howland–Tokelau island chain to the east. Water depths are generally 5000–6000 m, and the deep-water substrate ranges from siliceous ooze in the north to calcareous clay in the south.

There is considerable information available on manganese nodule abundance and grade in the basin (Exon, 1982a) but little information about nodule shapes, sizes and surface textures. Nodules are present at 45% of the 52 deepwater stations in the basin. Abundance data is limited to two stations but the maximum nodule abundance of 56·4 kg/m^2 is very high. Chemical analyses from 14 stations show an average Mn/Fe ratio of 1·08 and an average grade of 0·89% Ni plus Cu plus Co. The maximum Mn/Fe ratio is 3·0 and the maximum grade is 1·58%.

Favourable characteristics of the basin for the occurrence of economically significant nodule fields include the water depth and the substrate type. Although the known nodule occurrences are very far from ore-grade, large areas of the basin have not been adequately sampled. Exon (1982a) suggested that further sampling was necessary to define the potential of the basin better. This should be concentrated south of the equator in the general vicinity of the Nova Canton Trough west of Gardner Island (i.e., on the southern margin of the equatorial zone of high plankton productivity).

7.3.3   East Central Pacific Basin

This basin is bounded by the Howland–Tokelau island chain to the west, the Manihiki Plateau and North Penrhyn Basin to the south, and the Line

J

Islands to the east. Water depths are less than 5000 m in the east and 5000–6000 m elsewhere. The deep-water substrate appears to be largely siliceous ooze in the north and calcareous clay in the south.

Information on nodule shape, size and surface texture comes largely from the results of three Japanese cruises (Mizuno and Moritani, 1977; Moritani, 1979; Mizuno and Nakao, in 1982). The Japanese work has shown that the most common shapes of nodules are either spheroidal, ellipsoidal or discoidal, or polylobate or intergrown aggregate forms. Nodule sizes vary greatly. Surface textures have proved to be the most important criteria for classification. Where nodules formed at the sediment–water interface, the buried parts are rough and the exposed parts smooth. Nodules which are dominantly smooth occur very widely and in high abundances, but are of fairly low grade. Nodules which are dominantly rough and presumably formed below the sediment surface are less widespread and are of lower abundance where they predominate but are of fairly high grade.

Nodule abundance and grade data from the whole basin are analysed by Exon (1982a). Nodules are present in 68% of 140 deep-water stations. The average abundance from 30 stations is 7·83 kg/m^2 and the maximum recorded is 31·6 kg/m^2. All abundances greater than 10 kg/m^2 lie in water depths of 5250–5720 m. Chemical analyses from 57 stations give an average Mn/Fe ratio of 1·89 and an average grade of 1·73% Ni plus Cu plus Co. The maximum Mn/Fe ratio is 6·00 and the maximum grade is 3·55%, both very high values. All grades higher than 1·3% lie in water depths of 4900–5600 m.

Favourable characteristics of the basin include water depth and substrate type. The high grade of many of the nodules indicates that the basin is of possible economic significance. Concentrations of greater than 10 kg/m^2 are coupled with grades greater than 2% Ni plus Cu plus Co at only one of the 25 stations at which both sets of information are available. Exon (1982a) gave high priority to further sampling beneath the zone of high plankton productivity just north and south of the equator, where the prospects of finding economically significant nodule fields are best.

7.3.4 North Penrhyn Basin

The North Penrhyn Basin is bounded by the Line Islands in the east, the Manihiki Plateau in the west and the East Central Pacific Basin to the north. To the south, a slight rise separates it from the South Penrhyn Basin. Water depths are generally 5000–5500 m and the deep-water substrate is calcareous ooze ranging to calcareous clay.

There are considerable manganese nodule abundance and grade data available in the basin (Exon, 1981, 1982a) and some information on nodule shapes and sizes. Nodules are present at 70% of the 30 deep-water stations. Most of the nodules are small to medium in size. Shapes are highly variable, but tabular, ellipsoidal and polynucleate nodules predominate.

The nodule abundance is known from 10 stations and is generally less than 2 kg/m^2, with a maximum in the far SE of 10·7 kg/m^2. Chemical analyses are available from 18 stations (Exon, 1982a) and show an average Mn/Fe ratio of 1·62 and an average grade of 1·32% Ni plus Cu plus Co. The maximum Mn/Fe ratio is very high, 6·0, and the maximum grade is 2·23%. All grades higher than 1·8% lie in water depths of 4850–5350 m.

Favourable characteristics of the basin include water depth, and sediment type in some areas. The grades of the nodules at several stations exceed the 2% regarded as of commercial significance. However, the abundance is nearly everywhere well below 10 kg/m^2 and the higher grades are coupled with low abundances (Exon, 1981). The basin does not therefore appear to have any economic potential for manganese nodules.

7.3.5 South Penrhyn Basin

This basin is bounded by French Polynesia to the east, the North Penrhyn Basin to the north, the Manihiki Plateau and the Samoa Basin to the west, and the Aitutaki ridge to the south. Water depths generally range from 4800 to 5700 m. The sediment type ranges from calcareous ooze above 4500 m to pelagic clay below 4800 m.

Pautot and Melguen (1979) showed that the pelagic clays below 5000 m are enriched in hydroxides and zeolites. In the deep parts of the basin, zeolites, volcanic glass and biogenic siliceous debris are common. Of these constituents, zeolites predominate in the north and siliceous debris in the south. The nature of these sediments suggests that sedimentation rates are very low in the deeper part of the basin, thus favouring the growth of manganese nodules.

Information on nodule shape, size and surface texture comes largely from Landmesser et al. (1976). Most nodules in the basin are small to medium in size. Shapes vary greatly, with spheroidal, ellipsoidal, discoidal or poly-nucleate nodules all dominant in places.

Nodule abundance and grade data from the whole basin are discussed by Exon (1981), and were updated by Exon (1983) to include the area east of 155°W. Nodules are present in 82% of the 97 stations occupied—one of these "stations" is, in fact, a small area sampled in detail by RV *Vityaz* (cf. Exon, 1981). The maximum abundances are generally higher than 10 kg/m^2 and these high abundances are confirmed by photographic evidence (cf. Landmesser *et al.*, 1976). Abundances exceeding 10 kg/m^2 are restricted to 4800–5300 m.

Chemical analyses from 48 samples show an average Mn/Fe ratio of 0·99 and an average grade of 0·99% Ni plus Cu plus Co. The maximum Mn/Fe ratio is 2·12 and maximum grade is 2·02%. The higher grades lie in water depths of 5000–5400 m.

Favourable characteristics of the basin include water depth and substrate type. Although nodule abundances are high, grades are generally low. Exon (1981) showed that higher grades are associated with low abundances and in only one case is a grade of more than 1·3% associated with an abundance of 6 kg/m^2. Despite the abundance of nodules in the basins, its economic potential as regards manganese nodules appears low.

7.3.6   Samoan Basin

This basin is bounded by the South Penrhyn Basin and Manihiki Plateau to the east, the Robbie Ridge to the north, the Tonga Trench to the west, and the SW Pacific Basin to the south. Water depths generally range from 5000 to 5600 m. The deep-water sediments are dominantly volcanogenic turbidites and ash deposits near the Samoan Islands and these have inhibited nodule formation (Meylan *et al.*, 1978; Exon, 1982b). However, in the SE part of the Samoan Basin silty clays are common (Meylan *et al.*, 1978). These are of pelagic origin and nodules are present.

Information on nodule shape, size and surface texture come largely from Meylan *et al.* (1978), who showed that 90% of the nodules are small. Spherical nodules and polynucleate nodules with microbotryoidal surface texture predominate.

Nodule abundance and grade data are discussed by Exon (1981). Nodules are present at 64% of the 28 stations from the Samoan Basin which were considered. The nodule abundance is known at only two stations, where it is 5·7 and 15·6 kg/m^2. However, the nodule population is known from 15 camera stations, and varies from high to low. Nodule analyses from nine stations shown an average Mn/Fe ratio of 1·13 and a very low average grade of 0·67% Ni plus Cu plus Co. The maximum nodule grade is just 1·15%. There is no systematic relationship apparent between grade and water depth.

Favourable characteristics of the SE part of the basin include the water depth and possibly the sediment type. However, the nodule grades are so low that one can safely deduce that the basin has no economic potential as far as manganese nodules are concerned.

7.3.7 SW Pacific Basin

This very large basin is bounded to the west by the Tonga–Kermadec Trench, New Zealand and the Campbell Plateau, to the south by the Pacific–Antarctic Ridge, to the east by the East Pacific Rise, and to the north by rises connecting the southern Cook Islands, Austral Islands and Pitcairn Island to Easter Island. Water depths are generally 4500–5500 m and the deep-water substrate is largely red clay.

Only the western part of the basin, investigated during a RV *Tangaroa* cruise for manganese nodules by Meylan *et al.* (1975) and Bäcker *et al.* (1976), is discussed here. Nodules were recovered from 46 of 53 *Tangaroa* stations. They are dominantly spheroidal, but ellipsoidal nodules dominate south of Rarotonga; most are of small to medium size, although large nodules predominate at a few stations. Granular surface textures prevail. The nodule abundance is known from only a few stations and ranges up to 35·2 kg/m^2 (Exon, 1981). Chemical analyses are available for 27 stations, and these show an average Mn/Fe ratio of about 0·9 and an average grade of about 1·1% Ni plus Cu plus Co. The maximum Mn/Fe ratio is about 1·15 and the maximum grade is about 1·5%.

Overall, high nodule abundances and low grades characterize the basin. One favourable characteristic is the water depth, but the basin does not appear to have any economic potential for manganese nodules.

7.4 Economic Potential

It is generally agreed that nodule fields with an average abundance of wet nodules exceeding 10 kg/m^2 and an average grade exceeding 2% Ni plus Cu plus Co have some economic potential. The mineable area of such a field should exceed 45 000 km^2 (Halbach, 1980). Table 7.2 compares average data for seven SW Pacific basins with those from the siliceous oozes of the equatorial NE Pacific region, which is considered the most prospective region anywhere in the world's oceans.

Average nodule abundances are moderate to high where known in the SW Pacific, except in the North Penrhyn Basin. Abundance data are therefore not discouraging overall. Average nodule grades (Ni plus Cu plus Co, %), on the other hand, are well down on average grades from the siliceous oozes of the equatorial NE Pacific. Only in the East Central Pacific Basin do average grades, at 1·73%, approach the 2% considered of economic significance. Exon (1983) concluded that, on available information, only this basin has any medium-term economic potential. Only there are nodules of high to very high grade (maximum 3·55% Ni plus Cu plus Co) found in the same general areas as nodules in moderate to high abundances (maximum 31·6 kg/m^2). However, in the East Central Pacific Basin as elsewhere, there is a tendency for an inverse relationship between nodule abundance and grade to exist.

Exon (1983) considered that, despite the extreme patchiness of data points in the East Central Pacific Basin, it was possible to define the most prospective zone as extending northward from 5°S to at least 6°N (although excluding the equatorial area from 1°S to 1°N where both abundances and grades were low). He correlated this zone with high plankton productivity, and the favourable prevailing water depths which mean that the surface sediment lies at or near the CCD. At these depths, extensive solution of calcareous plankton releases valuable metals to the sediment column and leads to relative enrichment of metal-bearing siliceous plankton in the sediment. The Melanesian Basin is the only other SW Pacific basin where such favourable conditions may apply but it is virtually unsampled.

7.5 Nodule Facies

At present, the only nodules in the SW Pacific to have received detailed study of distribution, morphology, mineralogy, composition and internal structure, allowing nodule facies to be defined, are those collected by RV *Tangaroa* in its cruises to the SW Pacific and Samoan Basins in 1974 and 1976. The following section is therefore based entirely on these deposits. The reader is referred to Meylan (1978) and Glasby *et al.* (1981) for a more detailed discussion.

According to Meylan (1978), manganese nodules from the SW Pacific can be geographically divided into at least two facies, with nodules in each facies being distinguished by similarities in chemical and physical properties. The facies and their characteristics were defined as follows.

(1) The Cook Island Facies—spheroidal and faceted spheroidal primary morphologies; rough microbotryoidal surface textures; an internal structure consisting of a thin (1–2 mm) ferromanganese oxide accretion crust overlying a massive burrowed ferromanganese oxide subcrust and a glassy volcanic nucleus in some state of alteration and replacement; Mn/Fe ratios near or below 1; low Ni and Cu contents; and δMnO_2 as the principal manganese oxide phase.

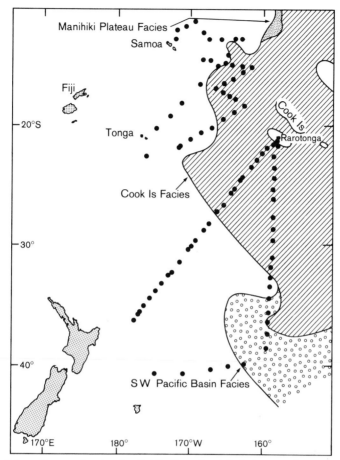

Fig. 7.2. Schematic map showing the distribution of nodule facies types in the SW Pacific.

(2) the SW Pacific Basin Facies—polynucleate primary morphology; microbotryoidal or rough botryoidal surface texture; an internal structure consisting of ferromanganese oxide layers interspersed with silicate micro-laminae, and no discrete nucleus; Mn/Fe ratios above 1; higher Ni and Cu (and lower Co) contents; and todorokite along with δMnO_2 as manganese oxide phases.

A schematic map showing the geographic distributions of these facies types (as well as the RV *Tangaroa* station positions) is given in Fig. 7.2.

In order to characterize the nodules more precisely, Meylan divided the *Tangaroa* cruise track into three arbitrary areas and determined the mean characteristics of the nodules in each of the areas. The mean size, shape and

Table 7.3. Nodule shape by geographic region

Shape[a]	Samoan Basin		SW of Rarotonga		South of Rarotonga[b]	
	Number	%	Number	%	Number	%
[S]	366	39	657	65	498	33
[E]	108	11	229	23	281	19
[D]	69	7	5	<1	222	15
[F]	44	5	62	6	61	4
[P]	291	31	9	1	189	13
[T]	19	2	34	3	18	1
[V]	46	5	3	<1	214	14
[B]	4	1	4	<1	3	<1

[a]Shapes are: spheroidal, ellipsoidal, discoidal, faceted, polynucleate, tubular, volcanic and biological respectively.
[b]Excluding Station G1007.

Table 7.4. Nodule size by geographic region

Size	Samoan Basin		SW of Rarotonga		South of Rarotonga[a]	
	Number	%	Number	%	Number	%
Small (<30 mm)	839	89	726	72	1174	78
Medium (30–60 mm)	96	10	256	26	296	18
Large (>60 mm)	12	1	21	2	56	4
Range (mm)	5–133		7–89		6–109	

[a]Excluding Station G1007.

chemical composition of the nodules from each of these areas are given in Tables 7.3–7.5. Whilst variations are apparent between areas, the overall similarities of the nodules are obvious.

Meylan found that almost all nodules in the SW Pacific have formed around glassy, volcanic megascopic nuclei by both accretion and replacement. Ferromanganese oxide layers have accreted upon the volcanic nuclei while the unstable glass in the nucleus has been altered to montmorillonite (and to a lesser extent phillipsite) and replaced by ferromanganese oxides. On a volumetric basis, it is considered that replacement has been a more significant process than accretion in the accumulation of the ferromanganese oxides in the SW Pacific.

The composition of the nodules appears to be related to their mineralogy. Nodules with higher Mn/Fe ratios (and higher Ni plus Cu contents) occur in nodules with higher todorokite/δMnO_2 ratios.

Table 7.5. Average composition of Pacific Ocean manganese nodules

	Samoan Basin[a]	Southern Penrhyn Basin[b]	Northern sector of SW Pacific Basin[c]	SE sector of SW Pacific Basin[d]	South Pacific[e]	NE equatorial Pacific (siliceous ooze)[f]	Entire Pacific[g]
Mn	15·6	14·8	16·7	21·6	15·9	22·4	19·3
Fe	17·0	15·8	21·1	12·7	14·2	8·2	11·7
Ni	0·34	0·27	0·40	0·91	0·55	1·16	0·66
Cu	0·21	0·14	0·22	0·37	0·29	1·02	0·39
Co	0·42	0·39	0·38	0·36	0·28	0·25	0·32

[a]Meylan (1978).
[b]From Landmesser et al. (1976).
[c]From Glasby et al. (1975).
[d]From Meylan and Goodell (1976).
[e]From Glasby (1976a).
[f]Data from Horn et al. (1972, 1973) presented by Glasby (1976a).
[g]From Bender (1970).

Macroscopic examination of fractured nodules from this region has revealed three basic structural zones in the nodules.

(1) A thin outer crust of accreted ferromanganese oxides, generally 0·5–2 mm thick.

(2) A thicker inner zone of oxides apparently formed by replacement of pre-existing material.

(3) A claystone or palagonite nucleus (Meylan, 1976).

This threefold division is most easily recognized in Cook Island Facies nodules and is analogous to the oxide crust–subcrust–nucleus structural zonation identified by Goodell *et al.* (1971) in most Southern Ocean nodules. For convenience, the terms outer layer, inner layer and nucleus will be used here for these three zones. The structural features of these nodules are quite different from those of the equatorial North Pacific nodules recently reported by Marchig and Halbach (1982).

Typically, the outer layers have a vague metallic lamination when viewed with the unaided eye. In SW Pacific Basin Facies nodules, lamination of the outer layer is more obvious because of the presence of thin intercalations of yellow to light brown fine-grained silicates, up to about 0·5 mm thick, and the transition from the outer to inner layers is gradual.

The inner layers, which may correspond to the mottled zone of Foster (1970), are composed of several structural elements, but throughout the region one inner layer is much like any other. The basic structural element of the inner layers is hard, massive, unlaminated (at least to the unaided eye) ferromanganese oxide, which has a vitreous black to dark red-brown luster on fresh fracture surfaces. In some nodules, the inner layer possesses a very faint metallic lamination immediately beneath the outer layer.

Interspersed within the inner layer are numerous tubular structures up to about 0·5 mm in diameter, apparently constructed by burrowing members of the benthic meiofauna. In many cases, the tubes are open and lined with dark orange to red-brown vitreous particles, probably silicates. In other cases, they are filled with yellow to red-brown finely crystalline material, or in some instances are so prevaded with ferromanganese oxides that only the barest outline remains.

The nodule nuclei consist (with the exception of an occasional shark's tooth) of volcanic material, probably hyaloclastite in various stages of alteration. Most commonly observed is buff-coloured, dull claystone, often with a blocky fracture. Manganese dendrites penetrating inward from the inner layer are a common feature, and ferromanganese oxide-lined microburrows represent another avenue of influx for replacement and filling material. Less commonly, the nuclei consist of waxy yellow-orange to vitreous red-brown palagonite, or dull, powdery Fe–Mn oxides. In many

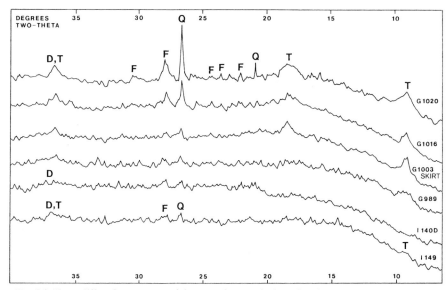

Fig. 7.3. X-ray diffraction patterns of the outer layers of selected nodules from the SW Pacific. T. stands for todorokite; D, δ-MnO$_2$; Q, quartz; F, feldspar. G1016 and G1020 from SW Pacific Basin Facies; others from Cook Island Facies. G1003 skirt represents second-generation growth around nodule equator.

instances, no well-defined nucleus exists, and either the inner layer extends to the nodule centre or else irregularly shaped pockets of nucleus material are scattered near the nodule centre.

Mineralogical differences between nodule facies and between structure zones also exist. X-ray diffraction analysis has shown that todorokite is generally poorly developed or lacking in the outer zone of Cook Island Facies nodules, whereas it is at least moderately developed in SW Pacific Basin Facies nodules (Fig. 7.3). As noted in other areas of the Pacific (cf. Usui, 1979), todorokite is most abundant in layers with a relatively high manganese content. δMnO$_2$ is generally only weakly developed in the nodule outer layers of both facies.

Silicate minerals are usually both more numerous and present in greater quantities in the nodule inner layers (Fig. 7.4). Phillipsite, which was not detected in any of the outer layers, appears to be weakly developed in many of the inner layers throughout the study area. Unexpectedly, todorokite is frequently better developed in the inner layers than in the outer layers, as well.

The claystone nuclei appear to consist primarily of montmorillonite (Fig. 7.5). Quartz is generally only weakly present in claystone and hyaloclastite

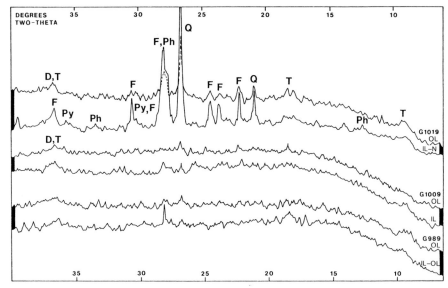

Fig. 7.4. X-Ray diffraction patterns of the outer and inner layers of selected nodules from the SW Pacific (Ol, outer layer; Il, inner layer; N, nucleus). Py stands for pyroxene; Ph, phillipsite; T, todorokite; D, δ-MnO$_2$; quartz; F, feldspar. G1019 from SW Pacific Basin Facies; others from Cook Island Facies.

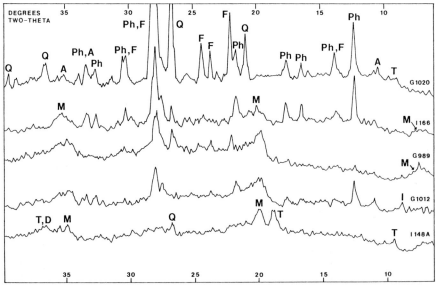

Fig. 7.5. X-Ray diffraction pattern of nuclei of selected nodules. Q stands for quartz; I, illite; A, amphibole; M, montmorillonite; P, phillipsite; F, feldspar; T, todorokite; D, δ-MnO$_2$. G1020 from SW Pacific Basin Facies; others from Cook Island Facies.

Table 7.6. Average structure zone compositions of managanese nodules by geographic area

Structure zone	Element	Samoan Basin	South and SW of Rarotonga, except stns. G1016–G1020	SE corner of Tangaroa area (stns. G1016–G1020)
Outer layer	Mn (%)	11·98	14·44	18·09
	Fe (%)	21·95	20·05	12·35
	Ni (ppm)	1498	2554	6254
	Cu (ppm)	1135	1428	2660
	Co (ppm)	4126	4430	2592
	Number of analyses	6	7	4
Inner layer	Mn (%)	16·42	15·26	14·46
(typical massive	Fe (%)	17·27	17·57	16·95
Fe-Mn oxides	Ni (ppm)	2487	3085	2453
with tubular	Cu (ppm)	1398	1905	1508
structures)	Co (ppm)	5241	4235	2955
	Number of analyses	2	7	1
Nucleus	Mn (%)	1·13	1·19	—
(tuff and	Fe (%)	12·90	10·21	—
claystone only)	Ni (ppm)	196	259	—
	Cu (ppm)	392	484	—
	Co (ppm)	258	209	—
	Number of analyses	1	3	0

nuclei, but feldspar has a moderate to strong X-ray diffraction expression in most such nuclei.

Chemical analysis has also revealed differences in the structure zones (Table 7.6). Supporting data can be found in Glasby *et al.* (1975) and Skornyakova (1976), although these authors did not discuss their findings. In Cook Island Facies nodules, the outer layers are depleted in manganese, nickel and copper and enriched in iron relative to the inner layers. Copper values as low as about 0·1% are not uncommon in outer layers. However, in the outer layers of SW Pacific Basin Facies nodules, manganese and its associated elements nickel and copper are enriched, and iron and cobalt are somewhat depleted, relative to the inner layers.

Only a few nodule nuclei have been analysed. Hyaloclastite, claystone and palagonite nuclei all have somewhat similar compositions. These compositions are, as would be expected, much more similar to that of sediment than to a ferromanganese oxide phase. However, wherever the volcanic nucleus has been noticeably infiltrated by ferromanganese oxides, contents of Mn, Fe, Ni, Cu and Co are all higher than in the unreplaced nucleus.

Based on the physical, mineralogical and chemical characteristics of the structural zones of SW Pacific nodules, it can be concluded that the nodules of this area have developed through both accretion and replacement processes. The outer layers which constitute the micro-botryoidal outermost shell have most probably accumulated primarily by precipitation of ferromanganese oxides from seawater (i.e., they have grown by accretion). This is in accordance with Marchig and Halbach's (1982) view that nodule laminations are precipitated from seawater. However, the inner layers which volumetrically constitute the bulk of most nodules have accumulated ferromanganese oxides through replacement of hyaloclastite and palagonite nuclei. Glasby (1978) has noted that nodules from the South Penrhyn Basin apparently have developed in a similar manner.

Replacement of nuclei in SW Pacific nodules has involved both physical and chemical processes. Physical replacement apparently has been accomplished by tiny benthic organisms which have burrowed into the fine-grained hyaloclastite nuclei, leaving tubular void spaces, often lined with detrital mineral grains. Harada and Nishida (1976) have observed a similar phenomenon in SW Pacific nodules. At first glance, these void spaces appear to be vesicles, but microscopic examination readily reveals a tubular aspect. Similar tubular structures can be seen in the outer accretion layer, and are probably analogous to the benthic foram tubes reported by Dugolinsky (1976).

Because the tubular structures occur principally in the ferromanganese oxide zones (outer and inner layers), the possibility that they are a constructional rather than replacement phenomenon must be considered. Several lines of evidence argue against the strictly constructional idea. Unfilled burrows of a size similar to those common in nodule inner layers have been found in semi-consolidated tuff fragments from the Samoan Basin. These fragments were partly or completely encrusted with an iron-rich oxide crust and may represent proto-nodules. The inner layers of most nodules have a massive aspect and lack concentric banding such as would be expected by accretion growth. In some cases, no obvious hyaloclastite or palagonite nucleus occupies the centre of a nodule. Instead, the inner layer may comprise the entire nodule beneath the outer layer. However, pockets of hyaloclastite in various stages of discoloration are often discernible as remnants of a nucleus in such cases. Additional evidence that tubular structures are part of a replacement inner layer includes the fact that inner layer re-entrants often invade the hyaloclastite nucleus. Manganese-lined and dendrite-surrounded burrows are also found in nuclei which are otherwise unaffected by ferromanganese oxides.

Replacement of nucleus material is also chemical in nature. This is obviously facilitated by the presence of tubular structures, as well as internal cracks which permit the movement of ions in interstitial fluids through the nodule (cf. Sorem and Fewkes, 1977). These avenues of element flux serve not only as conduits but also as chambers for crystal growth. Manganese-lined fractures have been noted by Fewkes (1976) among others. In SW Pacific nodules, internal open spaces appear to have been initially filled by detrital and authigenic silicates with metal oxide accumulations coming later in many instances. In the portions of hyaloclastite nuclei adjacent to the inner layers, the appearance of manganese dendrites seems to represent the first sign of replacement.

As replacement of the original nucleus proceeded, the nodules experienced chemical and mineralogical changes (cf. Burns and Burns, 1978a,b,c). The oxide replacement phase appears to be more manganese- than iron-rich and therefore evolving nodules tend to become more enriched in todorokite rather than δMnO_2, and the increasing Mn/Fe ratio is accompanied by higher values for nickel and copper.

Nodules from the SW Pacific "Cook Islands Facies" are quite distinct in size, shape, surface texture, internal structure, mineralogy and composition relative to those of the equatorial North Pacific between the Clarion and Clipperton Fracture Zones which appear at present to be the most promising from a commercial standpoint. It is now considered that this difference arises from variations in the biological productivity of the overlying surface waters. According to this hypothesis, siliceous organisms in the surface waters incorporate trace metals, sink to the seafloor and decompose releasing their trace metals (such as Ni, Cu and Zn) for incorporation in the nodules. In the high-productivity zone of the equatorial Pacific, this process is well defined and results in the formation of nodules enriched in Mn, Ni, Cu and Zn in which these elements are derived from both the overlying seawater and the underlying pore waters. The sediments are dominantly siliceous ooze. By contrast, nodules from the SW Pacific (Cook Island Facies) come from a region of extremely low biological productivity and are found on red clay sediments containing only low proportions of siliceous tests. In addition, the organic carbon content of sediments from this region is uniformly low (<0·25%) (Premuzic et al., 1982). Nodules from these areas are formed dominantly from seawater and do not have this additional source of trace metals (Ni, Cu and Zn) derived from the sediment column. As a result, the nodules from the SW Pacific (Cook Island Facies) contain much lower contents of Ni, Cu and Zn than those of the equatorial North Pacific. Whilst these nodules occur in high abundance (due to the low sedimentation rates), they are therefore unlikely to be of economic significance because they are characterized by low grades (Ni plus Cu contents).

Glasby and Thijssen (1982) have recently proposed that this release of divalent transition metal ions (Ni^{2+}, Cu^{2+}, and Zn^{2+}) into the pore waters of sediments from the high-productivity zone not only results in the higher contents of these elements in equatorial North Pacific nodules but also results in the stabilization of todorokite relative to δMnO_2. The nodules accrete faster because of the additional supply of elements from the underside and have different compositions on the upper and lower surfaces of the nodules. This different rate of growth on the upper and lower surfaces (due to deposition of elements from seawater and pore water) also influences the nodule morphology. The importance of the biogenic theory in controlling all aspects of nodule characteristics (morphology, composition, mineralogy and growth rate) is therefore emphasized.

The difference in characteristics between the nodules from the SW Pacific Basin Facies and the Cook Island Facies suggests that a similar situation could exist at southern latitudes. As the belt of circumpolar high productivity is approached, the same factors leading to high Mn/Fe ratios, high Ni, Cu and Zn contents and the formation of todorokite as the principal manganese oxide phase should prevail (cf. Meylan and Goodell, 1976), again being related to the high siliceous content of the sediments (cf. Edmond et al., 1979a; Nelson and Gordon, 1982). A comparison of the Cook Island and Kiribati nodules showed this to be the case in the Western Pacific (Exon, 1983). However, this is clearly not always the case, as shown by the work of Ostwald and Frazer (1973), and the reasons for this are not understood (cf. Glasby, 1976a).

8

The Origin and Distribution of Ore Minerals within the Oceanic Crust

M. A. MORRISON

Department of Geological Sciences, University of Birmingham, Birmingham, UK

R. N. THOMPSON

Department of Geology, Imperial College of Science and Technology, London, UK

8.1 Introduction

The past few years have seen a flood of new observations on the tectonic, volcanic and hydrothermal processes operating at mid-ocean ridges (Macdonald, 1983). Much of this can be attributed to the increasing development and use of deeply towed instrument packages, multibeam bathymetric mapping, ocean-bottom instruments and submersibles. This has led to the discovery of active hydrothermal systems and massive sulphide deposits at several localities on spreading centres and seamounts in the East Pacific [e.g., Franchetau *et al.* (1979), Lonsdale *et al.* (1980, 1982), Hekinian *et al.* (1980, 1983), Normark *et al.* (1982) and Malahoff (1982)] and fuelled a steadily growing interest in the economic potential of ocean crust mineralization [e.g., Duane (1982) and Welling (1982)]. More than 50 hydrothermal deposits have now been recorded on the ocean floor (Rona, 1984) and more are constantly being discovered as a result of continuing exploration [e.g., Boulegue *et al.* (1984)].

Sedimentation and Mineral Deposits
in the Southwestern Pacific Ocean
ISBN 0-12-195870-1

In contrast, our information on sub-seafloor mineralization remains scanty and inversely proportion to depth. Consequently discussions of oceanic mineralization draw heavily on analogies with ophiolite complexes [e.g., Cann (1980) and Sawkins (1984)] despite the continuing uncertainty as to the relationship of ophiolites to "normal" ocean crust [e.g., Miyashiro (1973), Malpas (1978), Nicolas and Violette (1982), McCulloch and Cameron (1983) and Dick and Bullen (1984)]. The purpose of this article is to review the available data on the composition, origin and distribution of ore-minerals within the ocean crust. Ridge-crest hydrothermal systems and surface deposits have recently been reviewed by Mottl (1983) and Rona (1984) and will not be discussed in any detail here except in so far as they provide constraints on the operation of sub-seafloor processes.

8.1.1 The nature of the ocean crust

Our current knowledge of the ocean crust stems from two main sources: (1) indirect studies using geophysical techniques such as seismic reflection and refraction, gravity, heat flow and magnetics; and (2) direct investigations on material recovered by drilling or dredging. The total seismically determined thickness of the crust varies from 5 to 7 km. Generalized models of the structure depict sediments [layer (1)], overlying basalts [layer (2)], which in turn overlie layer (3), consisting of gabbroic and more or less altered cumulate and ultramafic rocks (Fox *et al.*, 1973; Christensen and Salisbury, 1975).

Layer (2) is variable in thickness averaging 2·5 km. It can be divided into (2A) "fresh" pillow lavas and flows with intercalated sediments and breccia zones; (2B) predominantly altered pillows and flows; and (2C) intrusive dykes (Ewing and Houtz, 1979; Anderson *et al.*, 1982; Bratt and Purdy, 1984). In the North Atlantic, layer (2A) is 1·5 km thick at the ridge-crest and thins to 100 m as the crust ages to around 60 m.y. In the East Pacific it is 0·7 km thick at the ridge-crest and thins to 100 m by about 30 m.y. The distinction between seismic layers (2A) and (2B) generally disappears in crust older than 40 m.y., apparently as a result of increasing amounts of weathering and diagenesis with age (Ewing and Houtz, 1979; Muehlenbachs, 1980).

The nature of layer (3) is less well understood. The upper part, (3A), is relatively uniform in thickness ($\sim$3 km) and seismic velocity away from the ridge-crest but its presence below the ridge is a matter of debate (Rosendahl, 1976; Lewis, 1983). Petrological arguments imply that basaltic magma chambers persist at a shallow level beneath the ridge-crest and that the bulk of layer three consists of gabbros and related cumulates which have crystallized from the magmas [e.g., Cann (1974)]. The lower portion, (3B), increases in thickness away from the ridge and clearly grows with time at the expense of the underlying asthenosphere, either as a result of off-axis intrusion (Christensen and Salisbury, 1975) or, by underplating of hydrated and altered mantle material (Clague and Straley, 1977).

The generalized model outlined above provides a useful framework within which to discuss ocean crust mineralization. It should be emphasized that it represents a considerable oversimplification. Drilling of the upper crust has demonstrated substantial lateral heterogeneity, even between adjacent holes [e.g., Aumento and Melson (1977)]. Recent studies emphasize the lack of evidence for abrupt velocity discontinuities within layers (2) and (3) and the variability of the structure along the ridge [e.g. Lewis (1978) and Bratt and Purdy (1984)].

8.1.2 Sample distribution

Direct sampling of the ocean crust is possible by means of drilling or dredging. The two methods are complementary. Drilling provides the only means of obtaining a stratigraphic section but is only possible in areas with sufficient sediment thickness (≥ 100 m) to "spud in" the drill. In contrast, dredging is limited to regions devoid of sediment cover and has been carried out mainly on ridge-crests or fracture zones. The probability of exposure of lower crust is maximized in regions of intense tectonism such as active fracture zones where sections exceeding 5 km in thickness may be exposed.

The Deep Sea Drilling Project (DSDP) has had only limited success in sampling the igneous portion of the ocean crust. By completion of leg 92, in April 1983, several hundred holes had been drilled at 602 sites across the world's major ocean basins. Igneous basement was reached and penetrated at 239 sites but only exceeded 100 m at 28 of these. Average core recovery approximates to 20%. The most extensive reference section to date was recovered from hole 504B on the Costa Rica Rift where a total penetration of 1075·5 m was achieved during legs 69, 70 and 83. The upper part of the hole transected pillow lavas, flows and breccia horizons corresponding to layers (2A) and (2B). The core showed increasing alteration with depth culminating in a mineralized stockwork at 850 m subbottom. Below this the hole terminated in relatively massive, intrusive units of layer (2C) (Anderson *et al.*, 1982). Layer (3) has never been penetrated during the DSDP programme but gabbroic and peridotitic units, tectonically emplaced at a high level in the crust, were sampled during legs 37 and 45 (Aumento and Melson, 1977; Aumento and Rabinowitz, 1978).

Dredging from young crustal sections has recovered a wealth of material, a considerable portion of which must be from layer (3): basalts, metabasalts, gabbros, metagabbros, amphibolites, peridotites, serpentinites and mylonitized and brecciated material. Constraints on its interpretation are imposed by the uncertain temporal and spatial relations and the possibility that material exposed in some fracture zones is not typical of "normal" ocean

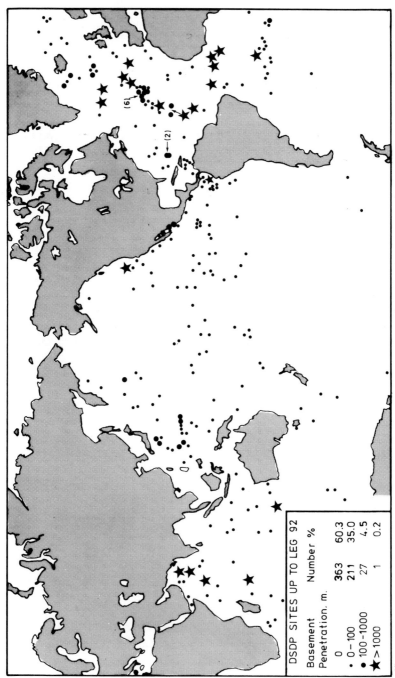

DSDP SITES UP TO LEG 92		
Basement Penetration. m.	Number	%
0	363	60.3
· 0–100	211	35.0
● 100–1000	27	4.5
★ >1000	1	0.2

Fig. 8.1. Distribution of DSDP sites (up to leg 92) at which oceanic basement was penetrated, together with the location of a few key dredge sites at which hydrothermally altered or mineralized samples were recovered (✫). [Data sources: Melson and Thompson (1971). Humphris and Thompson (1978a), and references therein; *DSDP Initial Reports*, volumes 1–76 and *Initial Core Description*. volumes 77–92.

crust (Fox *et al.*, 1973; Bonatti and Honnorez, 1976). A few key dredge sites from which hydrothermally altered or mineralized samples were recovered are shown in Fig. 8.1 together with the DSDP sites at which basement was penetrated. Examination of this shows that the distribution of sample sites is strongly biased towards slow-spreading ridges, particularly the North Atlantic. This creates problems in integrating the data on oceanic mineralization since most of the information on hot springs and associated vent deposits comes from intermediate- to fast-spreading ridge segments in the Pacific [e.g., Edmond *et al.* (1982) and Rona (1984)].

The patchy distribution of sample sites is hardly surprising since many of the original surveys were not undertaken primarily for petrological purposes. This in turn has meant that few mineralogical and chemical studies have been carried out on the *same* samples, making integration of the data difficult. To augment this rather disparate data base, probe analyses were made of sulphides and associated oxides in samples that had already been well characterized chemically and the results will be presented below (Tables 8.1 and 8.2). A final complication that must be borne in mind when assessing data for ocean floor rocks is that virtually *no* material has been recovered that is free from the effects of retrograde reactions or weathering.

Table 8.1. Microprobe analyses of magmatic sulphides in DSDP samples 483 21-347-51, leg 65, East Pacific Rise[a]

Element wt%	1	2	3	4	5	6	7
S	34·15	34·21	35·02	34·13	34·02	35·82	33·68
Fe	56·34	56·53	57·46	57·08	52·80	56·79	45·03
Cu	2·58	2·31	1·36	1·65	7·44	4·41	18·65
Ni	4·07	4·79	4·60	3·55	4·36	3·50	2·11
Si	0·12	—	—	—	0·16	—	0·10
Total	97·26	97·84	98·44	96·41	98·78	100·52	99·57

Elements wt%	8	9	10	11	12	13	14
S	35·29	57·02	51·84	51·50	52·21	53·08	53·55
Fe	58·04	46·42	46·62	47·74	46·07	46·80	46·43
Cu	2·48	—	—	—	—	—	—
Ni	4·15	—	—	—	—	—	—
Si	—	—	—	0·14	0·45	—	—
Total	99·86	98·49	98·46	97·38	98·73	99·88	99·98

[a]Also sought but not found were Mn, Co and Zn. Samples 1–8, globule within fresh glass selvedge: 1–3 centre, 4 halfway to rim, 5–8 rim. Samples 9–14, vesicle sulphides: 9, 10 centre of large mass filling vesicle, 11 three-quarters of way to rim and 12 rim of same sulphide mass; 13, 14 centre of separate vesicle sulphide.

Table 8.2. Microprobe analyses of ore minerals in metabasalts from the Mid-Atlantic Ridge at 22°S[a]

| | AII-60 1-142A | | AII-60 2-142B | | | | |
Element (wt%)	1	2	1	2	3	4	5
S	33·88	34·46	33·78	34·25	33·96	33·85	46·73
Fe	29·55	30·11	29·94	30·42	30·05	30·16	41·27
Cu	34·47	34·42	33·96	34·84	34·04	34·43	—
Ni	—	—	—	—	—	—	—
Si	—	0·15	—	0·16	0·11	—	2·08
Total	97·90	99·14	97·68	99·67	98·16	98·44	90·08

| | AII-60 2-142C | | | | | | | | | | | | | |
Element (wt%)	1	2	3	4	5	6	7	8	9	10	11	12	13	14
S	34·25	33·98	34·56	34·41	47·70	34·30	46·61	51·54	50·14	52·54	52·16	0·20	21·79	0·15
Fe	29·81	29·72	30·31	30·27	46·03	29·77	45·11	44·97	44·11	46·11	45·78	51·93	49·90	43·57
Cu	34·11	34·35	34·73	35·37	—	34·39	—	—	—	—	—	0·58	1·43	3·89
Ni	—	—	0·14	—	0·51	0·32	—	—	—	—	—	0·56	—	—
Si	0·12	0·13	0·16	0·18	0·12	—	1·06	0·49	0·76	0·10	0·34	2·08	0·86	6·82
Total	98·29	98·36	99·76	100·23	94·36	98·78	92·78	97·00	95·01	98·75	98·28	55·35	73·98	54·43

Table 8.2. (*continued*)

	AII-60 2-143												
Element (wt%)	1	2	3	4	5	6	7	8	9	10	11	12	13
S	52·74	53·09	53·16	54·17	53·44	52·42	52·46	52·77	52·86	52·44	52·73	52·80	51·71
Fe	46·33	45·78	46·05	46·99	45·89	45·69	45·85	46·27	46·08	46·79	45·11	46·16	45·63
Cu	—	—	—	—	—	—	—	—	0·39	—	—	—	—
Ni	—	—	—	0·38	0·32	—	—	—	—	—	—	—	—
Si	—	—	—	—	—	—	—	—	—	—	0·11	0·12	—
Total	99·07	98·87	99·21	101·54	99·65	98·11	98·31	99·04	99·23	98·23	98·95	99·08	97·34

[a]Also sought but not found were Mn, Co and Zn. Sample 1-142A: 1, 2, irregular sulphides filling quartz–epidote vein. Sample 2-142B: 1–4, irregular sulphides filling vein; 5, small equant porphyroblast in basalt matrix near vein. Sample 2-142C: 1–4, elongate masses with reddened oxidized rims filling centre of quartz–epidote vein; 5, small grain associated with chlorite at edge of vein and postdated by 1–4; 6, sulphide filling small vein; 7–11, partially oxidized suphide porphyroblasts in fabric of lava adjacent to vein; 12, highly oxidized sulphide in fabric of lava; 13, oxidized rim of 5; 14, red rim to vein sulphides. Sample 2-143: 1–7, all from the same epidote vein; 1, 2, core and 3, rim of large sulphide mass; 4, rim and 5, core of large euhedral sulphide; 8–13, core and rim analyses of three large sulphide grains in the fabric of the lava.

In several instances the successive re-equilibrations to lower temperatures that must have occurred may have obscured evidence for earlier phases of metal segregation and ore-mineral formation.

Despite the fragmentary nature of the data base, three distinct groups of ore minerals can be recognized on the basis of composition, paragenesis and stable isotope chemistry: (1) primary magmatic minerals—spinels and Fe–Cu–Ni sulphides; (2) pyrite or marcasite, associated with smectite–carbonate alteration due to low-temperature ($<250°C$) interaction with seawater, and (3) Cu–Fe sulphide veins, associated with pyrrhotite or pyrite disseminated within hydrothermally altered rocks. The first two assemblages are commonly encountered in basalts recovered from layer (2) and the third occurs mainly in dredge hauls. These groups are described in more detail below and the extent to which they can be held to characterize the ocean crust discussed.

8.2   Primary Magmatic Minerals

8.2.1   Spinels

Euhedral spinel phenocrysts varying in size from 10 to 500 μm and in colour from reddish brown (Cr-rich) to amber (Al-rich) have been reported from many ocean-floor basalts. They are often, but not always, restricted in occurrence to the least fractionated basalts with $FeO^*/(FeO^* + MgO) <$ 0.575 and $Cr > 350$ ppm where, $FeO^* =$ total iron as FeO (Sigurdsson and Schilling, 1976; Graham et al., 1978; Dick and Bryan, 1978). Solitary microphenocrysts are relatively rare; usually they are attached to, or included within, associated phenocrysts of olivine or plagioclase. Rarer (Al-rich) xenocrystal spinels have been reported from several localities in the North Atlantic (Sigurdsson, 1977; Dick and Bullen, 1984).

Spinels in basic rocks are commonly classified as magnesiochromites ($Mg > Fe^{2+}$, $Cr > Al$) or, chromian spinel ($Mg > Fe^{2+}$, $Al > Cr$). The majority of spinels in ocean floor basic and ultrabasic rocks straddle the boundary between these two groups having $Cr/(Cr + Al)$ ratios between 0.4 and 0.55. They are referred to as magnesiochromites to distinguish them from the rarer chromian spinels with $Cr/(Cr + Al)$ ratios of 0.2–0.4 that occur within olivines in picritic lavas from the North Atlantic (Sigurdsson and Schilling, 1976). This is a purely arbitrary division made for descriptive purposes only as a complete range of solid solution exists for this part of the spinel series (see Fig. 8.2).

Small crystals in glassy or variolitic rocks and the core compositions of microphenocrysts are usually in equilibrium with associated olivines, and spinel chemistry appears to be a sensitive indicator of liquid composition

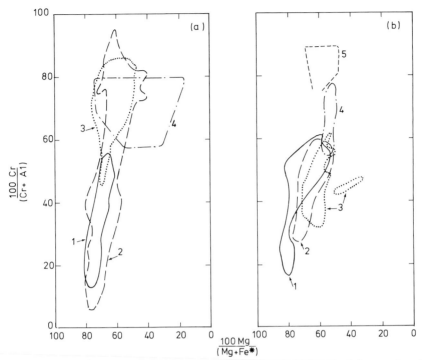

Fig. 8.2. Values of 100 Cr/(Cr + Al) versus 100 Mg/(Mg + Fe*) for spinels in oceanic and related rocks. (a) Fields: 1, abyssal peridotites; 2, alpine peridotites and ophiolites; 3, podiform chromites; 4, layered intrusions, (b) Fields: 1, Mid-Atlantic Ridge basalts from 30° to 40°N; 2, other mid-ocean ridge samples; 3, back-arc basin basalts; 4, ocean plateau basalts; 5, boninites. [Data sources: Frey *et al.* (1974), Graham *et al.* (1978), Dungan *et al.* (1978), Sigurdsson and Schilling (1976), Sigurdsson (1977), Prinz *et al.* (1976), Arai and Fuji (1978), Clarke and Loubat (1977), Thompson and Humphris (1980), Irvine (1967) and Dick and Bullen (1984).]

during the fractional crystallization of basic magmas (Sigurdsson and Schilling, 1976; Graham *et al.*, 1978; Dick and Bullen, 1984). Nevertheless, the zoning trends displayed by crystals in individual samples can be highly varied and large compositional ranges have been reported. The most common trends are increasing total Fe with a variable Cr/(Cr + Al), and increasing Cr/(Cr+ Al) with small variation in Mg/(Mg + Fe^{2+}), which are consistent with crystallization of olivine or olivine plus plagioclase, and consequent decreases in temperature and Mg/(Mg + Fe^{2+}) in the melt, but both types of trend can be encountered in the same samples (Sigurdsson, 1977; Dick and Bryan, 1978). Partially resorbed chromian spinels with overgrowths of magnesiochromite have also been recorded. Al-rich spinels are usually attributed to crystallization at elevated pressures [e.g., Bryan (1972), Frey *et al.* (1974), and Dick and Bryan (1978)]. To explain the diverse range of spinel compositions in some ocean-floor basalts and the probable presence of high-pressure xenocrysts, mixing of new magma with old in a high-level magma chamber beneath the ridge crest is usually invoked [e.g., Donaldson and Brown (1977), Dungan *et al.* (1978), and Dick and Bryan (1978)]. Dick and Bullen (1984) attribute abyssal dunites with spinels at the Cr-rich end of the compositional range to fractional crystallization of basalts at relatively shallow depths (<10 km) and equate them to the basal dunites of ophiolite complexes whilst abyssal dunites with a high-Al spinel are thought to represent residues precipitated at a higher pressure from ascending melts in the mantle.

Spinel composition fields for different occurrences are shown in Fig. 8.2. It is clear from this that there is considerable diversity amongst oceanic and related rocks. Back-arc basin basalts tend to have higher $Mg/(Mg + Fe^*)$ ratios at a given value of $Cr/(Cr + Al)$ than mid-ocean ridge basalts, whilst oceanic plateau basalts (Tokuyama and Batiza, 1981) and boninites have higher $Cr/(Cr + Al)$ at given values of $Mg/(Mg + Fe^*)$ than either back-arc basin or mid-ocean ridge basalts. Even amongst the latter there is a wide compositional range with spinels in basalts from the North Atlantic extending to more Fe- and Al-rich values than those from other spreading centres. There is considerable overlap of the spinel fields for abyssal basalts and peridotites, only limited overlap of these with the fields for Alpine-type peridotites and ophiolites and little, or no, overlap between spinels from mid-ocean ridge associations and those in podiform chromite ores or layered intrusions. Dick and Bullen (1984) stress the close similarity of spinel composition fields for genetically related basalts and peridotites in different associations and relate this to the degree of melting in the mantle source regions. They consider that many alpine-type periodotites and ophiolites formed in transitional environments, such as may exist during construction of young island arcs on ocean lithosphere or the earliest stages of arc or continental rifting or small aborted intra-arc rift basins. The more restricted range of spinel compositions in mid-ocean ridges clearly implies that any magmatic ore deposits formed in the oceanic lower crust or upper mantle will be similarly restricted and more Fe-rich and Cr-poor than those in podiform or layered chromite deposits currently exposed on land.

8.2.2 Sulphides

Fresh submarine basalts in which degassing has been inhibited by the pressure of the overlying water column contain considerably more sulphur than basalts erupted in subaerial or shallow (<200 m) water environments (Moore and Fabbi, 1971; Moore and Schilling, 1973). The sulphur in the basalts occurs in several forms: a discrete constituent in the basaltic glass, as subspherical globules within glass and interstitial grains in holocrystalline matrices, and as minute spherules embedded in vesicle walls.

8.2.2.1 Glass chemistry

Microprobe studies of basalt glasses have shown that they have sulphur contents typically in the range 800–1800 ppm (Fig. 8.3) and are at, or close to, sulphur saturation at near-liquidus temperatures. This is confirmed by the almost ubiquitous presence of sulphide globules within the glasses (Moore and Fabbi, 1971; Mathez, 1976; Mathez and Yeats, 1976; Czamanske and Moore, 1977). The modal abundance of these globules is low and Czamanske and Moore estimate that they contribute an average of only 12 ppm S to the bulk compositions and hence the glass compositions can be taken to reflect the magma compositions with respect to sulphur. Support for this was provided by Mathez (1976) who showed that glass inclusions in phenocrysts were similar compositionally to their matrix glasses or else depleted only in those elements entering the host phenocryst.

Sulphur solubility in mafic melts depends principally on fO_2, fS_2 and the Fe contents of the magmas as well as confining pressures (Katsura and Nagashima, 1974; Haughton et al., 1974). The role of temperature is difficult to evaluate. Haughton et al. concluded that at constant fO_2 and fS_2, a drop of 40°C could reduce sulphur solubility by a factor of two, but no detailed studies of the cooling of S-saturated liquids have been carried out. Figure 8.3

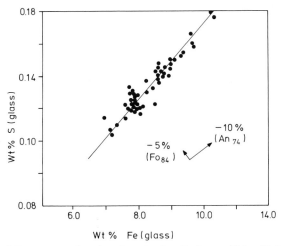

Fig. 8.3. S and Fe concentrations in submarine basalt glasses. [After Mathez (1979).]

shows the Fe and S concentrations in submarine basalt glasses. The data follow a trend similar to the experimental curve of Haughton *et al.* (1974) at 1200°C and 1 atm but are displaced to higher sulphur contents, presumably reflecting the greater solubility of sulphur in basalt liquids at higher pressures (Moore and Schilling, 1973). The good linear correlation demonstrates the strong dependence of sulphur contents on the iron content and implies that these elements probably complex in the silicate melt. This in turn implies that in sulphur-saturated magmas the Fe content will control the gas composition with respect to fO_2 and fS_2 (Mathez, 1976). Both Mathez and Czamanske and Moore (1977) considered fO_2 and fS_2 were subordinate to Fe content in controlling S solubility, in which case crystal fractionation may well be the dominant process responsible for maintaining differentiating melts at sulphur saturation.

8.2.2.2 *Sulphides*

Sulphide globules ranging in diameter from 2 to 100 μm are common in submarine basalt glasses and have been the subject of numerous studies [e.g., Kanehira *et al.* (1973), Mathez and Yeats (1976), Czamanske and Moore (1977), and Puchelt and Hubberten (1980)]. Most globules are spherical, but oblate and irregularly rounded ones also occur. They are frequently encountered in glass inclusions or embayments in phenocrysts. Sulphides also occur as small globules concentrated in and around skeletal oxides in quenched glasses or as angular, interstitial grains dispersed through holocrystalline groundmasses. In massive, more coarse-grained flows a complete transition is seen from globules in the quenched margins to irregular, apparently recrystallized grains in the flow interiors (Mathez, 1980).

The sulphides consist of monosulphide solid solution (MSS, Fe–Ni sulphide), intermediate solid solution (ISS, Cu–Fe sulphide) and occasional grains of exsolved pentlandite. In some cases inclusions or intergrowths of magnetite have been observed. More frequently, although a discrete oxide phase can be detected, the individual grains are too small (often <1 μm) for positive identification (Mathez, 1976). Experimental studies have shown that under most conditions, immiscible sulphide melts contain some dissolved oxygen which may form either wustite or magnetite, and be preserved as magnetite in disseminated sulphide droplets or sulphide ores (Naldrett, 1969; Shimazaki and Clark, 1973). The estimated magnetite content of ocean-floor sulphides ranges from 2 to 4% by volume, equivalent to about 1% dissolved oxygen in the sulphide liquid (Mathez, 1976; Czamanske and Moore, 1977).

The globules exhibit considerable internal heterogeneity which testifies to rapid subsolidus re-equilibration during cooling. Czamanske and Moore (1977) constructed the following cooling history for globules from the FAMOUS area of the Mid-Atlantic Ridge by comparison with experimental studies in the Fe–Ni–Cu–S–O system (Kullerud *et al.*, 1969; Craig and Kullerud, 1969; Naldrett, 1969), which appears to be of general application. Crystallization of MSS began at about 1100°C as the host glasses were rapidly increasing in viscosity. Separation of minor Fe oxide was then followed by solidification of the residual Cu–Fe–S liquid at about 850°C. Coarse pentlandite began to exsolve at about 610°C and a second generation formed between 600 and 300°C. Continuation of cooling eventually caused the ISS to break down into two phases at around 550°C. The extent of re-equilibration depends on the duration of the cooling interval. Figure 8.4 shows the compositions of the Ni-rich and Cu-rich phases together with the bulk compositions of the globules in quenched glasses from several localities projected onto the 700°C isothermal Cu–Fe–S surface. Despite the fact that these glasses must have been quenched from magmatic temperatures to less than 500°C in some tens of seconds the phase relations indicate re-equilibration to temperatures of ≤700°C. Mathez (1980) showed the re-equilibration continued to temperatures of 300°C or less in massive flows.

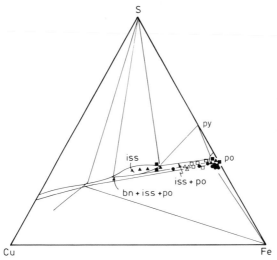

Fig. 8.4. Ni-rich and Cu-rich phases and estimated bulk compositions (open symbols) of sulphide globules projected into a generalized Cu–Fe–S system at 700°C. (●) East Pacific Ridge, (△, ▲) Mid-Atlantic Ridge, (□, ■) Nazca plate. Kullerud *et al.* (1969), Cabri (1973), po, pyrrhotite; py, pyrite; iss, intermediate solid solution; bn, bornite. The projections are made at constant S, that is, %Fe + %Cu = 100 − %S. [Data sources; Mathez and Yeats (1976), Czamanske and Moore (1977) and Table 8.1.]

Bulk sulphide compositions are difficult to determine because of the extensive exsolution but appear to be related to the compositions of the host magmas. Peridotites recovered during DSDP leg 37 contain sulphides with Ni plus Cu contents of 33·8 wt%, 2·8 wt% Co and Ni/Cu ratios of 3·3 (MacLean, 1977). Picritic basalts have sulphides with 26–20 wt% of Ni plus Cu and Ni/Cu ratios of 4·0 to 1·0, whilst in more evolved samples these values drop to 10 wt% and 0·5, respectively. Co, Mn and Zn occur only in trace quantities; the maxima reported are 0·3 wt% Co, 0·06 wt% Mn and 0·1 wt% Zn (MacLean, 1977; Mathez and Yeats, 1976; Czamanske and Moore, 1977). Judging from the sparse available data, Co enters the Ni-rich phase and Zn the Cu-rich phase.

These variations are in good agreement with experimentally determined distribution coefficients for transition metals between sulphide and silicate liquids. The partition coefficient D (where D = wt% metal in sulphide/wt% metal in silicate) obtained by Rajamani and Naldrett (1978) for a basaltic melt with 8·3% MgO was 274, 245 and 80 for Ni, Cu and Co, respectively at 1225°C. They further showed that in more basic melts (13·5% MgO) Cu becomes more sulphophile than Ni with D values of 333 and 231, respectively. Similar, but lower values of D were obtained by MacLean and Shimazaki (1976) for the synthetic system $FeO–FeS–SiO_2$ which gave partitioning ratios in the order Ni > Cu > Co > Fe > Zn for iron sulphide/iron silicate liquids. Zn showed a clear preference for the silicate liquid with D values of 0·5 or less.

The reported $\delta^{34}S$ values for these sulphides range from $-0·6‰$ to $+1·2‰$ for both globules and grains indicating that the sulphur is of mantle origin (Kanehira et al., 1973; Grinenko et al., 1975; Puchelt and Hubberten, 1980). This coupled with the textural and compositional variations clearly demonstrates that the sulphides are magmatic in origin resulting from immiscibility between silicate and sulphide liquids, that in many magmas a sulphide phase existed prior to eruption and that sulphide separation continued throughout emplacement. The close association of late-crystallizing oxides and a sulphide phase is to be expected since oxide separation will decrease iron concentrations in the residual silicate liquid and drive such liquids to FeS supersaturation (See Fig. 8.3).

8.2.2.3 Vesicle sulphides

Glassy basalt selvedges frequently contain vesicles with spherules decorating their walls. The shape, size and distribution of these wall decorations is extremely varied: some contain more than one generation, other vesicles have none (Moore and Schilling, 1973; Yeats and Mathez, 1976; Propach,

1978). Many of the spherules are Fe sulphides with high concentrations of Cu and Ni, and others appear to be oxidized and now consist of limonite accompanied by palagonitized glass. Accurate analyses are difficult to obtain but the available data suggest the spherule sulphides are richer in Fe and poorer in Cu and Ni than associated sulphides in the basalt glasses (Propach, 1978) (Table 8.1). Moore and Calk (1971) suggested that spherules formed by reaction of sulphur in the gas in the vesicles with metals in the silicate liquid. Their formation on the vesicle walls is attributed to the fact that S diffusion in the gas phase will be faster than in the surrounding melt. This explanation views vesicle spherules as disequilibrium features in response to rapid cooling, in contrast to the sulphide globules which are equilibrium features of silicate melts saturated with respect to sulphur.

8.3 Distribution of Magmatic Ores within the Ocean Crust

The present inadequate sampling of the ocean crust means that the extent of any mineralization in the lower portions must be mainly deduced from the crystallization histories and transition metal contents of the basalts. Whilst such ore bodies are unlikely to ever be an economic prospect they may provide an important source of transition metals for redistribution by deeply penetrating hydrothermal fluids.

The case for the fractionation and accumulation of spinel is well established. The textural and compositional relationships reviewed above demonstrate that spinel is a liquidus phase in many ocean-floor basalts and that the spinel compositions are systematically related to magma composition. Experimental studies have confirmed that spinel is an early crystallizing phase at high and low pressures [e.g., Fisk and Bence (1980)]. The Cr and Ni contents of ocean floor basalts range from 700 and 300 ppm, respectively, in primitive basalts to less than 50 ppm in more evolved samples [e.g., Frey et al. (1974), and Bougault (1977)]. Much of this variation can be attributed to the fractionation of olivine and spinel. Cumulate-textured gabbroic and peridotitic rocks have been recovered from a few localities on the ocean floor. The samples transected during DSDP leg 37 showed phase layering with concentrations of spinel and olivine at the base of the layers (Clarke and Loubat, 1977). Dick and Bullen (1984) demonstrated that spinels in abyssal dunites have a biomodal range of compositions that are consistent with both relatively high-pressure crystallization from ascending mantle melts and with relatively low-pressure crystallization possibly in a mid-ocean ridge magma chamber.

The extent to which sulphide accumulation may have also occurred is far less clear. Cu–Ni–Co–Fe sulphide ores of magmatic origin occur associated with dunitic and peridotitic rocks in the plutonic complex of the Troodos ophiolite (Panayiotou, 1980), but no comparable material has been reported from the ocean floor. Since the level of sulphur saturation in silicate liquids is related to the Fe content, the extent of any sulphide separation will be largely controlled by the effects of fractional crystallization on the Fe contents of the magmas.

Numerous chemical, mineralogical and experimental studies have shown that the low-pressure fractionation of ocean-floor basalts is dominated by olivine, or olivine plus plagioclase [e.g., Shido *et al.* (1971), Bryan *et al.* (1976) and Bender *et al.* (1978)]. The two vectors on Fig. 8.3 show the effects of the removal of these two phases on the Fe and S contents of the magmas. It is evident from this that, *provided the effects of decreasing temperature can be ignored,* the separation of mixtures of these two phases need not drive the liquids towards sulphur saturation. The addition of another Fe-bearing phase such as spinel or pyroxene, or the crystallization of plagioclase-free or plagioclase-poor assemblages would promote sulphide separation. Many authors have included clinopyroxene as a major fractionating phase in order to relate the compositions of associated basalts by crystallization models even though pyroxene is frequently absent, or only a minor component of the phenocryst assemblage [e.g., Bryan *et al.* (1976), Bougault (1977), Thompson *et al.* (1976) and Bryan (1979)]. The simplest way of resolving the conflict between the petrographic and chemical evidence is to invoke fractionation within the upper mantle at higher pressures (7–10 kbar) when clinopyroxene will also become a liquidus phase (Kushiro and Thompson, 1972; Bender *et al.*, 1978; Green *et al.*, 1979). At elevated pressures crystallization of pyroxene and plagioclase would immediately follow olivine. Subsequent ascent of the fractionated magmas to shallow levels would cause precipitation of olivine and plagioclase but not clinopyroxene, because of the expansion of the olivine plus plagioclase field relative to that of clinopyroxene (Kushiro and Thompson, 1972). Spinel and clinopyroxene xenocrysts that appear to be relics of high-pressure crystallization have been reported in ocean-floor basalts [e.g., Donaldson and Brown (1977) and Sigurdsson and Schilling (1976)]. High-pressure fractionation with or without subsequent mixing at low pressures has been invoked by many authors to explain the range of compositions in oceanic basalts [e.g., Flower (1981), O'Donnell and Presnall (1980), and Dick and Bullen (1984)]. Nearly all of these scenarios would be expected to lead to significant sulphide separation also.

Spinel, olivine, plagioclase and clinopyroxene all have crystal/liquid distribution coefficients for Cu of less than 1 (Gunn, 1971; Bougault and Hekinian, 1974; Bougault *et al.*, 1980). The only phase that can concentrate Cu is an immiscible sulphide liquid. Definitive evidence for sulphide separation would therefore be provided if successive basalts related by fractional crystallization showed decreasing Cu contents. Unfortunately, suites of demonstrably cogenetic ocean floor lavas are rare and Cu is not normally analysed for on a routine basis in petrogenetic studies.

Bougault *et al.* (1978) consider the Cu contents of abyssal basalts to be fairly uniform. Czamanske and Moore (1977), on the other hand, consider that significant depletion in Cu occurs during crystal fractionation. Using mass balance calculations for glasses from the FAMOUS area, they suggested that one-third of the Cu and commensurate amounts of S, Ni and Fe were removed from the parent magma by sulphide separation. Their results should be treated with caution as Cu and Ni distribution coefficients for sulphide/silicate liquids calculated from their model are too high. Nevertheless a survey of the available data reveals support for their model. Figure 8.5 shows the mean transitional metal abundances in groups of lavas recovered during DSDP legs 37 and 46 that appear to be related by simple fractional crystallization (Bougault *et al.*, 1978; Bougault, 1977). For each group the data were normalized to the abundances in the least evolved samples, which therefore plot along the line at unity. It can be seen from the diagram that fractionation in both groups is accompanied by progressive increases in Ti, V and Zn, smaller increases in Fe and Mn, and progressive *decreases* in Cr, Ni *and Cu*. Clearly, along these portions of the Mid-Atlantic Ridge at least, evolution of the magmas was accompanied by sulphide separation.

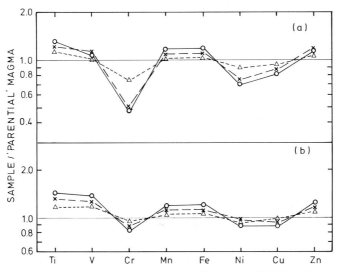

Fig. 8.5. Mean transition metal abundances in groups of lavas from DSDP legs 37 (a) and 46 (b). For details, see text.

K

The location and dimensions of magmatic ores within the ocean crust and uppermost mantle depends on the dimensions, depths and fluid dynamics of the axial magma chambers, which is a subject of considerable controversy (Macdonald, 1983). *If* the ophiolite analogy can be applied to mid-ocean ridges then chromite bodies (and sulphides) probably occur in the uppermost portion of the mantle beneath the ocean crust (Coleman, 1977; Sawkins, 1984), within the reach of hydrothermal circulation (Gregory and Taylor, 1981). Alternatively, should high-pressure crystallization processes prove to be dominant, then a considerable portion of the Cr, Ni and Cu budget of the oceanic lithosphere lies well beyond the reach of penetrating fluids.

8.4   Hydrothermal Minerals

Alteration of the oceanic crust by seawater has been the subject of intense study for several years. A wide variety of secondary minerals has been recognized and related to an almost equally diverse range of processes [e.g., Miyashiro *et al.* (1971), Bass (1976), Scott and Swanson (1976), Andrews *et al.* (1977), Robinson *et al.* (1977), and Cann (1979)]. Much of this diversity can be attributed to the fact that alteration was accomplished under open system conditions, and changes in redox conditions, pH and water/rock ratios as well as temperature can exert considerable influence on mineral stabilities.

Oxygen isotope data for ocean floor rocks is summarized in Fig. 8.6. Fresh, unaltered rocks, or primary minerals separated from altered rocks have $\delta^{18}O$ values of $5\cdot8 \pm 0\cdot3‰$ (Muehlenbachs and Clayton, 1972a; Pineau *et al.*, 1976). Most of the altered rocks recovered from the seafloor have been enriched in ^{18}O as a result of weathering or low-temperature hydrothermal alteration [e.g., Anderson and Lawrence (1976) and Gray *et al.* (1977)]. It should be emphasized that low temperature in this context could include alteration taking place at temperatures as high as 200°C or more (Stakes and O'Neill, 1982). H_2O and $\delta^{18}O$ are positively correlated in many of these samples, a feature best explained by mixing of fresh, anhydrous basalt with high-^{18}O, H_2O-rich secondary minerals formed from the basalts by reaction with seawater at low temperatures (Muehlenbachs and Clayton, 1972b; Stakes and O'Neill, 1977). In contrast, samples affected by alteration at high temperatures show either a decrease in $\delta^{18}O$, or else little change.

The principal minerals found in hydrothermally altered rocks are sulphides or mixtures of oxides and hydroxides, corresponding to alteration under relatively reducing or oxidizing conditions, respectively. In general,

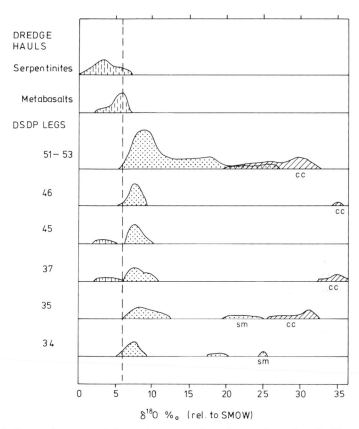

Fig. 8.6. Oxygen isotope variation in igneous rocks and secondary minerals. The vertical dashed line indicates the value for unaltered samples: sm, smectite; cc, calcite. [Data sources: Muehlenbachs and Clayton (1972b), Wenner and Taylor (1973), Muehlenbachs (1976, 1977, 1980), Anderson and Lawrence (1976), Gray *et al.* (1977), Hoernes *et al.* (1978) and Muehlenbachs and Hodges (1978).]

assemblages indicating deposition under relatively high sulphur fugacity and low oxygen fugacity are replaced by, or grade into, minerals deposited under conditions of relatively high fO_2 and locally high pCO_2 with time. Oxidative alteration is best developed in the upper portion of the crust and appears to decrease with depth as sulphide-bearing assemblages become more common. Nevertheless, oxidized and reduced assemblages that formed under both low and high temperatures have been recorded [e.g., Robinson *et al.* (1977) and Humphris and Thompson (1978a)].

K*

8.4.1 Low-temperature alteration

8.4.1.1 *Reduced assemblages*

The characteristic secondary mineral assemblage of low-temperature basalt alteration under sub to anoxic conditions is Mg smectite, Fe sulphide and associated minor carbonate. Alteration of the basalts is normally limited to replacement of olivine and interstitial glassy material by smectite and minor pyrite and primary magmatic sulphides by pyrite. Glassy pillow rinds may show palagonitization. Clinopyroxene, titanomagnetite and plagioclase are unaffected. Veins and vesicles are filled by smectite, sulphides and minor carbonate. The sequence of mineral deposition in the cavities is usually smectite, sulphide, carbonate [e.g., Bass (1977), Robinson *et al.* (1977) and Mathez (1980)].

The smectites are usually pale brown or green saponites characterized by high MgO (>20 wt%), low total iron (1–6 wt%) and very low K_2O (0·0–0·3 wt%). The minor carbonates associated with the smectites are aragonite and Mg-rich calcite. Picritic rocks may also contain dolomite and magnesite (Robinson *et al.*, 1977). Temperatures of 32–49°C were obtained by Lawrence *et al.* (1978), using oxygen isotope data, for saponites in DSDP leg 45 samples. These may well be minimal values as fibrous minerals such as smectites with a high surface area/volume ratio will be particularly prone to re-equilibration at lower temperatures. Saponites in ocean-floor samples probably form over a wide temperature range and also occur in association with higher temperature assemblages described below.

With progressively increasing temperature the saponites are joined by other Mg-rich layer silicates. The first to appear is talc, and it is followed by mixed-layer smectite–vermiculite–chlorite minerals (SVC) which may replace the saponites completely. In rocks containing saponite and talc the plagioclases are usually still unaltered, whilst the appearance of the SVC minerals usually coincides with the complete or partial replacement of plagioclase by Na–Ca zeolites and incipient alteration of the titanomagnetites to sphene. Clinopyroxene still remains pristine (Stakes and O'Neill, 1982; Cann, 1979). This assemblage is referred to as "zeolite-facies" by many authors [e.g., Cann (1979)]. This terminology, however, suffers from the problem that the index minerals, the zeolites, are relatively scarce and often found only as cavity fillers in vesicles or cross-cutting veins, which may indicate deposition during a later stage of alteration under different conditions. Examination of published descriptions reveals that rocks containing the assemblage saponite + SVC ± Talc ± chlorite and devoid of, or poor in, zeolites, have been reported from the Mid-Atlantic Ridge, the Indian Ridge, and the Nazca plate and occur over a wide interval of the site 504 drill hole on the Costa Rica Rift (Rozanova and Baturin, 1971; Shido *et al.*, 1974; Bass, 1976; Scott and Swanson, 1976; Anderson *et al.*, 1982). Zeolite-rich rocks, poor in Mg minerals, have been recovered from the Mid-Atlantic Ridge (Aumento *et al.*, 1971; Miyashiro *et al.*, 1971). The zeolite-rich group appear to be volumetrically less significant and zeolite veins only occur over a short interval in the site 504 drill hole. Oxygen isotope temperatures of 130–170°C have been obtained for saponite-talc-bearing pillow breccias by Stakes and O'Neill (1982). Comparison with active hydrothermal systems indicates temperatures may vary from 100 to 200°C (Kristmannsdottir, 1975).

The sulphides in all these samples consist of almost pure FeS and are usually described as pyrite. X-ray studies of DSDP leg 34 material revealed the presence of both marcasite and pyrite (Scott and Swanson, 1976; Bass, 1976), but whether or not marcasite also occurs in other similar assemblages is not known. "Disseminated-ore" type mineralization has been reported by Rozanova and Baturin (1971) in altered basalts containing saponite, SVC, talc, chlorite type alteration, whereas in the scarcer zeolite-rich samples sulphides appear to be rare or absent. The Fe sulphides replacing primary magmatic sulphides can contain up to 4·0 wt% Ni, whereas those in the veins and vesicles contain only Fe and S, indicating little mobility of Ni in the fluids at these temperatures. Co, Mn and Zn have not been detected (Humphris *et al.*, 1980a; Mathez, 1980). Reported $\delta^{34}S$ values for the separated sulphides vary widely from $-33\cdot4‰$ to $+23\cdot0‰$ (Field *et al.*, 1976; Krouse *et al.*, 1977). Krouse *et al.* noted a decrease in variation and in $\delta^{34}S$ depletion relative to unaltered basalts with depth. The data imply an abundant source of nonmagmatic sulphur, presumably seawater, but because the values range from strong $\delta^{34}S$ depletion relative to primary igneous values (0‰) to slight $\delta^{34}S$ enrichment relative to modern seawater (+21‰), they cannot be interpreted in terms of any simple equilibrium mixing model.

8.4.1.2 *Oxidized assemblages*

In many basalts the early smectite–sulphide depositional phase is replaced by, or grades into, a smectite–carbonate–oxide assemblage. The advent of oxidizing conditions is often marked by the appearance of yellow and red colourations and distinct alteration haloes with sharp, diffusion-controlled boundaries. These haloes may be up to 20 times as wide as the fractures they surround, testifying to the passage of large volumes of oxidizing fluids. Clastic material and microfossils that have been sluiced in from the seafloor and subsequently overgrown by smectites have been observed in some veins (Andrews *et al.*, 1977; Seyfried *et al.*, 1976). In other samples from lower in the crust, the transition may be more gradual (Humphris *et al.*, 1980b). The extent of oxidative alteration depends on permeability and exposure time on the seafloor and usually decreases with depth as sulphides become more abundant. In the site 504 hole, the zone of oxidative alteration extends to 300 m; in other deep holes it may reach 500 m plus (Robinson *et al.*, 1977; Mevel, 1980; Anderson *et al.*, 1982).

In extensively altered samples, saponite, remnant olivine and glass are all replaced by yellow and red smectites dusted with Fe oxides, plagioclase by mixtures of smectite, adularia and phillipsite, and titanomagnetite loses Fe and is oxidized to titanomaghemite. Leucoxene has been observed in oxidation haloes. Pillow rinds and hyaloclastite breccias are replaced by clays and oxides, or phillipsite and calcite. In less oxidized material smectite is the main secondary mineral deposited and sulphides are replaced by limonite. Veins and vesicles tend to show the following depositional sequence: smectite, phillipsite, occasional adularia, carbonates, oxides, sulphates (Andrews *et al.*, 1977; Humphris *et al.*, 1980a; Mevel, 1980; Morrison and Thompson, 1983).

The predominant smectite is a yellowish celadonite, rich in total iron (>20 wt%) and K_2O (4–9 wt%) and low in MgO (<6 wt%). Nontronitic clays are also fairly common but less so in extensively altered material. The often intense red colourations and very high iron contents reflect dusting with iron oxides. Vesicles may show a zonation from saponitic outer layers through varying shades of yellow to red inner layers reflecting deposition under conditions of steadily increasing fO_2. Occasionally an inner saponite core is seen separated from the red layer by a sharp boundary. This may be due to a cessation of seawater penetration because of either sediment deposition or plugging of the plumbing by secondary mineral precipitation (Robinson *et al.*, 1977; Pritchard *et al.*, 1979; Mevel, 1980).

The carbonates are Mg- and Fe-poor calcites. Rare manganoan calcite also occurs in extensively oxidized rocks either replacing plagioclase or as a late surface deposit (Humphris *et al.*, 1908a,b). The $\delta^{18}O$ values of the carbonates are high, ranging from +20‰ to +35‰ (Fig. 8.6). Calculated temperatures range from 0 to 30°C for most samples (Muehlenbachs, 1980; Juteau *et al.*, 1978). Stakes and O'Neill (1982) obtained temperatures of 65–85°C for calcite veins in saponite, talc-bearing pillow breccias. In all cases the calcite formation temperatures are lower than those for the earlier reduced assemblages. The $\delta^{13}C$ values for the carbonates cluster around zero, indicating that seawater is the predominant source of the carbon (Stakes and O'Neill, 1982). Sulphates are rare; the low-temperature varieties Mg–Al sulphate and gypsum have both been observed in late veins. Smectite–anhydrite veins occur in the saponite, SVC, talc-bearing portion of the site 504 drill hole and Anderson *et al.* (1982) report preliminary temperature determinations of 60–100°C for them.

In all these oxidized assemblages late-stage Fe and Mn oxides and hydroxides are common, but unfortunately probe analysis of them are not. Often they are described simply (hopefully?) as limonite and black Mn oxides. Identified varieties include geothite, haematite, todorokite and birnessite (Bass, 1976). These last two are principal constituents of the rapidly accumulating Mn crusts described from the TAG hydrothermal field in the Atlantic (Scott *et al.*, 1974), the Gulf of Aden (Cann *et al.*, 1977), the Mn-rich layers of the Galapagos mounds (Williams *et al.*, 1979), Mn oxide deposits from SW Pacific island arcs (Moorby *et al.*, 1984) and many others (Rona, 1984). Rare traces of native Cu have been identified in highly weathered breccias and vein debris from DSDP legs 37 and 53. The intimate association with clays and oxides suggest it is a product of low-temperature alteration rather than a residual phase from the basalts. Suggested mechanisms of formation are absorption of metals on colloidal particles from undersaturated solutions (Andrews *et al.*, 1977) or reduction of Cu in solution by reaction with Fe minerals (Humphris *et al.*, 1980a).

The accumulated data from DSDP drill holes shows that oxidative alteration begins soon after crustal formation and continues for several million years. Smectites, celadonites and analcites from DSDP hole 417A in the West Atlantic define a tightly constrained Rb–Sr isochron of 108 ± 3 m.y., identical to the 108 m.y. age of crust formation derived by palaeontological and magnetic anomaly correlation (Richardson *et al.*, 1980). Muehlenbachs (1980) showed that the pattern of $\delta^{18}O$ variation of DSDP basalts with age is consistent with a model of extensive circulation of cold seawater for the first 10 m.y., after which massive circulation gradually ceases except in locally permeable breccia horizons, where circulation can continue for up to 25–50 m.y. Scheidegger and Stakes (1980) have shown that the compositions of the low-temperature secondary minerals in DSDP cores vary systematically with age. Crust less than 15 m.y. old still contains a high proportion of Mg saponites; crust of intermediate age (15–50 m.y.) contains more Fe-rich minerals, mainly nontronite and celadonite; whilst crust of even greater age tends to contain adularia, K zeolites, montmorillonite and mica-like phyllosilicates rich in K and Al. This pervasive and long-lasting alteration of the upper portion of the crust must have obliterated evidence for early phases of alteration under reducing conditions except under favourable circumstances [e.g., Morrison and Thompson (1983)] and could cause the extent of hydrothermal alteration of the crust to be considerably underestimated.

8.5 High-temperature Alteration

Oceanic rocks affected by alteration at elevated temperatures contain a wide variety of secondary minerals. Material of this type is far scarcer than samples affected by low-temperature alteration (see Fig. 8.6) and detailed descriptions are even scarcer. The most commonly recovered group are the so-called "greenschist-facies" assemblages (Cann, 1979) characterized by chlorite, quartz, albite, sphene ± actinolite ± epidote in basic rocks and possibly by serpentine and talc in ultrabasic ones. Complete replacement of the igneous minerals is not uncommon. The most commonly encountered relict phase is clinopyroxene. Secondary mineral veins are numerous in these samples and several generations may be present.

Humphris and Thompson (1978a) showed that metabasalt "greenschist-facies" assemblages could be divided into two main groups, a reduced Mg-enriched group dominated by chlorite and an oxidized more Ca-rich group dominated by epidote. The division between the two groups is such that the chlorite-rich assemblages contain more than 15 modal percent chlorite and less than 15 modal percent of epidote. The chlorite-rich metabasalts are by far the most abundant group recovered from the seafloor and the rocks usually are altered pillow lavas. Epidote-rich rocks are far scarcer and may include altered dolerites as well as basalts. Many of the epidote-rich samples described by Humphris and Thompson (1978a) showed complete replacement of the primary minerals whereas the chlorite-rich pillows often had relatively unaltered material preserved in the cores of the pillows. In altered pillow lavas described as epidote-rich the epidote is frequently restricted to cross-cutting veins of quartz and epidote which may have formed under different conditions to the other minerals. Mottl (1983) has distinguished a further variety—quartz- and chlorite-rich greenstone breccias. These rocks are not predominantly metabasalts although they may contain basalt fragments; instead, the textural features indicate that they formed mainly as vein infillings. They consist of chlorite, quartz, pyrite, chalcopyrite and pyrrhotite. Unlike the chlorite-rich metabasalts they are Fe-rich, not Mg-rich, as the chlorites are iron-rich varieties.

"Amphibolite-facies" assemblages are characterized by hornblende and plagioclase in metagabbros and dolerites and by the appearance of tremolite in peridotites (Cann, 1979). Assemblages intermediate between "greenschist- and amphibolite-facies" assemblages have also been described. In all of these rocks, quartz and sulphides may be accessory phases but virtually no data on the extent of mineralization in hornblende–plagioclase rocks is available.

Sulphides are frequently found in samples affected by high-temperature alteration. Fe sulphides are common in Mg-rich metabasalts, both disseminated in the matrix and as cavity fillers in veins and vesicles, and scarce or absent in Ca-rich assemblages. "Disseminated-ore" type mineralization has only been reported from metabasalts poor, or devoid of, epidote (Bonatti *et al.*, 1976; Humphris *et al.*, 1978; Anderson *et al.*, 1982). Most of the sulphides are pyrite, but Bonatti *et al.* found both pyrrhotite and pyrite in metabasalts from the Mid-Atlantic Ridge. Zinc and copper sulphides are less abundant and have only been found in veins. Anderson *et al.* (1982) describe "stockwork-type" veins of sphalerite and chalcopyrite over an 18-m interval of the site 504 drill core. This is the only known occurrence of zinc sulphides from *within* the ocean crust, although zinc sulphides are debouching from active vents on the East Pacific Rise at 21°N (Haymon and Kastner, 1981; Styrt *et al.*, 1981). Chalcopyrite-pyrite veins cutting green-stones from the Mid-Atlantic Ridge have been reported by Bonatti *et al.* (1976) and Humphris and Thompson (1978a). Unlike the host basalts, the veins' assemblages frequently contain Ca-rich minerals such as epidote or zeolites (Bonatti *et al.*, 1976; Anderson *et al.*, 1982).

Only a handful of analyses have been reported for these sulphides. Accordingly, microprobe analyses were made of the ore-minerals in a subset of the chlorite-rich samples studied by Humphris and Thompson (1978a,b) from the Mid-Atlantic at 22°N. The results are presented in Table 8.2. The sulphides disseminated in the basalt matrix proved to be virtually pure pyrite, containing little or no Ni or Cu. The cross-cutting veins show the depositional sequence: chlorite, chalcopyrite or pyrite, quartz and epidote. The sulphides again contained little or no Ni. Manganese and zinc were looked for but not detected in all the suphides analysed, contrary to the suggestion by Humphris and Thompson (1978b) that zinc may have been deposited in the sulphide veins. In both these and the samples described by Bonatti *et al.* (1976), extensive late-stage oxidation has occurred. The veins in all the slides of one of the samples (AII-60 2-142) are surrounded by an oxidation halo extending for up to 4 mm beyond the vein margin. The sulphides in the groundmass and the vein show extensive oxidation predom-inantly to Fe-rich oxides and hydroxides (Table 8.2). Bonatti *et al.* also found Cu-rich oxides as well as Fe varieties in the rims of some sulphides.

Comparison with active geothermal systems indicates that temperatures greater than or equal to 230°C are needed for the formation of epidote, and >280°C for actinolite (Kristmannsdottir, 1975). Reported temperature determinations based on oxygen isotope data are: 125°C for antigorite in serpentinites (Wenner and Taylor, 1973); 200–300°C for quartz-epidote as-semblages (Muehlenbachs and Clayton, 1972b); and 280–380°C for altered peridotites from DSDP leg 45 (Hoernes *et al.*, 1978). Stakes and O'Neill (1982) obtained a fluid inclusion filling temperature at 275°C for fluids inside euhedral quartz crystals in a vein cutting metabasalt. Such tempera-tures are minima. From this they obtained the $\delta^{18}O$ values of the fluids from which the quartz precipitated, which proved to be +1·4‰, similar to the values measured for the hot springs on the EPR at 21°N (Craig *et al.*, 1980).

This result indicates that the high-temperature assemblages formed by interaction with seawater that had already undergone previous interaction with ocean crust. Accordingly, Stakes and O'Neill calculated temperatures using both $\delta^{18}O$ values of 0‰ (unaltered seawater) and 2‰. Their results give formation temperatures of 225–250°C for chlorite-rich greenstones. The highest temperatures they obtained were for the quartz–epidote veins cutting the greenstones which yielded temperatures of ≥350°C. The disparity in temperature between the host rocks and the veins stresses the dangers of mixing data from more than one hydrothermal event. Water/rock ratios calculated from elemental fluxes or $\delta^{18}O$ data for oceanic greenstones vary from 5:1 to greater than 50:1 [e.g., Humphris and Thompson (1978a)]. The shift in $\delta^{18}O$ values calculated for the fluids that deposited the veins by Stakes and O'Neill indicates water/rock ratios of 1·5:1 or less. Altered plagioclases from the Semail ephiolite indicate temperatures of at least 400–500°C and water/rock ratios of 0·15:1 (Gregory and Taylor, 1981); similar values are probably required for metagabbros recovered from the ocean floor.

8.6   Fractionation of Metals during Hydrothermal Alteration

Very few studies of the effects of hydrothermal alteration on the transition metal content of ocean floor basalts have been carried out. Investigations of low-temperature alteration have concentrated on changes across alteration haloes [e.g., Honnorez *et al.* (1978)] and no data is available for the earlier stages of alteration under reducing conditions. The only detailed study of high-temperature alteration is that of Humphris and her co-workers (Humphris and Thompson, 1978a,b; Humphris *et al.*, 1978) and no data is available for the "amphibolite-facies" assemblages. Transition element data for secondary mineral assemblages is equally scarce, being limited to Fe and Mn in most cases. This scanty data base can to some extent be supplemented by data from metalliferous sediments, ridge-crest hydrothermal systems and experimental studies, all of which have shed considerable light on the factors controlling metal transport in hydrothermal systems.

8.6.1   Experimental evidence

Recent experimental studies have duplicated many of the major element changes observed in hydrothermally altered basalts and accurately predicted many of the features of the recently discovered active hydrothermal vents. They have been less successful in reproducing the secondary mineral assemblages found in nature.

The most characteristic feature of all the experimental studies is the removal of Mg from seawater and its incorporation as $Mg(OH)_2$ into secondary minerals such as smectite, talc, chlorite and actinolite. The OH^- component is derived from the dissociation of seawater, the residual H^+ simultaneously causing a reduction in solution pH. This H^+ is subsequently consumed in hydrolysis reactions in which primary minerals are converted to secondary hydrated silicates. The loss of Mg is balanced by the removal of other elements from the basalts, notably Ca but also Si. Some Na is removed from seawater at low water/rock ratios (<5), but at higher ratios it is gained by the solutions. K is added to the rocks at low temperatures (<70°C) but at 150°C and above it is uniformly leached (Hajash, 1975; Mottl and Holland, 1978; Seyfried and Bischoff, 1981; Hajash and Chandler, 1981). The overall effect on the solution pH depends on the relative and absolute rates of these two types of reaction—those generating H^+ and those consuming it. The pH along with the water/rock ratio controls the concentration of metals in solution and the secondary minerals formed.

During experiments at low water/rock ratios (<50), i.e., rock-dominated conditions, Mg removal is essentially complete and hydrolysis reactions rapidly restore pH to near neutrality. Under these conditions the concentrations of metals in solution are low over a temperature range of 70–350°C, but start to rise at higher temperatures. In contrast, at higher water/rock ratios, i.e., under seawater-dominated conditions, the supply of Mg exceeds the amount the rock can accept, pH remains low and significant concentrations of Fe, Mn, Cu and Zn are maintained in solution even at the relatively low temperature of 150°C. V, Cr and Ni have not been found in solution in measurable amounts even at extreme conditions of low pH (Seyfried and Bischoff, 1977; Mottl and Seyfried, 1980; Hajash and Chandler, 1981; Seyfried and Mottl, 1982). Experiments at high water/rock ratios at temperatures less than 150°C have not been reported, but the general relationship between low pH and enhanced metal solubility can be expected to hold, although the absolute concentrations of metals in solution will be lower. All the studies have demonstrated that metal solubilities decrease with decreasing temperature, irrespective of water/rock ratio. The general pattern of solubility appears to be Mn > Fe > Zn > Cu > V, Cr and Ni, particularly at low temperatures. At extremely high temperatures, 350–500°C, H^+-generating reactions begin to predominate and metal concentrations in solution increase, even at low water/rock ratios. This appears to be due to the reduced solubility of calcium hydroxysilicates at higher temperatures. Ca thus begins to precipitate in secondary minerals such as epidote and generates H^+ through the removal of $Ca(OH)_2$ from solution (Hajash, 1975; Mottl et al., 1979).

Direct comparison of experimental and natural systems is complicated. In the experiments powdered samples are used, thereby eliminating the effects of permeability and surface area and flow geometry, lessening the effects of diffusion and therefore temperature, and maximizing the opportunity for less-reactive phases to participate in reactions. Water/rock ratios used in these experiments are the total mass of water divided by the total mass of rock in the system and are not strictly comparable to water/rock ratios calculated from stable isotope data. Basalt–seawater interaction in nature takes place under open, not closed conditions, hence true water/rock ratios are probably higher than those calculated by either experimental or isotopic approaches. The pH and reactivities of the solutions involved in natural systems will depend less on the total supply of components and more on the relative *rates* of supply versus reaction in the rock. Nevertheless, since the Mg concentrations in seawater-derived fluids largely control their acidity and the nature of the reactions taking place, the extent of Mg uptake in basalt can be used to a first approximation as a measure of the locally effective water/rock ratio during alteration. This approach has proved successful in explaining many of the mineralogical and chemical features of metabasalts (Seyfried and Mottl, 1982; Seyfried *et al.*, 1978).

During the early stages of low-temperature alteration Mg is taken up and deposited as smectite, whilst at higher temperatures the production of Mg-rich assemblages poor in Na- and Ca-rich minerals appears to predominate. Conditions during the production of these assemblages will have been "seawater-dominated" and considerable leaching of metals, mainly Fe, Mn, Zn and Cu, may have occurred. The H^+ generated will have been carried away together with any alkali and metal elements removed from the basalts to attack rocks elsewhere in the system. Samples containing both Mg- and Ca- or Na-rich minerals such as zeolites or epidote may have been exposed to both acid and neutral or alkaline conditions, with acid during the uptake of Mg and neutral subsequently. This would explain the tendency for zeolite and epidote to occur in veins cutting the earlier Mg-rich assemblages, the association of epidote with sulphides in veins but not in the host metabasalts, and the disparity in temperatures demonstrated by Stakes and O'Neill for the host metabasalt and epidote vein.

The scarcer, Ca-rich assemblages must have been produced under either "rock-dominated" conditions at near neutral pH, or by interaction with Ca-rich fluids that had lost their Mg at higher levels in the crust. Seyfried *et al.* (1978) invoked the first explanation to explain the formation of chlorite–albite–epidote–actinolite cores in altered pillow lavas with chlorite-rich rims, since the rims would be exposed to higher fluid fluxes. The latter mechanism is more likely to apply to the epidote-rich metabasalts described by Humphris and Thompson (1978a,b), since in many cases the rocks have been completely replaced by the Ca-rich assemblages and some are dolerites. Interaction of basalt with evolved Ca-rich, Mg-poor seawater means that Mg content will no longer provide a reliable guide to alteration conditions. If the fluids were evolved the alteration may have been accompanied by significant removal of metals in solution since the precipitation of $Ca(OH)_2$ would have maintained a low solution pH.

Evidence for high-temperature fluids that penetrate deep into the crust is provided by the frequent presence of metagabbros containing "amphibolite-facies" assemblages in dredge hauls. The available major element data indicate that the bulk changes in composition during the production of the Mg-rich "greenschist-facies" assemblages were greater than those during production of the Ca-rich asemblages, which in turn were greater than those involved in formation of the "amphibolite-facies" assemblages (Humphris and Thompson, 1978a; Bonatti et al., 1975). This is consistent with the sequential production of these assemblages under conditions of decreasing water/rock ratio and increasing depth of penetration and temperature of the fluids—presumably in the downflowing limb of a hydrothermal system.

Mottl (1983) produced a very similar model based on recalculating the experimental data into secondary mineral assemblages seen in recovered basalts using the data of Humphris and Thompson (1978a), since as noted above the experiments have successfully reproduced the major element transformations seen in nature but not the mineralogical ones. His model predicts the following sequence of assemblages with increasing temperature and decreasing water/rock ratio in the downflowing zone of an oceanic hydrothermal system: chlorite plus quartz; chlorite plus albite plus quartz; chlorite plus albite plus epidote plus actinolite plus quartz; chlorite plus albite plus epidote plus actinolite without quartz. Water/rock ratios were estimated by Mottl to be >50 for the first assemblage and <2 for the last. Despite the apparent good chemical fit of such models they should all be treated with caution, as most descriptions of altered rocks recovered from the seafloor fail to distinguish vein assemblages from those of the host rocks. The final group to consider is the unusual sulphide-bearing quartz chlorite breccias from the Mid-Atlantic Ridge described by Mottl (1983), which clearly do represent vein infillings. Their formation was attributed by Mottl to deposition from Mg-poor fluids in the *upwelling* limb of a hydrothermal system since they are rich in Fe, Mn, Zn, Cu and Si—all elements leached from basalts during the experimental studies.

8.6.2 Evidence from hydrothermally altered samples

The model outlined above indicates that the widespread Mg-rich assemblages were produced in the downflow region of a hydrothermal system.

Mn, Fe and Zn were probably mobilized during both low- and high-temperature alteration whilst Cu and Ni were probably leached during high-temperature alteration but were probably immobile during the formation of the saponite bearing assemblages. Verification of this is difficult because of the lack of data. Nevertheless, the occurrence of Fe sulphide veins in most of the Mg-rich metabasalts clearly indicates that Fe must have been mobile over a wide range of temperatures as the Fe content of seawater is too low ($<0\cdot1$ ppm; Hajash and Chandler, 1981) to provide an alternative source.

Figure 8.7 shows the Fe and Mn contents of secondary minerals culled from the literature. The diagram clearly shows that most secondary minerals, particularly the low-temperature smectites, have higher Fe/Mn ratios and lower Mn contents than unaltered basalts and basaltic glasses. The only minerals enriched in Mn relative to fresh basalts are the *late* Mn-rich carbonates deposited during the waning stages of low-temperature oxidative alteration (Humphris *et al.*, 1908a,b) and the metalliferous sediments formed as a result of effluent discharge from ridge-crest, hot-spring systems [e.g., Cronan (1973)]. This, coupled with the greater solubility of Mn

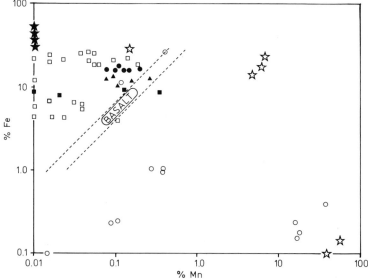

Fig. 8.7. Iron versus manganese for metalliferous sediments and secondary minerals in ocean basalts. (□) smectite, (○) carbonate, (●) chlorite, (■) epidote, (▲) amphibole, (⋆) sulphide, (✭) metalliferous sediment. The field for unaltered basalts and basaltic glasses is also indicated. [Data sources: Bostrom *et al.* (1969), Bonatti *et al.* (1972), Cronan (1973), Bonatti (1975), Bonatti *et al.* (1975), Robinson *et al.* (1977), Humphris and Thompson (1978a), Humphris *et al.* (1980a,b) and Morrison and Thompson (1983).]

relative to Fe, and the evidence for mobilization of iron noted above, clearly suggests that Mn is leached during both low- and high-temperature alteration and removed from the crust more efficiently than Fe. Data for the other transition elements are scarcer. Ni appears to be immobile during the alteration of primary magmatic sulphides and Cu, Ni and Zn have not been detected in cavity-filling sulphides in low-temperature assemblages. Smectites within altered basalts have been reported containing up to 500 ppm of Ni and 300 ppm of Cu. These values are close to those of unaltered basalts and basaltic glasses, suggesting that both of these elements were immobile during smectite formation. In contrast, up to 1700 ppm of Zn has been reported in saponites infilling veins, which clearly implies mobility of Zn at low-temperatures (Scott and Swanson, 1976; Robinson et al., 1977). The absence of Zn from the sulphides in these assemblages may simply reflect the frequent preference shown by this element for the silicate rather than the sulphide phase (MacLean and Shimazaki, 1976).

Humphris and Thompson (1978a,b) demonstrated that Fe, Mn, Cu and to a lesser extent Ni, but not V and Cr, were mobile during "greenschist-facies" alteration. Selected trace element data for some of their samples is shown in Fig. 8.8. Samples devoid of cross-cutting, sulphide-bearing veins are plotted in the upper diagram, and those with sulphide veins in the lower diagram. In each case the data for the hydrothermally altered pillow margins were normalized to the data for the relatively unaltered cores, except for those chlorite- and epidote-rich samples which were virtually completed altered and which were normalized to an average composition for fresh basalts from the same localities. Comparison with Fig. 8.5 shows that the trends are extremely different from those produced by igneous fractionation, especially in the shift of Fe and Mn values from near constant ratios. Nevertheless, Fig. 8.7 demonstrates that the change in Fe and Mn values is less extreme at high than at low temperatures. All the samples plotted in the upper diagram have clearly lost Cu, confirming that the epidote-rich assemblages were produced under conditions of low pH as postulated above. The lower diagram shows that much of this Cu appears to be redeposited in later veins. The other elements show no consistent trends. The magnitude of the changes can be estimated by comparison with the values for Ti, since this element has been shown to be immobile in numerous studies [Humphris and Thompson (1978a,b), and references therein]. The elements which have distributions most closely resembling that of Ti are Cr and Ni, which clearly implies immobility. Vanadium shows even less variation than Ti, whilst the Fe and Mn data appear to show evidence for both small losses or gains in different samples.

The data agree with the experimental predictions for mobility of Cu and immobility of Ni, Cr and V, but they clearly disagree with the prediction that

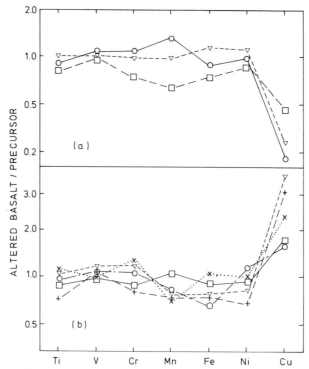

Fig. 8.8. Transition metal contents in hydrothermally altered basalts. (a) Samples devoid of sulphide-bearing cross-cutting veins. Solid line, chlorite-rich; dotted line, epidote-rich; dashed lines, chlorite-rich samples with groundmass pyrite. (b) Chlorite-rich samples with cross-cutting, sulphide-bearing veins. [Data from Humphris and Thompson (1978a,b).]

both Fe and Mn should be strongly leached from basalts during high-temperature alteration. Humphris and Thompson (1978a) showed that some chlorite- and epidote-rich samples were depleted in Fe and that it was predominantly pyrite-bearing metabasalts that appeared to show uptake of Fe. They postulated that this Fe-enrichment was due to the precipitation of sulphides from fluids that had gained Fe elsewhere in the system. This explanation cannot resolve the Mn problem since the sulphides contain no manganese (Fig. 8.7 and Table 8.2). The lack of any marked enrichment of Fe in the samples plotted in the lower part of Fig. 8.8 as compared to those in the upper part is also difficult to reconcile with this theory.

     One way of reconciling the experimental and geochemical evidence is to postulate that leaching of Fe and Mn under low-temperature conditions, during formation of the saponite-bearing assemblages, is strong enough to cause the solutions to approach saturation with these elements (and possibly with Zn?). The ubiquitous oxidative overprinting of the saponite-bearing metabasalts in the upper portions of the crust has caused their role in the hydrothermal alteration of the seafloor to be overlooked by many authors. The frequent presence of small amounts of Fe sulphide in these assemblages and the tendency for the abundance of sulphides to increase with depth and for vein sulphide deposition to immediately follow that of the Mg-rich minerals all demonstrate that the solutions must be close to Fe saturation under the prevailing conditions. The greater similarity in both Fe/Mn ratios and Fe and Mn contents of the "greenschist-facies" minerals to unaltered basalts as compared to the low-temperature minerals (Fig. 8.7) suggests that at higher temperatures the solutions are close to equilibrium with the rocks with regard to both Fe and Mn. Under these conditions, little other than minor, local redistribution of these elements will occur. *This in turn suggests that the net loss of Fe and Mn from the seafloor takes place during low-temperature alteration.* The lack of Zn data for metabasalts altered under varying conditions means that the extent to which a similar model may also be applied to this element cannot be ascertained. Such conditions will not apply to elements such as Cu, since no significant leaching occurs during earlier low-temperature interaction. Hydrothermal alteration is thus a highly selective process with different elements being gained by the solutions at different points along the fluid pathway. These processes cannot be reproduced in experimental studies since they are conducted under closed conditions with all the elements being made available for leaching at the same time.

Metalliferous sediments forming along mid-ocean ridges are enriched in Ni, Co, V, and Cr relative to pelagic sediments in addition to Mn, Fe, Zn and Cu. The data discussed above revealed little evidence for mobility of these elements at either high or low temperatures. An extension of the reasoning above suggests that they may therefore be mobilized at even higher temperatures and deeper levels in the crust. One possible (untestable) source is magmatic ores deep in the crust, and the other possibility is the gabbroic and cumulate rocks of layer (3). The few available analyses of metagabbros show that their major element compositions and CIPW norms overlap those of fresh basalts from the same localities and that the bulk compositional changes during "amphibolite-facies" alteration appear to be small (Bonatti et al., 1975). Comparison of these metagabbros with fresh basalts of equivalent Fe/Mg ratios [e.g., Frey et al. (1974)] shows that they have approximately one-half of their anticipated Co, Ni and V contents, and only one-tenth of their expected Cr and Cu contents. It seems highly likely that a considerable portion of the metal content of ridge-crest hydrothermal effluents must be leached from layer (3) or even deeper.

8.7   Distribution and Evolution of
Oceanic Hydrothermal Systems

Despite the certainty as to the existence of submarine hydrothermal activity there is little agreement on the form which the circulation takes. Attempts to model the systems are strongly limited by the paucity and non-uniform distribution of the data. At present the nature of relatively low-temperature circulation processes occurring on the ridge flanks is far better understood than that of those taking place at the ridge axis.

Some of the strongest evidence as to the scale and extent of hydrothermal circulation is provided by heat flow studies. Comparison of the measured conductive heat loss with that predicted from theoretical models have revealed a widespread negative anomaly that can only be explained by convective cooling of the crust by seawater circulation [e.g., Anderson (1972), Lister (1972) and Wolery and Sleep (1976)]. On a *regional* scale the *average* heat flow values are a function of seafloor age. There is a crestal zone characterized by exceptionally low and variable values, a transition zone within which heat flow values increase from values below to values in agreement with the predictions from thermal models, and a third zone where the heat flow very broadly agrees with the theoretical values. This third zone occurs in crust of different ages in different oceans: 4–6 m.y. in the Galapagos area, 10–15 m.y. on the East Pacific Rise, 40–60 m.y. in the Indian Ocean and 50–70 m.y. in the Atlantic (Anderson *et al.*, 1977). These ages correspond to the times required for the accumulation of 150–300 m of sediment indicating that circulation only ceases when a widespread blanket of impermeable sediment has formed (Pearson and Lister, 1973). Certainly penetration of the overlying sediments has caused hydrothermal circulation to recommence in more than one DSDP drill site [e.g., Duennebier and Blackington (1980) and Anderson *et al.* (1982)].

Detailed heat flow mapping has been carried out on two areas of the East Pacific where the sedimentation rates are high enough to enable measurements to be taken on relatively young crust: from 10 000 to 750 000 years old in the Galapagos area, and from 0·1 to 12·5 m.y. in age on the Juan de Fuca Ridge (Williams *et al.*, 1974; Corliss *et al.*, 1979; Davis *et al.*, 1980). These have demonstrated that circulation occurs on two different scales. In young crust (0·1 m.y. old), the scale of variability (distance between heat flow maxima) varies from 1 to a few kilometres. The heat flow maxima are related to faults which appear to act as discharges for warm water. In the Galapagos area these faults are capped by rows of hydrothermally deposited sediment mounds through which water at temperatures of 10–20°C is percolating (Corliss *et al.*, 1979; Williams *et al.*, 1979). If groundwater flow was restricted solely to the fault zones, then, since both the rising and sinking limbs of the circulation systems should be encountered, both elevated and depressed heat flow regions should be detected near the faults; but this was not the case. Instead, cellular convection by porous media flow appears to be occurring, implying a high permeability in the upper crust. The higher relative permeability and the greater crustal exposure in the fault zones combine to force the rising limb of the convection cell at localized discharge points in the manner predicted by Elder (1965). These small-scale variations in heat flow die out rapidly as the sediment thickness increases.

Davis *et al.* (1980) showed that larger scale variations (18–20 km) were present in all the profiles they studied. These appear to reflect regional horizontal circulation driven by cold water recharge at isolated basement outcrops and seamounts. The co-existence of two modes of convection even in relatively young crust affected by local venting means that the thickness of the permeable layer cannot be extracted from the heat flow data. The basement temperatures can however be calculated by extrapolation of the surface values, and these are low and usually less than 100°C, even in the youngest sites (Davis *et al.*, 1980). On other ridges, starved of sediment and open to the seafloor for even longer, the temperatures are likely to be even lower.

The evidence that widespread circulation can persist for up to 10^7 to 10^8 years in some cases shows that the uppermost crust must be highly permeable. This is borne out by in situ studies of DSDP drill holes. Geophysically the upper crust samples in DSDP site 504 consists of three distinct units— interpreted as seismic layers (2A), (2B) and (2C). The upper 100 m consists of rubbly pillow basalts, breccia zones and a few massive flows. The seismic velocities (for both P and S waves), density, thermal conductivity and electrical resistivity are all low. The estimated bulk porosity varies from 12–14% and the measured bulk permeability is as high as those of good oil producing sands ($\sim$60 millidarcies) and the fractures appear to be open and filled with seawater (Anderson *et al.*, 1982; Becker *et al.*, 1982). This zone behaved as an active aquifer after drilling had punctured the impermeable cap of sediments overlying the basement.

The second layer, (2B), from 100 to 650 m into the basement, consists of smaller pillows, more intense brecciation and a few massive flows. Towards the base of this layer the first dykes are encountered. This interval showed increasing gradients in seismic P and S wave velocities, electrical resistivity, and density. The estimated bulk porosity is 7–10% and the permeability decreases rapidly to only 4 md. This higher seismic velocity and lower permeability appear to be directly related to the deposition of secondary minerals in fractures and void spaces (Anderson and Zoback, 1982). Over the next 50 m of the section, considered to represent the transition from layer (2B) to (2C), all the physical properties change. The electrical resistivity, density and seismic velocity all rise, and the intensity of fracturing and hydrated mineral content all decrease. The estimated bulk porosity drops to less than 3% and the lower 750 m of the section has a permeability of only 0·1 md, a drop of nearly three orders of magnitude (Anderson et al., 1982; Becker et al., 1982). Recent electromagnetic experiments demonstrated a similar increase in electrical resistivity at a depth of ~1·4 km on the East Pacific Rise (EPR) at 21°N. This implies a similar strong decrease in fracture porosity and hence permeability to hydrothermal penetration (Young and Cox, 1981).

The primary porosity of layer (2C) at site 504 was already low before alteration reduced it further. The contrast between layers (2A) and (2B), however, is due to secondary minerals plugging the plumbing. The basalts in the drill core have been affected by several overlapping episodes of alteration at progressively lower temperatures. The minerals sealing the basalts in the upper part of the section are those deposited during low-temperature oxidative alteration, which was the last event to affect the crust before sedimentation stopped the circulation. As noted above, the accumulated data from DSDP drill holes indicates that this stage of alteration begins soon after crust formation and may continue in permeable horizons for up to 25–50 m.y. The correlation of the progressive isotopic and mineralogical changes with age shown by Muehlenbachs (1980) and Scheidegger and Stakes (1980) with the progressive disappearance of layer (2A) as a seismic entity with age is striking (Ewing and Houtz, 1979). This clearly implies that the disappearance of layer (2A) in older crust is due to the gradual sealing of fractures and voids by continuing low-temperature circulation in the upper crust. This process is unlikely to render the upper crust totally impermeable to hydrothermal penetration, since the heat flow and DSDP data clearly indicate that circulation is stopped by sediment deposition. A bore hole seismic experiment at site 504 showed a clear layer (2B)–(2C) break consistent with the other studies, but failed to detect the 100 m thick layer (2A), suggesting that such thin zones are beyond the resolution of both experimental and conventional seismic refraction techniques (Houtz and Ewing, 1976; Becker et al., 1982).

The lack of sediment cover on most ridge axes precludes both the taking of heat flow measurements and drilling, which cannot therefore be used to constrain the hydrothermal circulation at axial regions. Active hydrothermal vents have recently been discovered at several localities on the East Pacific Rise and in the Galapagos rift (Corliss et al., 1979; Macdonald et al., 1980; Rona, 1984). The exit temperatures and flow rates of the fluids vary from 10 to 20°C and a few centimetres per second in the Galapagos area to 350°C and 2·4 m/sec for the dramatic black smoker vents on the EPR at 21°N. The solutions exiting from the black smokers are precipitating Cu, Zn and Fe sulphides and have Fe/Mn ratios close to the average for ridge-crest metalliferous sediments. Studies of solution chemistry indicate that the lower temperatures in the Galapagos vents are due to rapid, shallow, sub-seafloor mixing of cold seawater with a hydrothermal fluid similar in composition and temperature to those spewing from the vents at 21°N. The hydrothermal fluids in both systems are enriched in Li, Rb, K, Ca, Si and Mn and depleted in Mg and Si relative to seawater. Estimates of water/rock ratios in the systems vary over a narrow range from 0·7 to 4·2 (Craig et al., 1980; Edmond et al., 1979b, 1982). These fluids are therefore complimentary to the epidote-rich "greenschist-facies" assemblages, but not to the Mg-rich metabasalts.

Chemical modelling indicates a wide slow recharge zone and rapid-flow discharge zones for these systems [e.g., Bischoff (1980)]. Quartz geothermometry for the Galapagos fluids indicates temperatures and pressure ranging from 375°C at 750 bars to 340°C at 1000 bars for the hydrothermal end-member, equivalent to a depth of ≥ 4 km for equilibration before uprise. A similar depth for the base of the upflow zone at 21°N is indicated by the similar fluid chemistry of the two systems (Edmond et al., 1979b, 1982; Mottl, 1983). The fluids must have risen rapidly from these depths without significant re-equilibration to lower temperatures or mixing with the fluids in the surrounding rocks. This, coupled with the narrow bore of the vents (~15 cm) and their linear arrangement along strike suggests that the discharges are located along faults (Macdonald et al., 1980). Isolation of the upflow zone from the surrounding highly permeable matrix may have been promoted by the early precipitation of a sulphide lining to the vent.

It was inferred above that the widespread saponite- and chlorite-bearing, Mg-rich assemblages were produced in the downflow zones of hydrothermal systems under seawater-dominated conditions. The requirement for a wide recharge zone for the vents thus explains the predominance of these Mg-rich assemblages in collections from the seafloor. Downwelling seawater will undergo successive interactions to produce the low- and high-temperature assemblages during the course of which it will become progressively depleted in Mg and enriched in K, Rb, Li, Ba, Ca and Si. The increasing temperature and decreasing permeability with depth will cause the conditions to progressively approach those of rock-dominated alteration, leading to the production of epidote-rich or "amphibolite-facies" assemblages. The corresponding changes in solution chemistry will cause the elemental compositions and $\delta^{18}O$ values to reflect low water/rock ratios similar to those of the vent fluids.

Converse et al. (1984) have shown that the heat loss from the hot springs at 21°N is equivalent to all of the heat released by the cooling oceanic crust between 0 and 540 km (18 m.y.) for a 1-km segment of the ridge crest, or all of the heat released by quick cooling of the lithosphere to depths of ~38 km along a 1-km ridge segment if the vents are steady-state features. This is highly unlikely as heat flow data away from the ridge and the close fit of the topography at 21°N to that predicted from conductive cooling models suggests that over a period of 10^4 yr or more the heat transported by the vent fluids is not significantly different from the predicted quantity. Balancing the heat losses over time requires the hot springs to be highly episodic, and in operation for about one-ninth to one-fifth of the time on any particular ridge crest segment (Macdonald et al., 1980; Converse et al., 1984). Estimated lifetimes for the vents, for thermal calculations, range from 10^5 to 10^2 yr. Inactive vents located just north of the active vents at 21°N appear to have been formed no more than 4×10^3 yr ago (Lalou and Brichet, 1980). This in turn implies that the magmatic activity driving the circulation is also episodic. Sleep (1975) has calculated that conductive heat losses alone are sufficient to preclude the development of a steady-state magma chamber at slow-spreading (<1 cm/yr, half rate) centres, and geophysical studies have failed to find such chambers (Macdonald, 1983). For faster spreading rates the situation is less clear. The available geophysical evidence appears to preclude the existence of a magma chamber under the northern Juan de Fuca Ridge (McClain and Lewis, 1982) but does indicate the existence of small (<4 km width) chambers beneath the EPR at 12° and 21°N (Filloux, 1982; Lewis and Garmany, 1982). Macdonald (1983) suggests that magma chambers are narrow, quasi-steady-state features, centred under the axial topographic high between transform faults and elongated parallel to the ridge axis. Mottl (1983) suggests that the best evidence for a magma chamber is the existence of active hydrothermal vents. Certainly, the vent systems discovered to date in the East Pacific do appear to be closely related to the topographic high (Ballard et al., 1984).

The thermal budgets of mid-ocean ridges have been extensively modelled by Lister (1972, 1974, 1977). In his model axial magma chambers are separated from the overlying permeable, hydrothermal convective regions by a cracking front underlain by a thin conductive boundary layer. The magma chamber limits the extent of penetrative convection, since water is unlikely to penetrate the upper boundary zone of the chamber. As the magma solidifies the water penetrates downwards through a propagating matrix of thermal contraction cracks. Rates of penetration of a few metres per year are predicted by the model. These rates are substantially faster than seafloor spreading and indicate that the entire crust down to the Moho can be cooled and penetrated within a few tens of thousands of years. The efficiency of cooling is such that even on the fastest-spreading ridges, magma chambers are unlikely to extend for more than a few kilometres beyond the axis. On slower-spreading ridges (<1 cm/yr, half rate), the period of intense hydrothermal activity is likely to be short-lived and highly episodic. Some residual circulation of cool water should persist however, driven by conductive heating from below.

The models of Lister have accurately predicted most of the features of both the recently discovered hydrothermal vents and the axial magma chambers currently being delineated by geophysical studies. In his latest version, Lister (1982) predicts temperatures of 350°C and flow rates of 0·5 m/sec for the upflow zones, in good agreement with the values obtained for the hot springs at 21°N. The depth to the base of this zone indicated by the geochemical results suggests that the magma chamber at 21°N may be partially solidified and undergoing active penetration by the hydrothermal fluids. This stage is not likely to last for more than a thousand years or so for a magma chamber of the size detected by Lewis and Garmany (1982), in good agreement with the age determinations on the nearby inactive vents. On slower-spreading ridges the lifetimes of individual systems must be even shorter. Only on very fast-spreading ridges (>5 cm/yr, half rate), are steady-state magma chambers likely to occur. In such cases the fluids are unlikely to penetrate below layer (2).

8.8   Distribution of Hydrothermal Minerals within the Crust

8.8.1   Mature ocean basins

The mineralogical and chemical evidence discussed above indicates that extensive leaching and redistribution of metals is most likely to occur during the intense, axial stage of hydrothermal circulation. The upflow zones of all the systems detected to date appear to be linear, ridge-crest parallel features, probably faults, and it is in these regions that ore deposits are most likely to occur. Deposition of minerals from solutions will take place either as a result of changing temperatures or pressures in the rising fluids, or as a result of mixing with low-temperature, relatively alkaline, oxidizing fluids such as normal seawater. The latter is the most likely mechanism given for the rapid upflow rates of the discharging solutions. Metals are precipitated from solution in reverse order of their uptake. The resulting precipitates formed as a result of both falling temperature and increasing fO_2 will be complex, and probable sequences are: Cu–Zn–Fe sulphides, Fe-rich silicates, oxides and hydroxides, Mn-rich oxides, hydroxides and carbonates and sulphates (Cronan et al., 1977; Haymon and Kastner, 1981).

Mixing with seawater can take place either after the fluids have debouched onto the seafloor as is currently happening at 21°N or within the upper part of the crust en route to the surface as appears to be happening in the Galapagos area (Edmond *et al.*, 1979b; Haymon and Kastner, 1981). Optimum conditions for sub-seafloor mixing will occur when the longest distances are being traversed by the rising fluids. Above steady-state magma chambers on fast-spreading ridges, or during the early stages of circulation above a short-lived magma chamber, the fluids are unlikely to penetrate below layer (2). As the phase of active cooling and penetration develops the upflow distances will become progressively longer, eventually stretching as far as the Moho. The thickness of the uppermost permeable portion of the crust layer (2A) is 1·5 km on the Mid-Atlantic Ridge, roughly twice that of layer (2A) at the East Pacific Rise. This suggests that hydrothermal fluids are more likely to spew onto the seafloor at fast-spreading ridges or during the early stages of hydrothermal activity. Much of the particulate material precipitating from the plumes at 21°N, together with the manganese and iron still held in solution, is being dissipated by the oceanic currents circulating at mid-depth in this area and thus contributing to the regional metalliferous sediments characteristic of the equatorial East Pacific (Edmond *et al.*, 1982). Slow-spreading ridges, where sub-seafloor mixing is more likely, can thus be considered as conservative with respect to hydrothermal ore minerals, whereas fast-spreading ridges are profligate. The regional distribution of metalliferous sediments in the oceans suggests that this distinction between slow- and fast-spreading systems may be general (Bostrom *et al.*, 1969).

Some estimate of the volumes of material produced by the hydrothermal systems can be made from the data on the black smoker fluids. The solutions contain roughly 100 ppm of Fe (Edmond *et al.*, 1982). The bulk of the particulate material in the plumes is pyrrhotite. If all of this is deposited around the immediate vent areas, then accumulation periods of only 2–3 yr are required to form a massive 15 million ton ore deposit such as Mavrouni in Cyprus (Mottl, 1983; Converse *et al.*, 1984). Since most of the material is dispersed by oceanic current, then extrapolation of the current rates of accumulation indicate that times of up to 0·5 m.y. are more realistic estimates of the times required to form such an ore body. Unfortunately the vents appear to be active for periods of a few tens of thousands of years. Even the relatively small volumes of material around the vents ($\sim$900 m^3; Haymon and Kastner, 1981) are only likely to be preserved if they are buried beneath subsequent lava flows or turbidite influxes, as they are rapidly degraded and oxidized by seawater once the hydrothermal activity ceases. The formation of massive deposits from such vent effluent therefore requires special and unusual conditions such as restricted ocean current circulation patterns or ejection of the fluids into a deep and sheltered basin as is currently happening to the fluids in the Red Sea (Rona, 1984). Such situations are unlikely to arise on fast-spreading ridges where the active hydrothermal systems appear to be located on topographic highs.

Despite the high rates of production at fast-spreading ridges, significant quantities of ore may only be preserved at slow-spreading centres where the deposition takes place in the subsurface environment. If precipitation occurs across a long mixing column with seawater then the ore minerals may form a widely scattered disseminated-ore type deposit of which the various mineralized samples recovered from the Mid-Atlantic Ridge and described above may be typical. Only if deposition is restricted to an interval of a few tens of metres is a massive body likely to form. The probable shorter lifetimes of magma chambers on slow-spreading as compared to fast-spreading ridges (Sleep, 1975; Lister, 1974; Macdonald, 1983) means that the hydrothermal activity is likely to be of a similar short duration; hence individual deposits in these areas are unlikely to exceed a million tons.

The mineralogical and chemical evidence reviewed above indicates that most of the leaching of Fe and Mn from the ocean crust takes place under seawater-dominated conditions during relatively low-temperature alteration. Iron-rich and Mn-rich deposits may thus be formed in the low-temperature systems operating on axial regions that are not underlain by magma chambers such as the northern Juan de Fuca Ridge (McClain and Lewis, 1982), or during the waning stages of hydrothermal activity above a "dying" magma chamber. Deposition will again be in fault-controlled upflow zones (Williams *et al.*, 1974; Davis *et al.*, 1980). The slower rates of production are more likely to yield volumetrically insignificant amounts of ore. Typical examples are probably the Galapagos mounds which consist of Fe-rich nontronitic clays and Fe and Mn oxides and hydroxides (Williams *et al.*, 1979). Locally such systems may be contributing to the formation of Mn-rich crusts and sediments on the flanks of spreading centres. Chemically and mineralogically they will be identical to the end products of deposition from a more vigorous hydrothermal system. The existence of such deposits alone need not necessarily indicate the presence of massive, high-temperature deposits below the surface.

8.8.2 Young ocean basins

The Red Sea is the most thoroughly explored region of oceanic mineralization in the world. About half the described sites of metal enriched deposits on slow-spreading centres are from its basins (Rona, 1984) and it is frequently quoted as a typical example of the type of hydrothermal activity and associated metallogenesis likely to develop during the initiation of spreading. Many of the characteristics of the Red Sea are uniquely favourable to the formation and concentration of hydrothermal deposits, including low to zero sediment input, restricted oceanic current circulation leading to reducing conditions in the basins, anoxic conditions in the basins preventing degradation of the metal-rich sediments, and adjacent evaporite beds that enhance the salinity of the hydrothermal fluids and thus their capacity to transport metals (Rona, 1984). Many of these conditions are unlikely to be present in other immature ocean basins.

Recent work in the Gulf of California has identified a further type of hydrothermal activity that may be important in young ocean or back-arc basins. The Guyamas basin in the centre of the Gulf of California is an active spreading centre formed during the last 3·5 m.y. It has a medium spreading rate (2·5–3 cm/yr), and high sedimentation rates, locally in excess of 1200 m/m.y., which has prevented extrusion of magma onto the seafloor and caused intercalation of layers (1) and (2) to occur. The resulting crustal structure is characterized by unusual amplitude magnetic anomalies and "diffuse" seismic properties similar to those seen in some back-arc basins (Saunders *et al.*, 1982; Lawver and Hawkins, 1978).

Drilling in the Guyamas basin during DSDP leg 64 recovered sequences of massive doleritic and gabbroic sills and turbiditic sediments. Einsele *et al.* (1980) have demonstrated that intrusion of the sills into the highly porous and unconsolidated sediments was accompanied by intense thermal alteration of the sediments, marked changes in interstitial water chemistry and the large-scale expulsion of heated pore fluids. Temperatures in some cases appear to have exceeded 300°C. This process appears not only to have created space for the invading magmas but also to have caused the formation of the hydrothermal deposits of talc, pyrrhotite and other sulphides observed around fault scarps on the floor of the basin (Lonsdale *et al.*, 1980).

Hydrothermal systems of this type differ in many ways from those operating on mid-ocean ridges open to seawater penetration. The metals are derived mainly from sediments and pore fluids, not the igneous basement. Studies of the Guyamas Basin sills showed they were less altered than most samples recovered from the upper part of the crust in other DSDP drill holes and appeared to have undergone little or no interaction with seawater. What little alteration did take place may have been under closed conditions (Morrison and Thompson, 1983). The decrease in sediment porosity associated with the sill emplacement prevents recharge of the system and the hydrothermal activity is a "one-off" event. New phases of activity are restricted to higher levels in the section as later intrusions rise until they meet unconsolidated sediments when they spread out to form sills.

The mouth of the Gulf of California is characterized by lower sedimentation rates, from 350–60 m/m.y., which permitted extrusion of the basalts onto the seafloor, resulting in a more typical basement stratigraphy that was sampled during DSDP legs 64 and 65. Nevertheless, Moorby *et al.* (1983) could find no evidence for the presence of an Fe-enriched basal sediment in this area of the type found elsewhere on the East Pacific Rise and rise flanks, suggesting that this area was relatively unaffected by hydrothermal alteration. Morrison and Thompson (1983) showed that the basalts had been affected by high-level hydrothermal activity, locally forming "greenschist-facies" assemblages, but that the circulation was exceptionally short-lived and choked by the rapid sediment influxes. Only at site 483 with the relatively low sedimentation rates of 63 m/m.y. were "normal" upper crustal oxidative assemblages found.

Saunders *et al.* (1982) have argued that a Guyamas Basin model is applicable to any area subjected to high sedimentation rates (>1000 m/m.y.), and thus may be important during the early stages of ocean basin or marginal basin development. Flanking such areas will be regions of intermediate sedimentation rate, similar to those at the mouth of the Gulf, which are likely to be barren and devoid of significant mineralization of any type. This in turn implies that studies of palaeosedimentation rates may provide a powerful tool for assessing the prospects for oceanic mineralization.

8.8.3 Other tectonic settings

Hydrothermal alteration and mineralization is not confined to any one tectonic setting in the oceans. The principal requirements for its operation are a magmatic heat source, seawater and permeable rocks. Experimental work has shown that the chemistry of the hydrothermal solutions is likely to be similar irrespective of the composition of the affected rock types. This is confirmed by the recent discoveries of hydrothermal deposits on seamounts (Lonsdale *et al.*, 1982) and in the SW Pacific arcs (Cronan, 1983; Exon and Cronan, 1983; Moorby *et al.*, 1984) similar to those on mid-ocean ridges. Nevertheless, the dimensions, geometries, lifetimes and flow characteristics of hydrothermal systems are critically dependent on the nature of the magma chambers supplying the driving energy.

This review has concentrated on mid-ocean ridges because of the recent advances in our understanding of magmatic and hydrothermal processes in these regions. Detailed evaluation of the ore-mineral potential of convergent margins must await more data on the tectonic, magmatic and chemical processes operating in these regions. Anticipated differences will be mainly in the size and distribution of any deposits rather than in their composition.

References

Adams, J. E. (1978). Unpublished Ph.D. thesis, Victoria University of Wellington. 109 pp.

Advisory Committee on Aggregates (1976). "Aggregates: The Way Ahead". U.K. Department of the Environment, H.M.S.O., London. 118 pp.

Alexandersson, G. and Klevebring, B.-I. (1978). "World Resources Energy Metals Minerals". de Gruyter, Berlin. 248 pp.

Alther, G. R. and Wyeth, R. K. (1981). *Environ. Geol.* **3**, 185–193.

Anderson, R. N. (1972). *Geol. Soc. Am. Bull.* **83**, 2947–2956.

Anderson, R. N., Honnorez, J., Becker, K., Adamson, A. C., Alt, J. C., Emmermann, R., Kempton, P. D., Kinoshita, H., Lavrene, C., Mottl, M. J. and Newark, R. L. (1982). *Nature.* **300**, 589–594.

Anderson, R. N. Langseth, M. G. and Sclater, J. G. (1977). *J. Geophys. Res.* **82**, 3391–3410.

Anderson, T. F. and Lawrence, J. R. (1976). "Initial Reports of the Deep Sea Drilling Project", Vol. 35, pp. 497–505. U.S. Gov. Print. Off., Washington, D.C.

Anderson, R. N. and Zoback, M. D. (1982). *J. Geophys. Res.* **87**, 2860–2868.

Andrews, A. J., Barnett, R. L., MacClement, B. A. E., Fyfe, W. S., Morrison, G., MacRae, N. D. and Starkey, J. (1977). "Initial Reports of the Deep Sea Drilling Project", Vol. 37, pp. 795–810. U.S. Gov. Print. Off., Washington, D.C.

Andrews, J. and Packham, G. *et al.* (1975). "Initial Reports of the Deep Sea Drilling Project", Vol. 30, U.S. Gov. Print. Off., Washington D.C. 735 pp.

Andrews, J. E., Friedrich, G., Pautot, G., Pluger, W., Renard, V., Melguen, M., Cronan, D., Craig, J., Hoffert, M., Stoffers, P., Shearme, S., Thijssen, T., Glasby, G., Le Notre, N. and Saget, P. (1983). *Mar. Geol.* **54**, 109–130.

Andrews, P. B. (1973). *N.Z. J. Geol. Geophys.* **16**, 793–830.

Andrews, P. B. and van der Lingen, G. J. (1969). *N.Z. J. Geol. Geophys.* **12**, 119–137.

Anglada, R., Froget, C. and Récy, J. (1975). *Sediment. Geol.* **14**, 301–317.

Anonymous (Undated). N.Z. Fishing Industry Board Economics and Marketing Division Rep., Vol. 33, 46 pp.

Anonymous (1969). *Ocean Ind.* **4**(8), 28.

Anonymous (1971). *Ocean Ind.* **6**(8), 7–9.

Anonymous (1974a). *Soil & Water* **10**(4), 43–45.

Anonymous (1974b). *Soil & Water* **11**(2), 4–6.

Anonymous (1974c). *Mining Mag.* **131**(1), 57.

Anonymous (1975). Mineral production and processing. DSIR Research 1975, pp. 17–38.

Anonymous (1979a). "Mineral Industries of Canada, Australia and Oceania", U.S. Bureau of Mines, U.S. Gov. Print. Off., Washington D.C. 67 pp.

Anonymous (1979b). *In* "Australia Mining Year Book, '79". Thomson Publications (Aust.), N.S.W., pp. 90, 92.

Anonymous (1981a). Report on the Inshore and Nearshore Resources Training Workshop, Suva, Fiji, 13–17 July, 1981. CCOP/SOPAC Technical Secretariat. 88 pp.

Anonymous (1981b). *Commonw. Geol. Liaison Off. Newsl.* **NL9**, 26.

Anonymous (1982a). *Proc. Concrete Coral Workshop, Rarotonga, Cook Islands, 31 March–2 April.* CCOP/SOPAC Rep. 32 pp.

Anonymous (1982b). *Wellington Evening Post* March 8, p. 9.

Anonymous (1982c). *Wellington Evening Post* June 30, p. 23.

Anonymous (1983a). *Cook Islands News* 22 July, pp. 1, 16.

Anonymous (1983b). *Cook Islands News* 8 July, pp. 1, 12.

Anonymous (1983c). *Geotimes* **28**(5), 16–19.

Anonymous (1984). *Town & Country Plann. Bull.* No. 17, 8 pp.

Aplin, A. C. (1983). Unpublished Ph.D thesis, University of London. 348 pp.

Aplin, A. and Cronan, D. S. (1985a). *Geochim. Cosmochim. Acta* **49**, 427–436.

Aplin, A. and Cronan, D. S. (1985b). *Geochim. Cosmochim. Acta* **49**, 437–451.

Arai, S. and Fujii, T. (1978). "Initial Reports of the Deep Sea Drilling Project", Vol. 45, pp. 587–594. U.S. Gov. Print. Off., Washington, D.C.

Archer, A. A. (1973). *Ocean Manage.* **1**, 5–40.

Archer, A. A. (1974). *Mining Mag.* **130**(3), 150–163.

Archer, A. A. (1976). *CCOP/SOPAC Tech. Bull.* No. 2, 21–38.

Archer, A. A. (1979). *In* "Manganese Nodules: Dimensions and Perspectives", pp. 71–81. D. Reidel Publishing Co., Dordrecht, Netherlands.

Arrhenius, G. and Bonatti, E. (1965). *Prog. Oceanogr.* **3**, 7–22.

Arrhenius, G., Cheung, K., Crane, S., Fisk, M., Frazer, J., Korkisch, J., Mellin, T., Nakao, S., Tsai, A. and Wolf, G. (1979). *Proc. Colloq. Int. CNRS* No. 289, 333–356.

Ash, R. P., Carney, J. N. and Macfarlane, A. (1978). New Hebrides Condominium Geological Survey, Regional Report, 49 pp.

Aubrey, D. G. (1981). *Oceans* **23**(4), 4–13.

Aumento, F., Loncarevic, B. D. and Ross, D. I. (1971). *Philos. Trans. R. Soc. Lond. Ser., A* **268**, 623–650.

Aumento, F., Melson, W. G. *et al.* (1977). "Initial Reports of the Deep Sea Drilling Project", Vol. 37, U.S. Gov. Print. Off., Washington, D.C. 1008 pp.

Aziz-Ur-Rahman and McDougall, I. (1973). *Geophys. J. R. Astron. Soc.* **33**(2), 141–155.

Bäcker, H., Glasby, G. P. and Meylan, M. A. (1976). *NZOI Oceanogr. Field Rep.* **6**, 88 pp.

Bakus, G. J. (1983). *Ocean Manage.* **8**, 305–316.

Ballard, R. D., Hekinian, R. and Franchetau, J. (1984). *Earth Planet. Sci. Lett.* **69**, 176–186.

Banner, A. H. (1974). *Proc. Second Int. Coral Reef Symp.* **2**, 685–702.

Bardsley, W. E. (1977). *N.Z. Geographer* **33**, 76–79.

Barrie, J. V. (1981). *Estuarine Coastal Shelf Sci.* **12**, 609–619.

Bartlett, T. (1980). *N.Z. Energy* **35**(11), 11, 13.

Bass, M. N. (1976). "Initial Reports of the Deep Sea Drilling Project", Vol. 34, pp. 393–432. U.S. Gov. Print. Off., Washington D.C.

Baturin, G. N. (1969). *Dokl. Acad. Sci. USSR, Earth Sci. Sect.* **189**, 227–230.

Baturin, G. N. and Dubinchuk, V. T. (1979). "Microstructures of Oceanic Phosphorites: An Atlas of Electron Micrographs". Nauka, Moscow (in Russian), 198 pp.

Baturin, G. N. and Pokryshkin, V. I. (1980). *Oceanology* **20**, 56–61.
Baturin, G. N., Merkulova, K. I. and Chalov, P. I. (1972). *Mar. Geol.* **13**, M37–M41.
Baubron, J. C., Buillon, J. H. and Récy, J. (1976). *Bull. Bur. Rech. Geol. Min.* 2° Ser. IV, No. 3, 165–175.
Beck, A. C. (1947). *N.Z. J. Sci. Technol.* **28B**, 307–313.
Becker, K., Von Herzen, R. P., Francis, T. J. G., Anderson, R. N., Honnorez, J., Adamson, A. C., Alt, J. C., Emmermann, R., Kempton, P. D., Kinoshita, H., Laverne, C., Mottl, M. J. and Newark, R. L. (1982). *Nature* **300**, 594–598.
Bender, J. F., Hodges, F. N. and Bence, A. E. (1978). *Earth Planet. Sci. Lett.* **41**, 277–302.
Bender, M. L. (1970). *In* "Encyclopedia of Geochemistry and Environmental Sciences" (R. W. Fairbridge, Ed.), pp. 673–677, van Nostrand-Reinhold Co., New York. 1321 pp.
Bentley, R. (1981). *N.Z. Eng.* **36**(10), 17, 20.
Bentz, F. P. (1974). *In* "The Geology of Continental Margins"(C. A. Burk and C. L. Drake, Eds.), pp. 537–547. Springer-Verlag, New York.
Berger, W. H., Adelseck, C. G. and Mayer, L. A. (1976). *J. Geophys. Res.* **81**, 2617–2627.
Bezrukov, P. L. (1973). *In* "Oceanography of the South Pacific 1972" (R. Fraser, Comp.), pp. 217–219. N.Z. National Committee for UNESCO, Wellington. 524 pp.
Bignell, R. D., Cronan, D. S. and Tooms, J. S. (1976). *Trans. Instn. Min. Metall. Sect. B* **85**, 273–278.
Bioresearches Ltd. (1974). A Preliminary Study of the Ecological Impact of Mining Gravel from sublittoral deposits off the East Coast of Northland. Bioresearches Ltd, Auckland. 25 pp. (unpublished report).
Birch, G. F. (1980). *SEPM Spec. Pub.* **29**, 79-100.
Bischoff, J. L. (1980). *Science* **207**, 1465–1469.
Bitoun, G. and Récy, J. (1982). *In* "Equipe de Géologie—Géophysique du centre ORSTOM de Noumea. Contribution à l'étude géodynamique du Sud-Ouest Pacific". Travaux et Documents de ORSTOM, No. 147, pp. 505–539.
Black, P. M. and Brothers, R. N. (1977). *Contrib. Mineral. Petrol.* **65**, 69–78.
Blake, D. H. and Miezitis, Y. (1967). *Bull.—Bur. Miner. Resour. Geol. Geophys. (Aust.)* No. 93, 56 pp.
Blaskett, K. S. and Hudson, S. B. (1965). *Proc. Eighth Commonw. Min. Metall. Congr.* **3**, 313–340.
Blow, W. H. (1969). *Proc. First Int. Conf. Planktonic Microfossils* **1**, 199–422.
Bohlen, W. F., Cundy, D. F. and Tramontano, J. M. (1979). *Estuarine Coastal Mar. Sci.* **9**, 699–711.
Boland, F. M. and Hamon, B. V. (1970). *Deep-Sea Res.* **17**, 777–794.
Bold, D. A. (1982). *Iron Steel Int.* Oct. pp. 243–254.
Bonatti, E. (1975). *Ann. Rev. Earth Planet. Sci.* **3**, 401–431.
Bonatti, E., Fisher, D. E., Joensuu, O., Rydell, M. S. and Beyth, M. (1972). *Econ. Geol.* **67**, 717–730.
Bonatti, E., Guerstein-Honnorez, B.-M. and Honnorez, J. (1976). *Econ. Geol.* **71**, 1515–1525.
Bonatti, E. and Honnorez, J. (1976). *J. Geophys. Res.* **81**, 4104–4116.
Bonatti, E., Honnorez, J., Kirst, P. and Radicati, F. (1975). *J. Geol.* **83**, 61–78.
Bonatti, E. and Nayudu, Y. R. (1965). *Am. J. Sci.* **263**, 17–39.
Bonhomme, M. (1979). Rap. dactyl. Centre Sédim. et Géoch. de la Surface, 4 pp.

Borchert, H. (1960). *Trans. Instn. Min. Metall.* **69**, 261–279.

Borchert, H. (1965). *In* "Chemical Oceanography", (J. P. Riley and G. Skirow, Eds.), pp. 159–204. Academic Press, London. 508 pp.

Border, S. and van Dyke, J. (1982). *Univ. Hawaii Law Rev.* **4**, 1–59.

Boström, K. and Peterson, M. N. A. (1966). *Econ. Geol.* **61**, 1258–1265.

Boström, K., Peterson, M. N. A., Joensuu, O. and Fisher, D. E. (1969). *J. Geophys. Res.* **74**, 3261–3270.

Bougault, H. (1977). "Initial Reports of the Deep Sea Drilling Project", Vol. 37, pp. 539–546. U.S. Gov. Print. Off., Washington, D.C.

Bougault, H. and Hekinian, R. (1974). *Earth Planet. Sci. Lett.*, **24**, 249–261.

Bougault, H., Cambon, P., Joron, L. and Treuil, M. (1978). "Initial Reports of the Deep Sea Drilling Project", Vol. 46, pp. 247–251. U.S. Gov. Print. Off., Washington, D.C.

Bougault, H., Joron, J. L. and Treuil, M. (1980). *Philos. Trans. R. Soc. Lond. Ser.*, A **297**, 203–213.

Boulegue, J., Perseil, E. A., Bernat, M., Dupre, B., Stouff, P. and Franchetau, J. (1984). *Earth Planet. Sci. Lett.* **70**, 249–259.

Bowler, J. M. (1963). *Proc. R. Soc. Vict.* **76**, 69–136.

Bratt, S. R. and Purdy, G. M. (1984). *J. Geophys. Res.* **89**, 6111–6125.

Bremner, J. M. (1980). *J. Geol. Soc.* **137**, 773–786.

Briggs, R. M., Lillie, A. R. and Brothers, R. N. (1978). *Bull. Bur. Rech. Géol. Min.* Sect. IV, No. 3, 171–189.

Brinkhuis, B. H. (1980). Marine Sciences Research Centre, State University of New York, Stony Brook, Spec. Rep. No. 34, 193 pp.

Brock, V. E., Jones, R. S. and Helfrich, P. (1965). *Hawaii Mar. Lab. Tech. Rep.* No. 5, 90 pp.

Brock, V. E., van Heukelem, W. and Helfrich, P. (1966). *Hawaii Inst. Mar. Biol. Tech. Rep.* No. 11, 56 pp.

Brothers, R. N. (1954a). *Trans. R. Soc. N.Z.* **82**, 677–694.

Brothers, R. N. (1954b). *N.Z. Geographer* **10**, 47–59.

Brothers, R. N. and Blake, M. C., Jr. (1973). *Tectonophysics* **17**, 337–358.

Brothers, R. N., Heming, R. F. and Hawke, M. M. (1980). *N.Z. J. Geol. Geophys.* **23**, 537–539.

Brown, G. A. (1971). *Underwater J.* **3**, 166–176.

Bryan, W. B. (1972). *Carnegie Inst. Wash. Yearb.* **71**, 396–403.

Bryan, W. B. (1979). *J. Petrol,* **20**, 293–325.

Bryan, W. B., Stice, G. D. and Ewart, A. (1972). *J. Geophys. Res.* **77**, 1566–1585.

Bryan, W. B., Thompson, G., Frey, F. A. and Dickey, J. S. (1976). *J. Geophys. Res.* **81**, 4285–4304.

Buchbinder, B. and Halley, R. B. (in press). *In* "Geology and Offshore Resources of the Pacific Island Arcs—Tonga Region" (D. W. Scholl and T. L. Vallier, Eds.) Circum–Pac. Counc. Energy Miner. Resour. Earth Science Series, Houston, Texas.

Buckenham, M. H. (1965). *Proceedings Eight Commonw. Min. Metall. Congr.* **7**, *Paper No. 215*, 10 pp.

Burk, C. A. (1973). *Bull.—Bur. Miner. Resour. Geol. Geophys. (Aust.)* No. 141, pp. 115–122.

Burnett, W. C. (1977). *Geol. Soc. Am. Bull.* **88**, 813–823.

Burnett, W. C. and Gomberg, D. N. (1977). *Sedimentol.* **24**, 291–302.

Burnett, W. C. and Veeh, H. H. (1977). *Geochim. Cosmochim. Acta* **41**, 755–764.

Burnett, W. C., Beers, M. J. and Roe, K. K. (1982). *Science* **215**, 1616–1618.

Burns, R. E., Andrews, J. E. *et al.* (1973). "Initial Reports of the Deep Sea Drilling Project", Vol. 21. U.S. Gov. Print. Off., Washington, D.C. 931 pp.

Burns, R. G. (1976). *CCOP/SOPAC Tech. Bull. No. 2*, 14–20.

Burns, V. M. (1979). *Min. Soc. Am. Short Course Notes* **6**, 347–380.

Burns, V. M. and Burns, R. G. (1978a). *Scanning Electron Microsc.* **1**, 245–252.

Burns, V. M. and Burns, R. G. (1978b). *Am. Mineral.* **63**, 827–831.

Burns, V. M. and Burns, R. G. (1978c). *Earth Planet. Sci. Lett.* **35**, 341–348.

Bushinskii, G. I. (1969). "Old Phosphorites of Asia and their Genesis". Israel Program for Scientific Translation, Jerusalem. 266 pp.

Cabri, L. J. (1973). *Econ. Geol.* **68**, 443–454.

Calvert, S. E. and Price, N. B. (1977). *Mar. Chem.* **5**, 43–74.

Campbell, J. F. and Hwang, D. J. (1982). *Pac. Sci.* **36**, 35–43.

Cann, J. R. (1974). *Geophys. J. R. Astron. Soc.* **39**, 169–187.

Cann, J. R. (1979). *In* "Deep Drilling Results in the Atlantic Ocean: Ocean Crust" (M. Talwani, G. C. Harrison and D. E. Hayes, Eds.), pp. 230–238. Maurice Ewing Series 2. Am. Geophys. Union, Washington, D.C.

Cann, J. R. (1980). *J. Geol. Soc. Lond.* **137**, 381–384.

Cann, J. R., Winter, C. K. and Pritchard, R. G. (1977). *Mineral. Mag.* **41**, 193–199.

Carney, J. N. (in press). Vanuatu Geological Survey, Regional Report.

Carney, J. N. and Macfarlane, A. (1978). *Bull. Aust. Soc. Explor. Geophysicists* **9**, 123–130.

Carney, J. N. and Macfarlane, A. (1979). New Hebrides Government Geological Survey, Regional Report, 71 pp.

Carney, J. N. and Macfarlane, A. (1982). *Tectonophysics* **87**, 147–175.

Carter, A. N. (1978). *Nature* **276**, 258–259.

Carter, L. (1975). *Mar. Geol.* **19**, 209–237.

Carter, L. (1980). *N.Z. J. Geol. Geophys.* **23**, 455–468.

Carter, L. (1981). *BP Recorder* No. 171, pp. 8–14.

Carter, L. and Heath, R. A. (1975). *N.Z. J. Mar. Freshwater Res.* **9**, 423–448.

Carter, L. and Herzer, R. H. (1979). *N.Z. Oceanogr. Inst. Mem.* No. 83, 33 pp.

Carter, L. and Ridgway, N. M. (1974). *NZOI Oceanogr. Summ.* No. 2, 7 pp.

CCOP/SOPAC (1980). *CCOP/SOPAC Tech. Bull. No. 3*, 285 pp.

CCOP/SOPAC Annual Report (1981).

CCOP/SOPAC Annual Report (1982).

CCOP/SOPAC Technical Secretariat (1977a). *Proc. Sixth Sess. CCOP/SOPAC* pp. 50.

CCOP/SOPAC Technical Secretariat (1977b). *Proc. Sixth Sess. CCOP/SOPAC* pp. 62.

Chansang, H., Boonyanate, P. and Charuchinda, M. (1981). *Proc. Fourth Int. Coral Reef Symp.* **1**, 129–136.

Chase, C. G. (1971). *Geol. Soc. Am. Bull.* **82**, 3087–3110.

Chave, K. and Mackenzie, F. T. (1961). *J. Geol.* **69**, 572–582.

Cherkis, N. Z. (1980). *CCOP/SOPAC Tech. Bull. No. 3*, 37–45.

Chester, R. and Aston, S. R. (1976). *In* "Chemical Oceanography", 2 ed., Vol. 6. (J. P. Riley and R. Chester, Eds.), pp. 281–390. Academic Press, London. 414 pp.

Chester, R. and Hughes, M. J. (1967). *Chem. Geol.* **2**, 249–262.

Chivas, A. R. and McDougall, I. (1978). *Econ. Geol.* **73**, 678–689.

Christensen, N. I. and Salisbury, M. M. (1975). *Rev. Geophys. Space Phys.* **13**, 57–86.

Clague, D. A. and Straley, P. F. (1977). *Geology* **5**, 133–136.

Clark, J. S. and Turner, R. C. (1955). *Can. Chem. Process.* **33**, 665–671.

Clarke, D. B. and Loubat, H. (1977). "Initial Reports of the Deep Sea Drilling Project", Vol. 37, pp. 847–855. U.S. Gov. Print. Off., Washington D.C.

Clarke, E. De C. (1912). *N.Z. Geol. Surv. Bull.* (N.S.) **14**, 58 pp.

Cole, J. W. Gill, J. B. and Woodhall, B. (in press). *In* "Geology and Offshore Resources of Pacific Island Arcs—Tonga Region" (D. W. Scholl and T. L. Vallier, Eds.) Circum-Pac. Counc. Energy Miner. Resour. Earth Science Series. Houston, Texas.

Cole, W. S. (1960). *Geol. Surv. Prof. Paper, (U.S.)* No. 374-A.

Coleman, P. J. (1969). New Hebrides Anglo-French Condominium, Annual Report of the Geological Survey for the year 1967, pp. 36–37.

Coleman, P. J. (1976). *CCOP/SOPAC Tech. Bull. No. 2*, 134–140.

Coleman, P. J., McGowran, B. and Ramsay, R. B. (1978). *Bull. Aust. Soc. Explor. Geophysicists* **9**(3), pp. 110–114.

Coleman, R. G. (1977). "Ophiolites: Ancient Oceanic Lithosphere". Springer-Verlag, New York, Berlin and Heidelberg, 229 pp.

Colley H. (1976). *Miner. Resour. Div. (Fiji) Mem.* No. 1, 123 pp.

Colley, H. (1978). *Courier* No. 49, 72–74.

Colley, H. and Greenbaum, D. (1980). *Econ. Geol.* **75**, 807–829.

Colley, H. and Warden, A. J. (1974). *Geol. Soc. Am. Bull.* **85**, 1635–1646.

Collot, J. Y., Daniel, J. and Burne, R. V. (1985). *Tectonophysics* **112**, 325–356.

Commercial Editor (1982). *Wellington Evening Post* Feb. 27, p. 17.

Commonwealth Secretariat (1976). Cook Islands Minerals Proposal for a Regime to Govern their Exploitation. London 42 pp.

Connelly, J. B. (1974). *J. Geol. Soc. Aust.* **21**, 459–469.

Concrete Research Association (1982). Proceedings of Concrete from Coral Workshop Rarotonga, Cook Islands, 31 March–2 April, 1982. New Zealand Concrete Research Association, Porirua. 32 pp.

Consolidated Gold Fields Australia Ltd and ARC Marine Ltd. (1980). Environmental Impact Statement Marine Aggregate. 3 vols. unpublished report.

Converse, D. R., Holland, H. D. and Edmond, J. M. (1984). *Earth Planet. Sci. Lett.* **69**, 159–175.

Cook, P. J. (1974). *Proc. Third Sess. CCOP/SOPAC*, pp. 75–85.

Cook, P. J. and Marshall, J. F. (1981). *Mar. Geol.* **41**, 201–221.

Cooper, P. F., MacNevin, A. A. and Winward, K. (1973). *Geol. Survey N.S.W.* Nos 38, 40 and 44, 80 pp.

Corliss, J. B., Dymond, J., Gordon, I. L., Edmond, J. M., Von Herzen, R. P., Ballard, R. D., Green, K., Williams, D., Bainbridge, A., Crane, K. and van Andel, Tj. H. (1979). *Science* **203**, 1073–1083.

Coulson, F. I. E. (1971). *Geol. Surv. Dep. Fiji, Bull.* No. 17, 49 pp.

Craig, H., Welhan, J. A., Kim, K., Poreda, R. and Lupton, J. E. (1980). *EOS* **61**, 992. (Abstr.).

Craig, J. R. and Kullerud, G. (1969). *In* "Magmatic Ore Deposits" (H. D. B. Wilson, Ed.). *Econ. Geol. Monogr.* No. 4, pp. 344–358.

Craig, P. M. (1975). Geological Map 1:50 000 Solomon Islands, Nendö Sheets EOI 1 and 2. Department of Geological Surveys, Honiara.

Craven, M. A. (1969). *Proc. Nat. Conf. Concrete Aggregates, Hamilton, N.Z.* pp. 7–16.

Cronan, D. S. (1969a). *Chem. Geol.* **5**, 99–106.

Cronan, D. S. (1969b). *Geochim. Cosmochim. Acta* **12**, 1562–1566.

Cronan, D. S. (1973). "Initial Reports of the Deep Sea Drilling Project", Vol. 16. pp. 601–604. U.S. Gov. Print. Off., Washington D.C.

Cronan, D. S. (1976). *Geol. Soc. Am. Bull.* **87**, 928–934.

Cronan, D. S. (1980). "Underwater Minerals". Academic Press, London. 362 pp.

Cronan, D. S. (1981). *In* "Report on the Inshore and Nearshore Resources Training Workshop", Suva, Fiji, 13–17 July, 1981. CCOP/SOPAC Technical Secretariat. pp. 31–32.

Cronan, D. S. (1983). *CCOP/SOPAC Tech. Bull. No. 4*, 55 pp.

Cronan, D. S. (1984). *South Pac. Mar. Geol. Notes* **3**(1), 1–17.

Cronan, D. S., Glasby, G. P., Halunen, J., Collen, J. D., Knedler, K. E., Johnston, J. H., Cooper, J., Landmesser, C. W. and Wingfield, R. T. (1981). *South Pac. Mar. Geol. Notes* **2**(2), 25–35.

Cronan, D. S., Glasby, G. P., Moorby, S. A., Thomson, J., Knedler, K. E. and McDougall, J. C. (1982). *Nature* **298**, 456–458.

Cronan, D. S., Moorby, S. A., Glasby, G. P., Knedler, K. E., Thomson, J. and Hodkinson, R. A. (1984). *In* "Marginal Basin Geology". *Spec. Publ.—Geol. Soc. Lond.* No. 16, pp. 137–149.

Cronan, D. S., Smith, P. A. and Bignell, R. D. (1977). *In* "Volcanic Processes in Ore Genesis". *Spec. Publ.—Geol. Soc. Lond.* No. 7, p. 80.

Cronan, D. S. and Thompson, B. (1978). *Trans. Instn. Min. Metall., Sect. B.* **87**, 87–90.

Cronan, D. S. and Tooms, J. S. (1969). *Deep-Sea Res.* **16**, 335–359.

Cruickshank, M. J. (1974). *In* "The Geology of Continental Margins" (C. A. Burk and C. L. Drake, Eds.), pp. 965–1000. Springer-Verlag, New York.

Cruickshank, M. J. and Hess, H. D. (1976). *Oceanus* **19**(10), 32–44.

Cullen, D. J. (1962). *N.Z. J. Geol. Geophys.* **5**, 309–313.

Cullen, D. J. (1967a). *Palaeogeogr., Palaeoclimatol., Palaeoecol.*, **3**, 289–298.

Cullen, D. J. (1967b). *N.Z. J. Mar. Freshwater Res.* **1**, 399–406.

Cullen, D. J. (1969). *J. Geophys. Res.* **74**, 4213–4220

Cullen, D. J. (1970). *Palaeogeogr., Palaeoclimatol., Palaeoecol.*, **7**, 13–20.

Cullen, D. J. (1975). *NZOI Oceanogr. Summary* **8**, 6 pp.

Cullen, D. J. (1978a). *NZOI Oceanogr. Field Rep.* **12**, 29 pp.

Cullen, D. J. (1978b). *Mar. Geol.* **28**, M67-M76.

Cullen, D. J. (1979a). *N.Z. Agric. Sci.* **13**, 85–91.

Cullen, D. J. (1979b). *In* "Proterozoic-Cambrian Phosphorites" (P. J. Cook and J. H. Shergold, Eds.), Australian National University Press, Canberra, 54 pp.

Cullen, D. J. (1980). *S.E.P.M. Spec. Pub.* **29**, 139–148.

Cullen, D. J., and Burnett, W. C. (1986). *Mar. Geol.* (in press).

Cullen, D. J. and Singleton, R. J. (1977). *NZOI Oceanogr. Field Rep.* **10**, 24 pp.

Cumberland, K. B. (1968). "Southwest Pacific". Whitcombe and Tombs Ltd., Christchurch. 423 pp.

Czamanske, G. K. and Moore, J. G. (1977). *Geol. Soc. Am. Bull.* **88**, 587–599.

Dabb, G. (1981). *In* "Pacific Islands Yearbook" (J. Carter, Ed.), pp. 21–31. 14th ed. Pacific Publications, Sydney. 559 pp.

D'Addario, G. W., Dow, D. B. and Swoboda, R. (1975). Geology of Papua New Guinea 1:2 500 000. Bureau of Mineral Resources, Canberra.

Dahl, A. L. (1977). *Proc. Third Int. Coral Reef Symp.* **2**, 571–575.

Dahm, J. and Healy, T. R. (1980). A Study of Dredge Spoil Dispersion off the entrance to Tauranga Harbour. Report to the Bay of Plenty Harbour Board, Tauranga. 64 pp.

Daniel, J. and Katz, H. R. (1981). *Geo-Mar. Lett.*, **1**, 213–219.

Daniel, J., Dugas, F., Dupont, J., Jouannic, C., Launay, J. and Monzier, M. (1976). *Cah. ORSTOM, Ser. Géol.* **8**(1), 95–101.

Davey, F. J. (1980). *N.Z. J. Geol. Geophys.* **23**, 533–536.
Davey, F. J. (1982). *Tectonophysics* **87**, 185–241.
Davies, H. L. (1985). Mineral Potential of the Southwest Pacific Islands. UNEP Regional Seas Reports and Studies, No. 69, 129–142.
Davies, P. J. and Marshall, J. F. (1972). *Bur. Min. Resour. (Aust.) Rec. 1972–1973* 13 pp. (unpublished).
Davis, E. E. and Lister, C. R. B. (1977). *J. Geophys. Res.* **82**, 4845–4860.
Davis, E. E., Lister, C. R. B., Wade, U. S. and Hyndman, R. D. (1980). *J. Geophys. Res.* **85**, 299–310.
de Broin, C. E., Aubertin, F. and Ravenne, C. (1977). *In* "Geodynamics in South-West Pacific", Symposium International, Noumea, New Caledonia, 27 August–2 September, 1976, pp. 37–50. Editions Technip, Paris.
Defossez, M., Monget, J. M., and Roux, P. (1980). *In* "Geology and Geochemistry of Manganese" (I. M. Varentsov and Gy. Grasselly, Eds.), Vol. 1, pp. 413–442. Hungarian Academy of Science, Budapest. 463 pp.
de Groot, S. J. (1979a). *Ocean Manage.* **5**, 211–232.
de Groot, S. J. (1979b). *Ocean Manage.* **5**, 233–249.
de Lacy, H. (1977). *N.Z. Farmer* **98**(21), 21–23.
Delegation of Papua New Guinea, The (1981). *Proc. Seventeenth Sess. CCOP.*, pp. 494–495.
Dick, H. J. B. and Bryan, W. B. (1978). "Initial Reports of the Deep Sea Drilling Project", Vol. 46, pp. 215–226. U.S. Gov. Print. Off., Washington D.C.
Dick, H. J. B. and Bullen, T. (1984). *Contrib. Mineral Petrol.* **86**, 54–76.
Dietz, R. S., Emery, K. O. and Shepard, F. P. (1942). *Geol. Soc. Am. Bull.* **53**, 815–848.
Donaldson, C. H. and Brown, R. W. (1977). *Earth Planet. Sci. Lett.* **37**, 81–89.
Dow, D. P. (1977). *Bull.—Bur. Miner. Resour. Geol. Geophys. (Aust.)* No. 201, 41 pp.
Down, C. G., and Stocks, J. (1976). U.K. Departments of the Environment and Transport Research Report No. 21.
Down, C. G. and Stocks, J. (1977). "Environmental Impact of Mining". Applied Science Publishers Ltd., London. 371 pp.
Downey, J. F. (1935). "Gold-mines of the Hauraki District, New Zealand". N.Z. Gov. Printer, Wellington. 305 pp.
Driessen, A. (1984). *In* "Australian Mineral Industry Annual Review for 1982", pp. 213–216. Australian Government Publishing Service, Canberra. 312 pp.
D.S.I.R. (1981). "Report of the Department of Scientific and Industrial Research for the Year Ended 31 March, 1981". N.Z. Govt. Printer, Wellington. 64 pp.
Duane, D. B. (1982). *Mar. Technol. Soc. J.* **16**, 87–90.
Dubois, J. (1969). *Ann. Géophys.*, **25**(4), 923–972.
Dubois, J., Ravenne, C., Aubertin, A., Louis, J., Guillaume, R., Launay, J. and Montadert, L. (1974). *In* "The Geology of Continental Margins" (C. A. Burk and C. L. Drake, Eds.), pp. 521–535. Springer-Verlag, New York.
Duennebier, F. and Blackington, G. (1980). *Nature* **284**, 338–340.
Dugas, F., Carney, J. N., Cassignol, C., Jezek, P. A. and Monzier, M. (1977). *In* "Geodynamics in South-West Pacific", Symposium International, Noumea, New Caledonia, 27 August–2 September, 1976, pp. 105–116. Editions Technip, Paris.
Dugolinsky, B. K. (1976). Unpublished Ph.D. thesis, University of Hawaii. 228 pp.
Duncan, J. F. and Metson, J. B. (1982). *N.Z. J. Sci.* **25**, 111–116.
Dungan, M. A., Long, P. E. and Rhodes, J. M. (1978). "Initial Reports of the Deep Sea Drilling Project", Vol. 45, pp. 461–477. U.S. Gov. Print. Off., Washington D.C.

Dupont, J., Launay, J., Ravenne, C. and de Broin, C. E. (1975). *C. R. Acad. Sci., Ser. D* **281**, 605–608.

Eade, J. V. (1971). Tonga bathymetry. New Zealand Oceanographic Institute Chart, Oceanic Series 1:1 000 000, Wellington.

Eade, J. V. (1972). Ha'apai bathymetry. New Zealand Oceanographic Institute Chart, Island Series 1:200 000, Wellington.

Eade, J. V. (1978). *Proc. Seventh Sess. CCOP/SOPAC*, pp. 51–53.

Eade, J. V. (1979). *Proc. Eighth Sess. CCOP/SOPAC*, pp. 93–96.

Eade, J. V. (1980). CCOP/SOPAC Cruise Rep. 39 (unpublished).

Eade, J. V., Kitikeaho, F. and Soakai, S. (1978). *Proc. Seventh Sess. CCOP/SOPAC*, pp. 67.

Eden, R. A. and Smith, R. (1983). Ministry of Energy and Mineral Resources, Mineral Resources Department, Rep. 50, Suva.

Eden, R. A. and Smith, R. (1984). Ministry of Lands, Energy and Mineral Resources, Mineral Resources Department, Suva.

Edmond, J. M., Jacobs, S. S., Gordon, A. L., Mantyla, A. W. and Weiss, R. F. (1979a). *J. Geophys. Res.* **84**, 7809–7826.

Edmond, J. M., Measures, C., McDuff, R. E., Chan, L. M., Collier, R., Grant, B., Gordon, L. I. and Corliss, J. B. (1979b). *Earth Planet. Sci. Lett.* **46**, 1–18.

Edmond, J. M., Von Damm, K. L., McDuff, R. E. and Measures, C. I. (1982). *Nature* **297**, 187–191.

Edwards, A. R. (1973). "Initial Reports of the Deep Sea Drilling Project", Vol. 21, pp. 641–661. U.S. Gov. Print. Off., Washington, D.C.

Edwards, A. R. (1975). "Initial Reports of the Deep Sea Drilling Project", Vol. 30, pp. 667–684. U.S. Gov. Print. Off., Washington, D.C.

Einsele, G., Geiskes, J. M., Curray, J., Moore, D. M., Aguayo, E., Aubry, M. P., Fornari, D., Guerrero, J., Kastner, M., Kelts, K., Lyle, M., Matoba, V., Molina-Cruz, A., Niemitz, J., Rueda, J., Saunders, A., Schrader, M., Simoniet, B. and Vacquier, V. (1980). *Nature* **283**, 441–445.

Eiseman, F. B. (1982). *Oceans* **15**(6), 36–37.

Elder, J. W., (1965). *Mongr. Ser. Am. Geophys. Union* **8**, 211–239.

Ellen, P. E. (1958). *N.Z. Concrete Construction* March 12, pp. 18–26.

Emelyanov, E. M. (1971). *In* "The Geology of the East Atlantic Continental Margin" (F. M. Delaney, Ed.), pp. 99–103. Institute of Geological Sciences, London.

Emery, K. O. (1975). *Technol. Rev.* **77**(3), 31–33.

Ewart, A. and Bryan, W. B. (1972). *Geol. Soc. Am. Bull.* **83**, 3281–3298.

Ewart, A. and Bryan, W. B. (1973). *In* "The Western Pacific: Island Arcs, Marginal Seas, Geochemistry" (P. J. Coleman, Ed.), pp. 503–522. University of Western Australia Press.

Ewart, A., Brothers, R. N. and Mateen, A. (1977). *J. Volcanol. Geotherm. Res.* **2**, 205–250.

Ewing, J. and Houtz, R. (1979). *In* "Deep Drilling Results in the Atlantic Ocean: Ocean Crust" (M. Talwani, C. G. Harrison and D. E. Hayes, Eds.), pp. 1–14. Maurice Ewing Series 2, Am. Geophys. Union, Washington, D.C.

Exon, N. F. (1979). *Rep.—Bur. Miner. Resour. Geol. Geophys. (Aust.)* No. 1979/62, 8 pp.

Exon, N. F. (1981). *South Pac. Mar. Geol. Notes* **2**(4), 47–65.

Exon, N. F. (1982a). *South Pac. Mar. Geol. Notes* **2**(6), 77–102.

Exon, N. F. (1982b). *South Pac. Mar. Geol. Notes* **2**(7), 103–120

Exon, N. F. (1983). *Mar. Min.* **4**, 79–107.

L

Exon, N. F. and Cronan, D. S. (1983). *Mar. Geol.* **52**, M43–M52.

Exon, N. F., Moreton, D. and Hicks, G. (1980). *BMR J. Aust. Geol. Geophys.* **5**, 67–68.

Exon, N. F., Stewart, W. D., Sandy, M. J. and Tiffin, D. L. (in press). *BMR J. Aust. Geol. Geophys.*

Exon, N. F. and Tiffin, D. L. (1982). *Trans., Third Circum-Pac. Energy Miner. Resour. Conf., Honolulu, August 22–28, 1982.* pp. 623–630.

Falvey, D. A. (1975). *Bull. Aust. Soc. Explor. Geophysicists* **6**, 47–49.

Falvey, D. A. (1978). *Bull. Aust. Soc. Explor. Geophysicists* **9**, 117–123.

Fewkes, R. H. (1976). Unpublished Ph.D thesis, Washington State University. 169 pp.

Field, C. W., Dymond, J. R., Heath, G. R., Corliss, J. B. and Dasch, E. J. (1976). "Initial Reports of the Deep Sea Drilling Project", Vol. 34, pp. 381–384. U.S. Gov. Print. Off., Washington, D.C.

Field, M. (1981). *Wellington Evening Post* July 4, p. 1.

Filloux, J. H. (1982). *J. Geophys. Res.* **87**, 8364–8378.

Finch, J. (1947). *N.Z. J. Sci. Technol.* **29B**, 36–51.

Fischer, A. G. (1969). *Geol. Soc. Am. Bull.* **80**, 549–552.

Fisher, N. H., Ed. (1973). *Bull—Bur. Miner. Resour. Geol. Geophys. (Aust.)* No. 141, 225 pp.

Fisher, R. L. and Engel, C. G. (1969). *Geol. Soc. Am. Bull.,* **80**, 1373–1378.

Fisk, M. R. and Bence, A. E. (1980). *Earth Planet. Sci. Lett.* **48**, 113–123.

Fleming, C. A. (1946). *N.Z. J. Sci. Technol. Sect. B* **27**, 347–365.

Fleming, C. A. (1968). *J., Geol. Soc. Lond.* **125**, 125–170.

Flower, M. F. J. (1981). *J. Geol. Soc. Lond.* **138**, 695–712.

Fordyce, R. E. and Cullen, D. J. (1979). NZOI Rec. **4**, 45–53.

Foster, A. R. (1970). Unpublished M.S. thesis, Washington State University, 169 pp.

Fox, P. J., Schreiber, E. and Peterson, J. J. (1973). *J. Geophys. Res.* **78**, 5155–5172.

Franchetau, J., Needham, H. D., Choukroune, P., Juteau, T., Seguret, M., Ballard, R. D., Fox, P. J., Normark, W., Carranza, A., Cordoba, D., Guerrero, J., Rangin, C., Bougault, M., Cambon, P. and Hekinian, R. (1979). *Nature* **277**, 523–528.

Frazer, J. Z. (1977). *Mar. Min.* **1**, 103–123.

Frazer, J. Z. (1980). *In* "Deep Sea Mining" (J. T. Kildow, Ed.), pp. 41–83. MIT Press, Cambridge, Massachusetts. 243 pp.

Frazer, J. Z. and Fisk, M. B. (1980). *Scripps Inst. Oceanogr. Rep.* SIO 80–16, 117 pp.

Frazer, J. Z. and Fisk, M. B. (1981). *Deep-Sea Res.* **28A**, 1533–1551.

Frey, F. A., Bryan, W. B., and Thompson, G. (1974). *J. Geophys. Res.* **79**, 5507–5527.

Friedrich, G. (1974). *Erzmetall.* **27**, 350–353.

Friedrich, G. (1976). *CCOP/SOPAC Tech. Bull. No. 2,* 39–53.

Friedrich G., Glasby, G. P., Plüger, W. L. and Thijssen, T. (1981). Inter Ocean '81 Dusseldorf IO 81-302/01, pp. 72–81.

Friedrich, G., Glasby, G. P., Thijssen, T. and Plüger, W. L. (1983). *Mar. Min.* **4**, 167–253.

Furkert, F. W. (1947). *Trans. R. Soc. N.Z.* **76**, 373–402.

Galvin, P., Ed. (1906). "The New Zealand Mining Handbook". N.Z. Mines Dep., Wellington. 589 pp.

Galtier, L. (1984). *Proc. Inter. Sem. Offshore Miner. Resour. 2nd, GERMINAL*, pp. 29–45.

Game, P. M. (1970). *Bull. Br. Mus. (Nat. Hist.), Mineral.* **2**(5), 223–284.

Garner, D. M. (1957). *N.Z. Oceanogr. Inst. Mem.* No. 2, 18–27.

Gauss, G. A. (1980). *Proc. Ninth Sess. CCOP/SOPAC*, pp. 76–77.

Gauss, G. A. (1982). *South Pac. Mar. Geol. Notes* **2**(9), 131–153.

Gauss, G. A., Eade, J. and Lewis, K. (1983). *South Pac. Mar. Geol. Notes* **2**(10), 155–184.

Gayman, W. (1978). *Ocean Manage.* **4**, 51–104.

George, G. W. P., Ed. (1978). "Australia's Offshore Resources: Implication of the 200-mile Zone". Australian Academy of Science, Canberra. 143 pp.

Gibb, J. G. (1974). Technical Note on Commercial Dredging of Offshore Deposits. Water and Soil Division, N.Z. Ministry of Works. 6 pp. (unpublished).

Gibb, J. G. (1977). N.Z. Water & Soil Technical Publication No. 5, N.Z. Ministry of Works, 16 pp.

Gibb, J. G. (1978). *N.Z. J. Mar. Freshwater Res.* **12**, 429–456.

Gibb, J. G. (1979a). Unpublished Ph.D thesis, Victoria University of Wellington. 217 pp.

Gibb, J. G. (1979b). *Soil & Water* **15**(3), 18–19.

Gibb, J. G. (1981). N.Z. Water & Soil Technical Publication No. 21, N.Z. Ministry of Works, 63 pp.

Gibb, J. G. (1982). *Proc. Eleventh N.Z. Geogr. Conf.*, pp. 129–134.

Gibb, J. G. (1983). *N.Z. Eng.* **38**(1), 15–19.

Gibb, J. G. (1984). *In* "Natural Hazards in New Zealand" (I. Speden and M. J. Crozier, Eds.), pp. 134–158. New Zealand National Commission for UNESCO, Wellington. 500 pp.

Gibb, J. G. and Adams, J. (1982). *N.Z. J. Geol. Geophys.* **25**, 335–352.

Gill, J. B. (1976a). *Geology* **4**, 123–126.

Gill, J. B. (1976b). *Geol. Soc. Am. Bull.* **87**, 1384–1395.

Gill, J. B. and McDougall, I. (1973). *Nature* **241**, 176–180.

Gillie, R. D. (1979). Unpublished Ph.D. thesis, University of Canterbury. 331 pp.

Gilson, R. (1980). "The Cook Islands 1820–1950". Victoria University Press, Wellington. 242 pp.

Glasby, G. P. (1972). *Hawaii Inst. Geophys. Rep.* HIG–72–23, pp. 59–82.

Glasby, G. P. (1976a). *N.Z. J. Geol. Geophys.* **19**, 707–736.

Glasby, G. P. (1976b). *N.Z. J. Geol. Geophys.* **19**, 771–790.

Glasby, G. P. (1978). *South Pac. Mar. Geol. Notes* **1**(7), 71–80.

Glasby, G. P. (1980). *Chem. Geol.* **31**, 347–361.

Glasby, G. P. (1981). *South Pac. Mar. Geol. Notes* **2**(3), 37–46.

Glasby, G. P. (1982a). *Mar. Min.* **3**, 231–270.

Glasby, G. P. (1982b). *Alpha* No. 20, 4 pp.

Glasby, G. P. (1982c). *N.Z. Int. Rev.* **7**(5), 26–28.

Glasby, G. P. (1983). *N.Z. Geographer* **39**, 3–11.

Glasby, G. P., Bäcker, H. and Meylan, M. A. (1975). *Erzmetall* **28**, 340–342.

Glasby, G. P., Friedrich, G., Plüger, W. L., Thijssen, T. and Kunzendorf, H. (1983a). *NZOI Oceangr. Summ.* No. 22, 10 pp.

Glasby, G. P., Hunt, J. L., Rankin, P. C. and Darwin, J. H. (1979). N.Z. Soil Bureau Scientific Report No. 36, 127 pp.

Glasby, G. P. and Katz, H. R. Eds. (1976). *CCOP/SOPAC Tech. Bull.* No. 2, 165 pp.

Glasby, G. P., Keays, R. R. and Rankin, P. C. (1978). *Geochem. J.* **12**, 229–243.

Glasby, G. P. and Lawrence, P. (1974a). *N.Z. Oceangr. Inst. Chart, Misc. Ser.* No. 33.

Glasby, G. P. and Lawrence, P. (1974b). *N.Z. Oceangr. Inst. Chart, Misc, Ser.* No. 34.

Glasby, G. P. and Lawrence, P. (1974c). *N.Z. Oceangr. Inst. Chart, Misc. Ser.* No. 35.

Glasby, G. P. and Lawrence, P. (1974d). *N.Z. Oceangr. Inst. Chart, Misc. Ser.* No. 36.

Glasby, G. P. and Lawrence, P. (1974e). *N.Z. Oceangr. Inst. Chart, Misc. Ser.* No. 37.

Glasby, G. P. and Lawrence, P. (1974f). *N.Z. Oceangr. Inst. Chart, Misc. Ser.* No. 38.

Glasby, G. P. and Lawrence, P. (1980). *N.Z. Oceangr. Inst. Chart, Misc. Ser.* No. 40.

Glasby, G. P., Meylan, M. A., Margolis, S. V. and Bäcker, H. (1981). *In* "Geology and Geochemistry of Manganese"(I. M. Varentsov and Gy. Grasselly, Eds.), Vol. 3, pp. 137–183. Hungarian Academy of Sciences, Budapest. 357 pp.

Glasby, G. P. and Read, A. J. (1976). *In* "Handbook of Strata-Bound and Stratiform Ore Deposits" (K. H. Wolf, Ed.), Vol. 7. Elsevier, Amsterdam. 656 pp.

Glasby, G. P., Stoffers, P., Sioulas, A., Thijssen, T. and Friedrich, G. (1982). *Geo-Mar. Lett.* **2**, 47–53.

Glasby, G. P. and Summerhayes, C. P. (1975). *N.Z. J. Geol. Geophys.* **18**, 477–490.

Glasby, G. P. and Thijssen, T. (1982). *Neues Jahrb. Mineral.* **145**, 291–307.

Glasby, G. P., Thijssen, T., Plüger, W. L., Friedrich, G., Stoffers, P., Frenzel, G., Andrews, J. E. and Roonwal, G. S. (1983b). *Hawaii Inst. Geophys. Rep.* No. 83–1, 112 pp.

Glasby, G. P., Tooms, J. S. and Howarth, R. J. (1974). *N.Z. J. Sci.* **17**, 387–407.

Gocht, W. and Wolf, A. (1982). *In* "Training Programme for the Management and Conservation of Marine Resources" (Class A Ocean Mining). Research Institute for International Technical and Economic Co-operation of the Aachen Technical University (RWTH). 223 pp.

Gomez, E. D. (1980). *In* "Marine and Coastal Processes in the Pacific: Ecological Aspects of Coastal Zone Management". Papers presented at a Unesco Seminar held at Motupore Island Research Centre, University of Papua, New Guinea, 14–17 July, 1980. 251 pp.

Gomez, E. D. (1983). *Ocean Manage.* **8**, 281–295.

Gomez, E. D., Alcala, A. C. and San Diego, A. C. (1981). *Proc. Fourth Int. Coral Reef Symp.* **1**, 275–282.

Goodell, H. G., Meylan, M. A. and Grant, B. (1971). *Antarct. Res. Ser.* **15**, 27–92.

Gordon, A. L. (1975). *Deep-Sea Res.* **22**, 357–377.

Gorton, M. P. (1977). *Geochim. Cosmochim. Acta* **41**, 1257–1270.

Gow, A. J. (1967). *N.Z. J. Geol. Geophys.* **10**, 675–695.

Graham, A. L., Symes, R. F., Bevan, J. C., and Din, V. K. (1978). "Initial Reports of the Deep Sea Drilling Project", Vol. 45, pp. 581–586. U.S. Gov. Print Off., Washington, D.C.

Graham, I. J. and Watson, J. L. (1980). *N.Z. J. Geol. Geophys.* **23**, 447–454.

Grange, K. (1985). *N.Z. Dive* **7**(4), 24–26.

Grange, K. R., Singleton, R. J., Richardson, J. R., Hill, P. J. and Main, W. deL. (1981). *N.Z. J. Zool.* **8**, 209–227.

Grant–Taylor, T. L. (1976). *N.Z. Geol. Surv. Rep.* M52, 75 pp.

Gray, J., Cumming, G. L. and Lambert, R. St. J. (1977). "Initial Reports of the Deep Sea Drilling Project", Vol. 37, pp. 607–612. U.S. Gov. Print. Off., Washington, D.C.

Green, D. (1970). Marine Mineral Development. Geol. Survey of Fiji (unpublished report).

Green, D. M., Hibberson, W. O. and Jacques, A. L. (1979). In "The Earth, its Origin, Structure and Evolution" (M. W. McElhinney, Ed.), pp. 265–299. Academic Press, London.

Greenbaum, D. (1980). Commonw. Geol. Liaison Office Spec. Liason Rep. CGLO SLR 3/1, pp. 71–88.

Greenbaum, D., Mallick, D. I. J. and Radford, N. W. (1975). New Hebrides Condominium Geological Survey, Regional Report, 46 pp.

Greene, H. G. and Wong, F. L. (1983). Hydrocarbon resource studies in the Southwest Pacific, 1982. U.S. Geol. Surv. Open-File Rep. No. 83–293, 24 pp.

Gregory, R. T. and Taylor, H. P., Jnr. (1981). J. Geophys. Res. 86, 2737–2755.

Greig, D. A. (1982). Unpublished M.Sc. thesis, Auckland University. 138 pp.

Griffiths, G. A. (1979). Nature 282, 61–63.

Griffiths, G. A. (1981). Water Resour. Bull. 17, 662–671.

Grigg, R. W. (1974). Proc. Second Int. Symp. Coral Reefs, Great Barrier Reef, Aust. 2, 235–240.

Grigg, R. W. (1976). Sea Grant Tech. Rep. (Univ. Hawaii), No. 77–03, pp. 1–48.

Grigg, R. W. (1977). "Hawaii's Precious Corals". An Island Heritage Book, Norfolk Island. 64 pp.

Grigg, R. W. (1979). In "Literature Review and Synthesis of Information on Pacific Island Ecosystems". Paper 6, Fish and Wildlife Service Report FWS/OBS–79/35, U.S. Department of the Interior, Washington, D.C.

Grigg, R. W. (1981). Proc. Fourth Int. Coral Reef Symp. 1, 243–246.

Grigg, R. W. and Bayer, F. M. (1976). Pac. Sci. 30, 167–175.

Grigg, R. W. and Eade, J. W. (1981). In "Report on the Inshore and Nearshore Resources Training Workshop", Suva, Fiji, 13–17 July, 1981, pp. 13–18. CCOP/SOPAC Technical Secretariat, Suva.

Grim, P. J. (1969). J. Geophys. Res. 74, 3933–3934.

Grinenko, V. A., Dmitriev, L. V., Migdisov, A. A. and Sharas'kin, A. Ya. (1975). Geochem. Int. 12, 132–137.

Gross, M. G., Barnard, W. D., Bokuniewicz, H., Gunnerson, C. G., Nichols, M. N., Saila, S. B. and Windom, H. L. (1979). In "Proceedings of a Workshop on Scientific Problems Relating to Ocean Pollution", Estes Park, Colorado, July 10–14, 1978 (E. D. Goldberg, Ed.), pp. 35–50. U.S. Department of Commerce. 225 pp.

Guillon, J. H. (1975). Mem. ORSTOM No. 76, pp. 11–120.

Gunn, B. M. (1971). Chem. Geol. 8, 1–13.

Hackman, B. D. (1973). In "The Western Pacific" (P. J. Coleman, Ed.), pp. 179–191. University of Western Australia Press.

Hackman, B. D. (1980). Overseas Mem. Inst. Geol. Sci. No. 6, 115 pp.

Hails, J. R. (1969). Proc. R. Soc. N.S.W. 102, 21–39.

Hails, J. R. (1974). Earth–Sci. Rev. 10, 171–202.

Hails, J. R. (1975). J. Geol. Soc. Lond. 131, 1–5.

Hajash, A. (1975). Contrib. Mineral. Petrol. 53, 205–226.

Hajash, A. and Chandler, G. W. (1981). Contrib. Mineral. Petrol. 78, 240–255.

Halbach, P. (1980). Geojournal 4, 407–422.

Halbach, P., Manheim, F. T. and Otten, P. (1982). Erzmetall 35, 447–453.

Halunen, A. J. (1978). *Bull. Am. Assoc. Petrol. Geol.* **62**, 219 (abstract).

Hamill, P. F. and Ballance, P. F. (1985). *N.Z. J. Geol. Geophys.* **28**, 503–511.

Hamilton, G. S. A. (1967). *N.Z. Concrete Construction* July 12, pp. 118–122.

Hamilton, W. (1979). *U.S. Geol. Surv. Prof. Pap.* No. 1078, 345 pp.

Hamon, B. V. and Tranter, D. J. (1971). *Aust. Nat. Hist.* **17**, 129–133.

Harada, K. and Nishida, S. (1976). *Nature* **260**, 770–771.

Harray, K. G. and Healy, J. R. (1978). *N.Z. J. Mar. Freshwater Res.* **12**, 99–107.

Haughton, D. R., Roeder, P. L. and Skinner, B. J. (1974). *Econ. Geol.* **69**, 451–467.

Hawkins, J. W. Jr. (1974). *In* "The Geology of Continental Margins" (C. A. Burk and C. L. Drake, Eds.), pp. 505–520. Springer-Verlag, New York.

Hawkins, J. W. Jr. (1976). *Earth Planet. Sci. Lett.* **28**, 283–297.

Hawkins, J. W. Jr. (1977). Petrologic and Geochemical Characteristics of Marginal Basin Basalts. American Geophysical Union Monograph (Maurice Ewing Series), pp. 355–365.

Hay, R. F., Mutch, A. R. and Watters, W. A. (1970). *N.Z. Geol. Surv. Bull.* (N.S.) No. 83, 86 pp.

Hayes, D. E. and Frakes, L. A. (1975). "Initial Reports of the Deep Sea Drilling Project", Vol. 28, pp. 909–942. U.S. Gov. Print. Off., Washington D.C.

Hayes, D. E. and Ringis, J. (1973). *Nature* **243**, 454–458.

Haymon, R. M. and Kastner, M. (1981). *Earth Planet. Sci. Lett.* **53**, 363–381.

Healing, R. A., Frazer, J. Z. and Archer, A. A. (1979). *In* "Manganese Nodules: Dimensions and Perspectives", pp. 37–58. D. Reidel Publishing Co., Dordrecht, Netherlands. 194 pp.

Healy, T. (1977). *N.Z. Geographer* **33**, 90–92.

Healy, T. (1980a). *In* "The Land Our Future, Essays on Land Use and Conservation in New Zealand" (A. G. Anderson, Ed.), pp. 239–260. Longman Paul, N.Z.

Healy, T. (1980b). *Soil & Water* **16**(4), 12–14.

Healy, T. (1981). *Soil & Water* **17**(3), 22–24.

Healy, T. (1982a). *Proc. Eleventh N.Z. Geogr. Conf.* pp. 135–139.

Healy, T. (1982b). *Proc. Eleventh N.Z. Geogr. Conf.* pp. 167–172.

Healy, T. R., Harray, K. G. and Richmond, B. (1977). University of Waikato Department of Earth Sciences, Occasional Report No. 3, 64 pp.

Heath, R. A. (1975). *N.Z. Oceangr. Inst. Mem.* **55**, 80 pp.

Heath, R. A. (1979). *N.Z. J. Geol. Geophys.* **22**, 259–266.

Heath, R. A. (1981a). *NZOI Oceanogr. Summary* **18**, 15 pp.

Heath, R. A. (1981b). *N.Z. J. Geol. Geophys.* **24**, 361–372.

Heathershaw, A. D., Carr, A. P. and Blackley, M. W. L. (1981). Inst. of Oceanogr. Sci. Rep. No. 118, 67 pp.

Heezen, B. C., Glass, B. and Menard, H. W. (1966). *Deep-Sea Res.* **13**, 445–458.

Hekinian, R., Fevrier, M., Bischoff, J. L., Picot, P. and Shanks, W. C. (1980). *Science* **207**, 1433–1444.

Hekinian, R., Fevrier, M., Avedik, F., Cambon, P., Charlou, J. L., Needham, H. D., Raillard, J., Boulegue, J., Merlivat, L., Moinet, A., Manganini, S. and Lange, J. (1983). *Science* **219**, 1321–1324.

Herbich, J. B. (1981a). *In* "Marine Environmental Pollution, 2" (R. A. Geyer, Ed.), pp. 227–240. Elsevier, Amsterdam. 574 pp.

Herbich, J. B. (1981b). *In* "Marine Environmental Pollution, 2" (R. A. Geyer, Ed.), pp. 241–260. Elsevier, Amsterdam. 574 pp.

Hilde, T. W., Uyeda, S. and Kroenke, L. (1977). *Tectonophysics* **38**, 145–165.

Hindle, W. H. (1976). *Bull. Miner. Resour. Div. Fiji* No. 1.

Hirst, J. A. and Kennedy, E. M. (1962). *Geol. Surv. Dep., Suva, Fiji. Econ. Invest.* No. 1, 8 pp.

Hodkinson, R. A. (1985). Unpublished Ph.D. thesis, University of London (in preparation).

Hoernes, S., Friedrichsen, M. and Schock, H. H. (1978). "Initial Reports of the Deep Sea Drilling Project", Vol. 45, pp. 541–549. U.S. Gov. Print. Off., Washington, D.C.

Hoffert, M. (1980). Unpublished D.Sc. thesis, University of Strasbourg. 231 pp.

Hoffert, M., Karpoff, A. M., Schaaf, A. and Pautot, G. (1979). *Proc. Colloq. Int. C.N.R.S.* No. 289, 101–112.

Hoffmeister, J. E. (1932). Bernice P. Bishop Museum Bulletin (Honolulu), Vol. 96. 93 pp.

Hohnen, P. D. (1978). *Bull.—Bur. Miner. Resour. Geol. Geophys. (Aust.)*, No. 194. 39 pp.

Hollett, K. J. and Moberly, R. (1982). *Environ. Geol.* **4**, 31–42.

Hollister, C. D., Johnson, D. A. and Lonsdale, P. F. (1974). *J. Geol.* **82**, 275–300.

Holmes, R. (1980). *Proc. Ninth Sess. CCOP/SOPAC* pp. 68–69.

Holmes, R. W. (1919). *Proc. N.Z. Soc. Civ. Eng.* **5**, 74–141.

Honnorez, J., Bohlke, J. K. and Honnorez-Guerstein, B. M. (1978). "Initial Reports of the Deep Sea Drilling Project", Vol. 46, pp. 299–329. U.S. Gov. Print. Off., Washington, D.C.

Honza, E., Keene, J. B. and Shipboard Scientists (1984). Cruise Report, M/S *Natsushima*, December 4, 1983–January 5, 1984. Rep 84/11 (PRAG 6), PNG Geological Survey, Department of Minerals and Energy, Port Moresby.

Horn, D. R., Delach, M. N. and Horn, B. M. (1973). *Int. Decade Ocean Explor. Tech. Rep.* No. 3, 51 pp.

Horn, D. R., Horn, B. M. and Delach, M. N. (1972). *Int. Decade Ocean Explor. Tech, Rep.* No. 1, 78 pp.

Horvitz, L. (1980). *CCOP/SOPAC Tech. Bull. No. 3*, 261–271.

Houtz, R. and Ewing, J. (1976). *J. Geophys. Res* **81**, 2490–2498.

Houtz, R. E. and Phillips, K. A. (1963). *Geol. Surv. Dep. Suva, Fiji, Econ. Rep.* No. 1, 36 pp.

Howorth, R. (1982a). *Proc. Eleventh Sess. CCOP/SOPAC*, pp. 122–125.

Howorth, R. (1982b). *CCOP/SOPAC Tech. Rep.* No. 22.

Howorth, R. (1982c). *CCOP/SOPAC Tech. Rep.* No. 24.

Howorth, R. (1982d). *CCOP/SOPAC Tech. Rep.* No. 25.

Howorth, R. (1983a). *CCOP/SOPAC Tech. Rep.* No. 29.

Howorth, R. (1983b). *CCOP/SOPAC Tech. Rep.* No. 31.

Hughes, G. W. (1979). Ririo, Choiseul Geological Map Sheet CH 4, 1: 50 000. Geological Survey Division, Ministry of Natural Resources, Honiara, Solomon Islands.

Hughes, G. W. and Turner, C. C. (1976). Solomon Islands Geological Survey Bulletin 2, Government Printer, Honiara, 80 pp.

Hughes, G. W. and Turner, C. C. (1977). *Geol. Soc. Am. Bull.* **88**, 412–424.

Hume, T. M. and Harris, T. F. W. (1981). *Water & Soil Misc. Publ.* No. 28, 63 pp.

Humphris, S. E., Morrison, M. A. and Thompson, R. N. (1978). *Chem. Geol.* **23**, 125–137.

Humphris, S. E. and Thompson, G. (1978a). *Geochim. Cosmochim. Acta.* **42**, 107–125.

Humphris, S. E. and Thompson, G. (1978b). *Geochim. Cosmochim. Acta.* **42**, 127–136.

Humphris, S. E., Thompson, R. N. and Marriner, G. F. (1980a). "Initial Reports of the Deep Sea Drilling Project", Vols. 51–53, pp. 1201–1217. U.S. Gov. Print. Off. Washington, D.C.

Humphris, S. E., Melson, W. G. and Thompson, R. N. (1980b). "Initial Reports of the Deep Sea Drilling Project", Vol. 54, pp. 773–787. U.S. Gov. Print. Off. Washington, D.C.

Hussong, D. M., Uyeda, S., et al. (1981). "Initial Reports of the Deep Sea Drilling Project", Vol. 60, 929 pp. U.S. Gov. Print. Off. Washington D.C.

Hutchison, A. J. H. (1973). N.Z. Eng. 28(8), 217–224.

Hutton, C. O. (1945). N.Z. J. Sci. Technol. Sect. B 27, 15–16.

Immel, R. and Osmond, J. K. (1976). Chem. Geol. 18, 263–272.

Ingamells, C. O. (1981). Geochim. Cosmochim. Acta 45, 1209–1216.

Ingram, J. (1983). N.Z. Eng. 38(9), 23, 25, 27.

Intergovernmental Oceanographic Commission (1975). Intergovernmental Oceanographic Commission Workshop, Report No. 6, 4 pp. & 9 annexes.

Intergovernmental Oceanographic Commission (1980). Intergovernmental Oceanographic Commission Workshop, Report No. 27, 9 pp. & 9 annexes.

Intergovernmental Oceanographic Commission (1983). Intergovernmental Oceanographic Commission Workshop, Report No. 35, 6 pp. & 6 annexes.

Irvine, T. N. (1967). Can. J. Earth Sci. 4, 71–103.

Ives, D. (1984). Pet. Gaz. Summer, pp. 4–8.

Jacques, A. L. (1980). Admiralty Islands Papua New Guinea; Explanatory Notes 1:250 000 Geological Series. Sheet SA/55–10, SA/55–11. Department of Minerals and Energy, Geological Survey of Papua New Guinea, Port Moresby, 25 pp.

Jacques, A. L. and Webb, A. W. (1975). Rep—Geol. Surv. Papua New Guinea 75/5, 8 pp.

Johannes, R. E. (1975). In "Tropical Marine Pollution" (E. F. J. Wood and R. E. Johannes, Eds.), pp. 13–15, Elsevier, Amsterdam, 192 pp.

Johannes, R. E. (1978). Ann. Rev. Ecol. Syst. 9, 349–364.

Johnson, D. A. (1974). Mar. Geol. 17, 71–78.

Johnson, R. W. (1979). BMR J. Aust. Geol. Geophys. 4, 181–207.

Johnson, R. W., Wallace, D. W. and Ellis, D. J. (1976). In "Volcanism in Australasia" (R. W. Johnson, Ed.), pp. 297–316. Elsevier, Amsterdam.

Johnston, E. B. (1980). Soil & Water. 16(1), 16–18.

Johnston, J. H. and Glasby, G. P. (1978). Geochem. J. 12, 153–164.

Jones, G. (1981). Aust. J. Mar. Freshwater Res. 32, 369–377.

Jones, G. and Candy, S. (1981). Aust. J. Mar. Freshwater Res. 32, 379–398.

Jones, H. A. (1980). AMI Q. 13, 1–14.

Jones, H. A. and Davies, P. J. (1979). Mar. Geol. 30, 243–268.

Jones, J. G. (1967). Geol. Soc. Am. Bull. 78, 1281–1288.

Jones, M. R. and Stevens, A. W. (1983). Queensland Gov. Min. J. 84, 5–11.

Jouannic, C. and Thompson, R. M. (1983). CCOP/SOPAC Tech. Bull. No. 5.

Jouannic, C., Taylor, F. W. and Bloom, A. L. (1982). In "Contribution a l'étude géodynamique du Sud-Ouest Pacific". Travaux et Documents de l'ORSTOM, No. 147, pp. 223–246.

Judd, B. T. and Palmer, E. R. (1973). Proc. Australas. Inst. Min. Metall. 247, 23–33.

Juteau, T., Bingol, F., Noack, Y., Whitechurch, H., Hoffert, M., Wirrmann, D. and Courtois, C. (1978). "Initial Reports of the Deep Sea Drilling Project", Vol. 45, pp. 613–645. U.S. Gov. Print. Off., Washington, D.C.

Kanehira, K., Yui, D., Sakai, H. and Sasakai, A. (1973). Geochem. J. 7, 89–96.

Kaplin, P. A. (1981). N.Z. Geographer 37, 3–12.

Karig, D. E. (1970). *J. Geophys. Res.* **75**, 239–254.

Katsura, T. and Nagashima, S. (1974). *Geochim. Cosmochim. Acta.* **38**, 517–531.

Katz, H. R. (1974). *In* "The Geology of Continental Margins" (C. A. Burk and C. L. Drake, Eds.), pp, 549–565. Springer-Verlag, New York.

Katz, H. R. (1976a). *CCOP/SOPAC Tech. Bull. No. 2*, 153–165.

Katz, H. R. (1976b). *Am. Assoc. Pet. Geol. Bull.* **60**, 1947–1956.

Katz, H. R. (1977). *In* "Geodynamics in South-West Pacific", Symposium International, Noumea, New Caledonia, 27 August–2 September, 1976, pp. 165–166. Editions Technip, Paris.

Katz, H. R. (1978a). *Geol. Soc. Am. Bull.* **89**, 1118–1119.

Katz, H. R. (1978b). *Am. Assoc Pet. Geol. Bull.* **62**, 1900–1905.

Katz, H. R. (1979). *Am. Assoc. Pet. Geol. Bull.* **63**, 1680–1688.

Katz, H. R. (1980). *CCOP/SOPAC Tech. Bull. No. 3*, 59–75.

Katz, H. R. (1981a). *CCOP/SOPAC Tech. Rep. No. 12.*

Katz, H. R. (1981b). *Am. Assoc. Pet. Geol. Bull.* **65**, 2254–2259.

Katz, H. R. (1982). *Trans. Third Circum-Pac. Energy Miner. Resour. Conf., Honolulu, August 22–28* pp. 181–189.

Katz, H. R. (1983). *Am. Assoc. Pet. Geol. Bull.* **67**, 1689–1694.

Katz, H. R. (1984). *Geol. Soc. Aust., Abstr.* No. 12, 294–296.

Katz, H. R. (In press). *In* "Geology and Offshore Resources of Pacific Island Arcs—Vanuata Region" (H. G. Greene and F. L. Wong, Eds.), Earth Science Series, Circum-Pac. Coun. Energy Miner. Resour.

Katz, H. R. and Daniel, J. (1981). Structural map of the New Hebrides island arc. Tenth Session of CCOP/SOPAC, Port Vila, Vanuatu, 6–14 October, 1981. Technical Secretariat CCOP/SOPAC, Suva, Fiji. (Unpublished.)

Katz, H. R. and Glasby, G. P. (1979). *South Pac. Mar. Geol. Notes* **1**(9), 95–110.

Kear, D. (1965a). *N.Z. Geol. Surv. Rep.* No. 2, 25 pp.

Kear, D. (1965b). *Proc. Eighth Commonw. Min. Metall. Congr.* **7**, *Paper No. 219*, 10 pp.

Kear, D. (1967). *N.Z. DSIR Inf. Ser.* No. 63, pp. 63–110.

Kear, D. (1979). Geology of Ironsand Resources of New Zealand. N.Z. DSIR. 164 pp. (unpublished report).

Kear, D. and Hunt, J. L. (1969). *Proc. Nat. Conf. Concrete Aggregates, Hamilton, N.Z.* pp. 17–25

Kear, D. and Wood, B. L. (1959). *N.Z. Geol. Surv. Bull.* **63**, 92 pp.

Kennedy, B., Daniels, N. H. and Marshall, T. (1967). *N.Z. J. Sci.* **10**, 701–720.

Kennett, J. P., Burns, R. E., Andrews, J. E., Churkin, M., Davies, T. A., Dumitrica, P., Edwards, A. R., Galehouse, J. S., Packham, G. H. and van der Lingen, G. (1972). *Nature, Phys. Sci.* **239**, 51–55.

Kent, G. (1980). "The Politics of Pacific Island Fisheries". Westview Press, Boulder, Colorado. 191 pp.

Kent, P. (1980). "Minerals from the Marine Environment". Edward Arnold, London, 88 pp.

Kilpatrick, J. B. and Hassal, D. C. (1981). *N.Z. J. Bot.* **19**, 285–297.

Kirk, R. M. (1977a). Survey of Inner Shelf Sediments from the Vicinity of the Harbour. Report to the Engineer's Department, Timaru Harbour Board. 23 pp.

Kirk, R. M. (1977b). *In* "Environment '77 Coastal Zone Workshop" (G. A. Knox, Ed.), pp. 47–54. Environmental Centre (Canterbury) Inc., Christchurch. 142 pp.

Kirk, R. M. (1978). Disposal of Dredge-spoil at Tarakohe and Coastal Erosion at Pohara Beach. Morris and Wilson Consulting Engineers Ltd. 43 pp.

Kirk, R. M. (1980). A review of New Zealand research on the physical aspects of coastal and estuarine environments. Paper presented at the New Zealand Marine Sciences Society.

Kirk, R. M. and Hewson, P. A. (1978). *Proc. Conf. Erosion Assess. Control N.Z.*, pp. 93–101.

Kirk, R. M., Owens, I. F. and Kelk, J. G. (1977). *Third Aust. Congr. Coastal Ocean Engineering, Melbourne, 18–21 April* pp. 240–244.

Kitt, W. (1981). *In* "Chemistry in a Young Country" (P. P. Williams, Ed.), pp. 185–194. N.Z. Institute of Chemistry Inc., Christchurch.

Knight, C. L., Ed. (1975). "Economic Geology of Australia and Papua New Guinea, 1. Metals", Australasian Institute of Mining and Metallurgy, Monograph Series No. 5, Parkville, Victoria, 1126 pp.

Knight, C. L., Ed. (1976). "Economic Geology of Australia and Papua New Guinea", Australasian Institute of Mining and Metallurgy, Monograph Series No. 8, Parkville, Victoria, 423 pp.

Kolodny, Y. and Kaplan, I. R. (1970). *Geochim. Cosmochim. Acta.* **34**, 3–24.

Krause, D. C. (1965). *Geol. Soc. Am. Bull.* **76**, 27–42.

Kress, A. G. and Veeh, H. H. (1980). *Mar. Geol.* **36**, 143–157.

Krishnaswami, S., Mangini, A., Thomas, J. H., Sharma, P., Cochran, J. K., Turekian, K. K. and Parker, P. D. (1982). *Earth Planet. Sci. Lett.* **59**, 217–234.

Kristmannsdottir, H. (1975). *In* "Proceedings of the Second United Nations Symposium on the Development and Use of Geothermal Resources", I, pp. 441–445. U.S. Gov. Print. Off., Washington, D.C.

Kroenke, L., Jouannic, C. and Woodward, P. (1983). Bathymetry of the Southwest Pacific, CCOP/SOPAC, Suva, Fiji.

Kroenke, L., Moberly, R., Winterer, E. L. and Heath, G. R. (1971). "Initial Reports of the Deep Sea Drilling Project", Vol. 7, pp. 1161–1226. U.S. Gov. Print. Off., Washington, D.C.

Kroenke, L., Scott, R. *et al.* (1980). "Initial Reports of the Deep Sea Drilling Project", Vol. 59, U.S. Gov. Print. Off., Washington, D.C. 820 pp.

Kroenke, L. and Tongilava, S. L. (1975). *South Pac. Mar. Geol. Notes* **1**(2), 9–15.

Kroenke, L. W. (1972). Unpublished Ph.D. thesis, Hawaii Institute of Geophysics, University of Hawaii, 119 pp.

Kroenke, L. W. and Bardsley, E., Eds. (1975). *CCOP/SOPAC Tech. Bull. No. 1*, 91 pp.

Krouse, H. R., Brown, H. M. and Farquharson, R. B. (1977). "Initial Reports of the Deep Sea Drilling Project", Vol. 37, pp. 621–624. U.S. Gov. Print. Off., Washington, D.C.

Kudrass, H.-R. and Cullen, D. J. (1982). *Geol. Jahrb. Reihe. D* No. 51, pp. 3–41.

Kuhn, G. G., Baker, E. D. and Campen, C. (1980). *Shore & Beach* **48**(4), 9–13.

Kullerud, G., Yund, R. A. and Moh, G. H. (1969). *In* "Magmatic Ore Deposits" (H. D. B. Wilson, Ed.). *Econ. Geol. Monogr.* No. 4, pp. 323–343.

Kushiro, I. and Thompson, R. N. (1972). *Carnegie Inst. Wash. Yearbook* **71**, 403–406.

Ladd, H. S. (1970). *U.S. Geol. Surv. Prof. Pap.* No. 640–C, 12 pp. + 5 plates..

Ladd, H. S., and Hoffmeister, J. E. (1945). Bernice P. Bishop Museum Bull. **181**, 392 pp.

Lallier-Verges, E. and Clinard, C. (1983). *Mar. Geol.* **52**, 267–280.

Lalou, C. and Brichet, E. (1980). *C. R. Acad. Sci. Ser. D* **290**, 819–822.

Landmesser, C. W., Andrews, J. E. and Packham, G. H. (1975). "Initial Reports of the Deep Sea Drilling Project", Vol. 30, pp. 647–662. U.S. Gov. Print. Off., Washington, D.C.

Landmesser, C. W., Kroenke, L. W., Glasby, G. P., Sawtell, G. H., Kingan, S., Utanga, E., Utanga, A. and Cowan, G. (1976). *South Pac. Mar. Geol. Notes* **1**(3), 17–39.

Landsea Minerals Ltd. (1975). Feasibility of Offshore Dredging along the East Coast of Northland. 48 pp. (unpublished report.)

Lapouille, A. (1978). *Bull. Aust. Soc. Exp. Geophysicists* **9**, 130–133.

Lapouille, A. (1982). *In* "Contribution a l'étude géodynamique du Sud-Ouest Pacific". Travaux et Documents de l'ORSTOM, No. 147, pp. 409–438. August–2 September, 1976, pp. 51–61. Editions Technip, Paris.

Larue, B. M., Collot, J. Y. and Malahoff, A. (1980). *CCOP/SOPAC Tech. Bull. No. 3*, 77–83.

Larue, B. M., Daniel, J., Jouannic, C. and Récy, J. (1977). *In* "Geodynamics in South-West Pacific", Symposium International, Noumea, New Caledonia, 27 August–2 September, 1976, pp. 51–56. Editions Technip, Paris.

Lauder, B. T. (1983). *N.Z. DSIR Ind. Process. Div. Rep.* No. IPD/TSO/2014, 15 pp. & appendix.

Launay, J., Dupont, J., Lapouille, A., Ravenne, C. and de Broin, C. E. (1977). *In* "Geodynamics in South-West Pacific", Symposium International, Noumea, New Caledonia, 27 August–2 September, 1976, pp. 155–163. Editions Technip, Paris.

Launay, J., Dupont, J. and Lapouille, A. (1982). *South Pac. Mar. Geol. Notes* **2**(8), 121–130.

Lawrence, J. R., Drever, J. J. and Kastner, M. (1978). "Initial Reports of the Deep Sea Drilling Project", Vol. 45, pp. 609–612. U.S. Govt. Print Off., Washington, D.C.

Lawver, L. A. and Hawkins, J. W. (1978). *Tectonophysics* **45**, 323–329.

Lensen, G. J. (1978). *In* "The Geology of New Zealand" (R. P. Suggate, G. R. Stevens and M. T. Te Punga, Eds.), Vol. 2, pp. 482–488. N.Z. Govt. Printer, Wellington.

Lewis, B. T. R. (1978). *Ann. Rev. Earth Planet. Sci.* **6**, 377–404.

Lewis, B. T. R. (1983). *J. Geophys. Res.* **88**, 3348–3354.

Lewis, B. T. R. and Garmany, J. D. (1982). *J. Geophys. Res.* **87**, 8417–8425.

Lewis, K. B. (1974). *Earth-Sci. Rev.* **10**, 37–71.

Lewis, K. B. (1979). *Mar. Geol.* **31**, 31–43.

Lewis, K. B. (1982). *NZOI Rec.* **4**(9), 121–133.

Lewis, K. B. and Eade, J. V. (1974). *NZOI Oceanogr. Summ.* No. 6, 8 pp.

Lewis, K. B., Utanga, A. T., Hill, P. J. and Kingan, S. C. (1980). *South Pacif. Mar. Geol. Notes* **2**(1), 1–23.

Liddy, J. C. (1972). *Mining Mag.* **126**(3), 197–203.

Lillie, A. R. (1970). *N.Z. J. Geol. Geophys.* **13**, 72–116.

Lillie, A. R. and Brothers, R. N. (1970). *N.Z. J. Geol. Geophys.* **13**, 145–183.

Lindner, A. W. (1972). *Aust. Pet. Explor. Assoc. J.* 62–68.

Lindner, A. W. (1975). *Proc. First Sess. CCOP/SOPAC*, pp. 77–84.

Lister, C. R. B. (1972). *Geophys. J. R. Astron. Soc.* **26**, 515–535.

Lister, C. R. B. (1974). *Geophys. J. R. Astron. Soc.* **39**, 465–509.

Lister, C. R. B. (1977). *Tectonophysics* **37**, 203–218.

Lister, C. R. B. (1982). *In* "The Dynamic Environment of the Ocean Floor" (K. A. Fanning and F. T. Manheim, Eds.), pp. 441–470. D. C. Heath, Lexington, Massachussets.

Lister, J. J. (1891). *Geol. Soc. Lond. Q. J.* **47**, 590–617.

Lock, J. (1984). *Geol. Soc. Austr., Abstr.* No. 12, 336–338.

Lonsdale, P. (1981). *Mar. Geol.* **43**, 153–193.

Lonsdale, P. F., Batiza, R. and Simkin, T. (1982). *Mar. Technol. Soc. J.* **16**, 54–61.
Lonsdale, P. F., Bischoff, J. L., Burns, V. M., Kastner, M. and Sweeney, R. E. (1980). *Earth Planet. Sci Lett.* **49**, 8–20.
Lonsdale, P. and Smith, S. M. (1980). *Mar. Geol.* **34**, M19–M24.
Lonsdale, P. and Spiess, F. N. (1977). *Mar. Geol.* **23**, 57–75.
Loughnan, F. C. and Craig, D. C. (1962). *Aust. J. Mar. Freshwater Res.* **13**, 48–56.
Ludwig, W. J. and Houtz, R. E. (1979). Isopach map of sediments in the Pacific Ocean Basin and marginal sea basins. American Association of Petroleum Geologists Chart.
Luyendyk, B. P., Macdonald, K. C. and Bryan, W. B. (1973). *Geol. Soc. Am. Bull.* **84**, 1125–1134.
Luyendyk, B. P., Bryan, W. B. and Jezek, P. A. (1974). *Geol. Soc. Am. Bull.* **85**, 1287–1300.
McAdam, G. D., Dall, R. E. A. and Marshall, T. (1969). *N.Z. J. Sci.* **12**, 649–686.
McClain, K. J. and Lewis, B. T. R. (1982). *J. Geophys. Res.* **87**, 8477–8490.
McClymont, B. (1982). N.Z. Commission for the Environment Newsletter, May, pp. 15–16.
McCulloch, M. T. and Cameron, W. E. (1983). *Geology* **11**, 727–731.
Macdonald, K. C. (1982a). *Ann. Rev. Earth Planet. Sci.* **10**, 155–190.
Macdonald, K. C. (1982b). *Mar. Technol. Soc. J.* **16**, 26–32.
Macdonald, K. C. (1983). *Rev. Geophys. Space Phys.* **21**, 1441–1454.
Macdonald, K. C., Becker, K., Spiess, F. N. and Ballard, R. D. (1980). *Earth. Planet. Sci. Lett.* **48**, 1–7.
McDougall, I. (1963). *Nature* **198**, 67.
McDougall, I. and van der Lingen, G. J. (1974). *Earth Planet. Sci. Lett.* **21**, 117–126.
McDougall, J. C. (1961). *N.Z. J. Geol. Geophys.* **4**, 283–300.
McDougall, J. C. and Brodie, J. W. (1967). *N.Z. Oceanogr. Inst. Mem.* No. 40, 56 pp.
McDougall, J. C. and Eade, J. V. (1981). *South Pac. Mar. Geol. Notes* **2**(5), 67–75.
MacKay, A. D., Gregg, P. E. H. and Syers, J. K. (1980). *N.Z. J. Agric. Res.* **23**, 441–449.
McKellar, J. B. (1975). *In* "Economic Geology of Australia and Papua New Guinea, 1. Metals" (C. L. Knight, Ed.), pp. 1055–1062. Australian Institute of Mining and Metallurgy Monograph Series No. 5, Parkville, Victoria, 1126 pp.
McKelvey, V. E., Wright, N. A. and Bowen, R. W. (1983). *U.S. Geol. Surv. Circ.*, No. 886, 55 pp.
McLean, R. F. (1978). *In* "Landform Evolution in Australasia" (J. L. Davies and M. A. J. Williams, Eds.), pp. 168–196. Australian National University Press, Canberra, 376 pp.
McLean, R. F. and Kirk, R. M. (1969). *N.Z. J. Geol. Geophys.* **12**, 138–155.
MacLean, W. H. (1977). "Initial Reports of the Deep Sea Drilling Project", Vol. 37, pp. 875–882. U.S. Gov. Print. Off., Washington, D.C.
MacLean, W. H. and Shimazaki, H. (1976). *Econ. Geol.* **71**, 1049–1057.
McPherson, R. I. (1978). *N.Z. Geol. Surv. Bull.* No. 87, 95 pp.
McTavish, R. A. (1966). *Micropaleontology* **12**, 1–36.
Maillet, P., Monzier, M., Selo, M. and Storzer, D. (1982). *In* "Contribution a l'étude géodynamique du Sud-Ouest Pacific". Travaux et Documents de l'ORSTOM, No. 147, pp. 441–458.
Malahoff, A. (1982). *Mar. Technol. Soc. J.* **16**, 39–45.
Malahoff, A., Stephen, R. H., Naughton, J. J., Keeling, D. L. and Richmond, R. H. (1982a). *Earth Planet. Sci. Lett.* **57**, 398–414.

Malahoff, A., Feden, R. H. and Fleming, H. S. (1982b). *J. Geophys. Res.* **87**, 4109–4125.

Mallick, D. I. J. and Greenbaum, D. (1977). New Hebrides Condominium Geological Survey, Regional Report, 84 pp.

Mallick, D. I. J. and Neef, G. (1974). New Hebrides Condominium Geological Survey, Regional Report, 103 pp.

Malpas, J. (1978). *Philos. Trans. R. Soc. Lond., Ser. A* **288**, 527–546.

Mammerickx, J., Smith, S. M., Taylor, T. L. and Chase, T. E. (1974). Bathymetry of the South Pacific map. Scripps Inst. of Oceanogr., University of California, La Jolla, California.

Marchig, V. and Halbach, P. (1982). *Tschermaks Mineral. Petrogr. Mitt.* **30**, 81–110.

Maritime Policy Branch (1980). Coastal Zone Management Seminar 1980, Marine Reserves, Vol. 3. N.Z. Ministry of Transport, Wellington.

Marshall, J. F. (1980). *Bull.—Bur. Miner. Resour. Geol. Geophys. (Aust.)* No. 207, 39 pp.

Marshall, J. F. and Cook, P. J. (1980). *J. Geol. Soc. Lond.* **137**, 765–771.

Marshall, T. and Finch, J. (1967). *N.Z. J. Sci.* **10**, 193–205.

Marshall, T. and Nicholson, D. S. (1955). *N.Z. J. Sci. Technol., Sect. B* **37**, 201–220.

Marshall, T., Suggate, R. P. and Nicholson, D. S. (1958). *N.Z. J. Geol. Geophys.* **1**, 318–324.

Martin, W. R. B. (1955). *N.Z. Eng.* **10**, 317–336.

Martin, W. R. B. (1956). *Nature* **178**, 1476.

Martin, W. R. B. (1961). *N.Z. J. Geol. Geophys.* **4**, 256–263.

Martin, W. R. B. and Long, A. M. (1960). *N.Z. J. Geol. Geophys.* **3**, 400–409.

Mason, B. (1945). *N.Z. J. Sci. Technol., Sect. B* **26**, 227–238.

Mathez, E. A. (1976). *J. Geophys. Res.* **81**, 4269–4276.

Mathez, E. A. (1980). "Initial Reports of the Deep Sea Drilling Project", Vols. 51–53, pp. 1069–1085. U.S. Gov. Print. Off., Washington, D.C.

Mathez, E. A. and Yeats, R. S. (1976). "Initial Reports of the Deep Sea Drilling Project", Vol. 34, pp. 363–373. U.S. Gov. Print. Off., Washington, D.C.

Matthews, E. R. (1982). Unpublished Ph.D. thesis, Victoria University of Wellington, 324 pp.

Melson, W. G. and Rabinowitz *et al.* (1978. "Initial Reports of the Deep Sea Drilling Project", Vol. 45, U.S. Gov. Print. Off., Washington, D.C. 717 pp.

Melson, W. G. and Thompson, G. (1971). *Philos. Trans. R. Soc. Lond., Ser. A* **268**, 423–441.

Menard, H. W. (1964). "Marine Geology of the Pacific". McGraw-Hill, New York, 271 pp.

Menard, H. W. and Frazer, J. Z. (1978). *Science* **199**, 969–971.

Mero, J. L. (1965). "The Mineral Resources of the Sea". Elsevier, Amsterdam, 312 pp.

Metson, J. B. (1980). Unpublished Ph.D. thesis, Victoria University of Wellington.

Mevel, C. (1980). "Initial Reports of the Deep Sea Drilling Project", Vols. 51–53, pp. 1299–1317. U.S. Gov. Print. Off., Washington, D.C.

Meylan, M. A. (1976). *CCOP/SOPAC Tech. Bull. No. 2*, pp. 92–98.

Meylan, M. A. (1978). Unpublished Ph.D. thesis, University of Hawaii. 312 pp.

Meylan, M. A., Bäcker, H. and Glasby, G. P. (1975). *NZOI Oceanogr. Field Rep.* No. 4, 24 pp.

Meylan, M. A., Glasby, G. P., McDougall, J. C. and Kumbalek, S. C. (1982). *N.Z. J. Geol. Geophys.* **25**, 437–458.

Meylan, M. A., Glasby, G. P., McDougall, J. C. and Singleton, R. J. (1978). *NZOI Oceanogr. Field Rep.* No. 11, 61 pp.

Meylan, M. A. and Goodell, H. G. (1976). *CCOP/SOPAC Tech. Bull. No. 2,* pp. 99–117.

Mineral Resources Committee, (1967). *N.Z. DSIR. Inf. Ser.* No. 63, pp. 43–61.

Mineral Resources Division (1981). Parliament of Fiji Parliamentary Paper No. 31 of 1981. 35 pp.

Mineral Resources Division, Fiji (1974). *Proc. Third Sess. CCOP/SOPAC,* pp. 85–91.

Mines Division (1979). Annual Returns of Production from Quarries and Mineral Production Statistics 1979. N.Z. Mines Division, Ministry of Energy.

Ministry of Transport (1980). Who Cares for the Coast? Pamphlet ISBN 0–477–06615–1. 30 pp. N.Z.

Mitchell, A. H. G. (1966). New Hebrides Condominium Geological Survey, Report No. 3. 41 pp.

Mitchell, A. H. G. (1970). *Sedimentol.* **41**, 201–243.

Mitchell, A. H. G. (1971). New Hebrides Condominium Geological Survey, Regional Report, 56 pp.

Mitchell, A. H. G. and Warden, A. J. (1971). *J. Geol. Soc. Lond.* **127**, 501–529.

Mitchell, J. K. (1982). *Ocean Yearb.* **3**, 258–319.

Miyashiro, A. (1973). *Earth Planet. Sci. Lett.* **19**, 218–224.

Miyashiro, A., Shido, F. and Ewing, M. (1971). *Philos. Trans. R. Soc. Lond., Ser. A* **268**, 589–604.

Mizuno, A. and Moritani, T., Eds. (1977). *Geol. Surv. Japan Cruise Rep.* No. 8, 217 pp.

Mizuno, A. and Nakao, S., Eds. (1982). *Geol. Surv. Japan Cruise Rep.* No. 18, 399 pp.

Monro, A. D. and Beavis, G. (1945). *N.Z. J. Sci. Technol., Sect. B* **27**, 237–241.

Monro, A. D. and Gibbs, H. S. (1938). *N.Z. J. Sci. Technol.* **19**, 523–526.

Monzier, M. (1975). Compagne Georstom II. Etude preliminaire enchantillons Ilons dragues ou preleves Comparaison des resultats concernant les encroutements et nodules polymetalliques avec ceux obtenus de la compagne Georstom I. ORSTOM, Noumea. 19 pp. (Unpublished report).

Monzier, M. (1976). *CCOP/SOPAC Tech. Bull. No. 2.* 124–128.

Monzier, M. and Missegue, F. (1977). Polymetallic nodules sampling in the Cook Islands Archipelago DANAIDES II and GEOTRANSIT II surveys. Preliminary report. ORSTOM-CNEXO, Noumea, New Caledonia (in French and English).

Moorby, S. A., Cronan, D. S. and Glasby, G. P. (1984). *Geochim. Cosmochim. Acta.* **48**, 433–442.

Moorby, S. A., Varnavas, S. P. and Cronan, D. S. (1983). "Initial Reports of the Deep Sea Drilling Project", Vol. 65, pp. 425–430. U.S. Gov. Print. Off., Washington, D.C.

Moore, J. G. and Calk, L. C. (1971). *Am. Mineral.* **56**, 476–488.

Moore, J. G. and Fabbi, B. P. (1971). *Contrib. Mineral. Petrol.* **33**, 118–127.

Moore, J. G. and Schilling, J. G. (1973). *Contrib. Mineral. Petrol.* **41**, 105–118.

Moore, J. R., Chairman (1975). Mining in the outer continental shelf and in the deep ocean. Panel on Operational Safety in Marine Mining, National Academy of Science, Washington, D.C. 119 pp.

Morgan, J. R. (1981). Marine Policy, Vol. 5, pp. 344–345 (book review).

Moritani, T., Ed. (1979). *Geol. Surv. Japan Cruise Rep.* No. 12, 256 pp.

Morley, I. W. (1981). "Black Sands: A History of the Mineral Sand Industry in Eastern Australia". University of Queensland Press, St. Lucia. 278 pp.

Morrison, M. A. and Thompson, R. N. (1983). "Initial Reports of the Deep Sea Drilling Project", Vol. 65, pp. 643–660. U.S. Gov. Print. Off., Washington, D.C.

Morton, J. (1981). *N.Z. Environ.* No. 29, pp. 34–35.

Mottl, M. J. (1983). *Geol. Soc. Am. Bull.* **94**, 161–180.

Mottl, M. J. and Holland, H. D. (1978). *Geochim. Cosmochim. Acta.* **42**, 1103–1115.

Mottl, M. J. and Seyfried, W. E. (1980). *In* "Sea Floor Spreading Centers: Hydrothermal Systems" (P. A. Rona and R. P. Lowell, Eds.), pp. 66–82. Dowden, Hutchinson and Ross, Stroudsburg, Pennsylvania.

Mottl, M. J., Holland, H. D. and Corr, R. F. (1979). *Geochim. Cosmochim. Acta.* **43**, 869–884.

Muehlenbachs, K. (1976). "Initial Reports of the Deep Sea Drilling Project", Vol. 34, pp. 377–440. U.S. Gov. Print. Off., Washington, D.C.

Muehlenbachs, K. (1977). *Can. J. Earth Sci.* **14**, 771–776.

Muehlenbachs, K. (1980). "Initial Reports of the Deep Sea Drilling Project", Vols. 51–53, pp. 1159–1167. U.S. Gov. Print. Off., Washington, D.C.

Muehlenbachs, K. and Clayton, R. N. (1972a). *Can. J. Earth Sci.* **9**, 172–184.

Muehlenbachs, K. and Clayton, R. N. (1972b). *Can. J. Earth Sci.* **9**, 471–478.

Muehlenbachs, K. and Hodges, F. N. (1978). "Initial Reports of the Deep Sea Drilling Project", Vol. 46, pp. 257–258. U.S. Gov. Print. Off., Washington, D.C.

Mulder, C. J. and Nieuwenhuizen, C. v. d. (1971). B.I.P.M. Report EP–42347, Open-File Tonga Government Petroleum Report No. 16.

Murray, J. (1902). *Queensland Geogr. J.* **19**, 691–711.

Murray, J. (1906). *Queensland Geogr. J. (N.S.)* **21**, 71–134.

Murray, J. and Renard, A. F. (1891). Rep. Sci. Results Explor. Voyage Challenger, 525 pp.

Naldrett, A. J. (1969). *J. Petrol.* **10**, 171–201.

Nayudu, Y. R. (1971). *Antarct. Res. Ser.* **15**, 247–282.

Neef, G. (1982). *Tectonophysics* **87**, 177–183.

Nelson, D. M. and Gordon, L. I. (1982). *Geochim. Cosmochim. Acta.* **46**, 491–501.

New Zealand Geological Survey (1981). *Alpha* No. 22, 4 pp.

New Zealand Steel Limited (Undated). Steel from Ironsands. Information Brochure. 20 pp.

Nichols, M. M. (1979). *In* "Ocean Dumping and Marine Pollution" (H. D. Palmer and M. G. Gross, Eds.), pp. 147–161. Dowden, Hutchinson and Ross, Stroudsburg, Pennsylvania, 268 pp.

Nicholson, D. S. (1967). *N.Z. J. Sci.* **10**, 447–456.

Nicholson, D. S. (1969). *N.Z. J. Sci.* **12**, 111–117.

Nicholson, D. S., Cornes, J. J. S. and Martin, W. R. B. (1958). *N.Z. J. Geol. Geophys.* **1**, 611–616.

Nicholson, D. S. and Fyfe, H. E. (1958). *N.Z. J. Geol. Geophys.* **1**, 617–634.

Nicholson, D. S., Shannon, W. T. and Marshall, T. (1966). *N.Z. J. Sci.* **9**, 586–598.

Nicolas, A. and Violette, J. F. (1982). *Tectonophysics* **81**, 319–339.

Noakes, L. C. and Jones, H. A. (1975). *In* "Economic Geology of Australia and Papua New Guinea, 1. Metals" (C. L. Knight. Ed.), pp. 1093–1104. Australasian Institute of Mining and Metallurgy, Monograph Series No. 5, Parkville, Victoria, 1126 pp.

Normark, W. R., Lupton, J. E., Murray, J. W., Koski, R. A., Clague, D. A., Morton, J. L., Delaney, J. R. and Johnson, H. P. (1982). *Mar. Technol. Soc. J.* **16**, 46–53.

Norris, R. M. (1964). *N.Z. Oceanogr. Inst. Mem. 26*, 39 pp.

Norris, R. M. (1978). *N.Z. Oceanogr. Inst. Mem. 81*, 28 pp.

O'Brien, D. J. and Marshall, T. (1965). *N.Z. J. Sci.* **8**, 3–15.

O'Brien, D. J. and Marshall, T. (1968). *N.Z. J. Sci.* **11**, 159–169.

O'Brien, G. W., Harris, J. R., Milnes, A. R. and Veeh, H. H. (1981). *Nature* **294**, 442–444.

O'Brien, G. W. and Veeh, H. H. (1980). *Nature* **288**, 690–692.

Oceanic Explorations Co. (1975). Oceanic Exploration Papua-New Guinea. Final Report: L'Etoile No. 1 Well, Offshore Bougainville Island. Open File Company Report, PNG Geological Survey, Port Moresby, Department of Minerals and Energy, Port Moresby.

O'Donnell, T. H. and Presnall, D. C. (1980). *Am. J. Sci.* **280A**, 845–868.

Ostwald, J. and Frazer, F. W. (1973). *Miner. Deposita* **8**, 303–311.

Our Own Correspondent (1980). *Mining Ann. Rev.* pp. 432–433.

Owen, R. M. (1977). *Mar. Min.* **1**, 85–102.

Packham, G. H. and Terrill, A. (1975). "Initial Reports of the Deep Sea Drilling Project", Vol. 30, pp. 617-633. U.S. Gov. Print. Off., Washington, D.C.

Page, R. W. and McDougall, I. (1972). *Econ. Geol.* **67**, 1065–1074.

Page, R. W. and Ryburn, R. J. (1977). *Pac. Geol.* **12**, 99–105.

Pain, C. F. (1976). *N.Z. J. Geol. Geophys.* **19**, 153–177.

Palmer, E. R. and Judd, B. (1973). *N.Z. Eng.* **28**(8), 227–233.

Paltech, Pty Ltd (1979). Age and Environmental Determinations of Six Samples from Maewo and Santo Islands, New Hebrides. Report 1979/17 to New Hebrides Geological Survey, Occ. 1/79.

Panayiotou, A. (1980). *Int. Ophiolite Symp., Cyprus* pp. 102–116. Geol. Surv. Dept., Repub. Cyprus.

Panel on Operational Safety in Marine Mining (1975). "Mining in the Outer Continental Shelf and in the Deep Ocean". National Academy of Sciences, Washington, D.C. 119 pp.

Paris, J. P. (1981). Mémoire du B.R.G.M., No. 113, 278 pp.

Paris, J. P., Andreieff, P. and Coudray, J. (1979). *C. R. Acad. Sci., Ser. D* **288**, 1659–1661.

Paris, J. P. and Bradshaw, J. D. (1977). *In* "Geodynamics in South-West Pacific", Symposium International, Noumea, New Caledonia, 27 August–2 September 1976, pp. 209–215. Editions Technip, Paris.

Paris, J. P. and Lille, R. (1977). *In* "Geodynamics in South-West Pacific", Symposium International, Noumea, New Caledonia, 27 August–2 September 1976, pp. 195–208. Editions Technip, Paris.

Pasho, D. W. (1976). *N.Z. Oceanogr. Inst. Mem. 77*, 27 pp.

Paterson, O. D. (1965). *Proc. Eighth Commonw. Min. Metall. Cong.* **2**, 336–342.

Pautot, G. and Melguen, M. (1976). *CCOP/SOPAC Tech. Bull. No. 2*, pp. 54–61.

Pautot, G. and Melguen, M. (1979). *In* "Marine Geology and Oceanography of the Pacific Manganese Nodule Province" (J. L. Bischoff and D. Z. Piper, Eds.), pp. 621–650. Plenum Press, New York.

Pautot, G., Hoffert, M., Karpoff, A. M. and Schaaf, A. (1979). *Proc. Colloq. Int. C.N.R.S.* No. 289, pp. 113–118.

Payne, R. R. and Conolly, J. R. (1972). *In* "Ferromanganese Deposits on the Ocean Floor" (D. R. Horn, Ed.), pp. 81–92. National Science Foundation, Washington, D.C. 293 pp.

Pearson, W. C. and Lister, C. R. B. (1973). *J. Geophys. Res.* **78**, 7786–7787.

Pemberton, D. G. (1979). *N.Z. Eng.* **34**(5), 98–101.

Pevear, D. R. (1966). *Econ. Geol.* **61**, 251–256.

Phillips, K. A., Comp. (1967). *N.Z. J. Geol. Geophys.* **10**, 1175–1203.

Pickrill, R. A. (1977). *N.Z. J. Geol. Geophys.* **20**, 1–16.

Pickrill, R. A. and Mitchell, J. S. (1979). *N.Z. J. Mar. Freshwater Res.* **13**, 501–520.

Pineau, F., Javoy, M., Hawkins, J. W. and Craig, H. (1976). *Earth Planet. Sci. Lett.* **28**, 299–307.

Piper, D. Z. and Fowler, B. (1980). *Nature* **286**, 880–883.

Piper, D. Z. and Williamson, M. E. (1977). *Mar. Geol.* **23**, 285–303.

Polunin, N. V. C. (1983). *Oceanogr. Mar. Biol. Ann. Rev.* **21**, 455–531.

Pomeyrol, R. (1951). Revue de L'Institut Français du Pétrole, Vol. 6, No. 8, pp. 271–282.

Porter, J. W., Woodley, J. D., Smith, G. J., Neigel, J. E., Battey, J. F. and Dallmeyer, D. G. (1981). *Nature* **294**, 249–250.

Pratt, R. M. (1971). *Southeast. Geol.* **13**, 19–38.

Premuzic, E. T., Benkovitz, C. M., Gaffney, J. S. and Walsh, J. J. (1982). *Org. Geochem.* **4**, 63–77.

Prescott, J. R. V. (1980) *Ocean Yearb.* **2**, 317–345.

Price, G. D. (1983). Final Report on West Coast Offshore Project—Northern Exploration Licences, New Zealand. CRA Exploration Pty Ltd, Report No. 11808. 52 pp., 2 Appendices & map section (proprietary report).

Price, G. D., Falconer, R. K. and Voon, Q. C. (1982). Final Report on Otago Offshore Project, New Zealand. CRA Exploration Pty Ltd, Report No. 11514. 88 pp., 6 Appendices & accompanying map volume (proprietary report).

Prinz, M., Keil, K., Green, J. A., Reid, A. M., Bonatti, E. and Honnorez, J. (1976). *J. Geophys. Res.* **81**, 4087–4103.

Pritchard, R. G., Cann, J. R. and Wood, D. A. (1979). "Initial Reports of the Deep Sea Drilling Project", Vol. 49, pp. 709–714. U.S. Gov. Print. Off., Washington, D.C.

Propach, G. (1978). "Initial Reports of the Deep Sea Drilling Project", Vol. 45, pp. 551–556. U.S. Gov. Print. Off., Washington, D.C.

Puchelt, H. and Hubberten, H. W. (1980). "Initial Reports of the Deep Sea Drilling Project", Vols. 51–53, pp. 1145–1148. U.S. Gov. Print. Off., Washington, D.C.

Rajamani, V. and Naldrett, A. J. (1978). *Econ. Geol.* **73**, 82–93.

Ramsay, W. R. H. (1978). *Bull. Aust. Soc. Explor. Geophysicists* **9**, 107–110.

Ramsay, W. R. H. (1982). *Tectonophysics* **87**, 109–126.

Rankin, P. C. and Glasby, G. P. (1979). In "Marine Geology and Oceanography of the Pacific Manganese Nodule Province" (J. L. Bischoff, and D. Z. Piper, Eds.), pp. 681–698. Plenum Press, New York.

Ravenne, C., Pascal, G., Dubois, J., Dugas, F. and Montadert, L. (1977a). In "Geodynamics in South-West Pacific", Symposium International, Noumea, New Caledonia, 27 August–2 September, 1976, pp. 63–77. Editions Technip, Paris.

Ravenne, C., de Broin, C. E., Dupont, J., Lapouille, A. and Launay, J. (1977b). In "Geodynamics in South-West Pacific", Symposium International, Noumea, New Caledonia, 27 August–2 September, 1976, pp. 145–154. Editions Technip, Paris.

Récy, J., Dubois, J., Daniel, J., Dupont, J. and Launay, J. (1977a). In "Geodynamics in South-West Pacific", Symposium International, Noumea, New Caledonia, 27 August–2 September, 1976, pp. 345–356. Editions Technip, Paris.

Récy, J., Missege, F. and Monzier, M. (1977b). Chemical Analysis Results about Metal Contents of Polymetallic Nodule Samples in the Cook Island Archipelago DANAIDES II and GEOTRANSIT II cruises. ORSTOM-CNEXO, Noumea, New Caledonia (unpublished manuscript) (in French and English).

Reed, J. J. and Hornibrook, N. de B. (1952). *N.Z. J. Sci. Technol.* **34B**, 173–188.

Reid, J. (1976). "Salt for New Zealand". Dominion Salt Ltd, Lake Grassmere, Marlborough, 48 pp.

Reid, S. J. (1981). *N.Z. J. Sci.* **24**, 51–58.

Renard, V. (1976). *25th Int. Geol. Congr.* **3**, 791 pp.

Richard, J. J. (1962). "Catalogue of the Active Volcanoes in the World, including Solfatara Fields, Part 13, Kermadec, Tonga and Samoa". International Association of Volcanology, Rome, 38 pp.

Richards, J. R., Cooper, J. A., Webb, A. W. and Coleman, P. J. (1966). *Nature* **211**, 1251–1252.

Richardson, J. R. (1985). Significance of steep rock wall faunas within fiord basins (in press).

Richardson, S. H., Hart, S. R. and Staudigel, H. (1980). *J. Geophys. Res.* **85**, 7195–7200.

Richmond, R. N. (1981). *U.S. Geol. Surv. Prof. Pap.* No. 1193, p. 201.

Rickard, M. J. (1966). *Geol. Surv. Mem.—Fiji, Dept.* No. 2, 81 pp.

Rickwood, F. K. (1968). *J. Aust. Pet. Explor. Assoc.* **8**, 51–61.

Ridge, J. D. (1976). "Annotated Bibliographies of Mineral Deposits in Africa, Asia (Exclusive of the USSR) and Australia". Pergamon Press, Oxford, 545 pp.

Riggs, S. (1979). *In* "Proterozoic-Cambrian Phosphorites" (P. J. Cook and J. H. Shergold, Eds.), pp. 63–64. Australian National University Press, Canberra.

Riley, P. B., Monro, I. S. and Schofield, J. C. (1985). *N.Z. J. Geol. Geophys.* **28**, 299–312.

Ripper, D. and Grund, R. (1969). The Geology of Permit 48 (Territory of Papua New Guinea). Continental Oil Company of Australia Ltd and New Guinea Cities Service Inc. Open-file Company rep., PNG Geological Survey. Department of Minerals and Energy, Port Moresby.

Ritchie, G. S. (1981). *The Dock & Harbour Authority* **62**, 163–166.

Ritchie, L. D. and Saul, P. J. (1974). Fisheries Management Division, N.Z. Ministry of Agriculture and Fisheries, Wellington. 36 pp. (unpublished report).

Ritchie, L. D. and Saul, P. J. (1975). Fisheries Management Division, N.Z. Ministry of Agriculture and Fisheries, Wellington. 22 pp. (unpublished report).

Robinson, G. P. (1969). New Hebrides Geological Survey, Regional Report, 77 pp.

Robinson, P. T., Flower, M. F. J., Schminke, H.-U. and Ohnmacht, W. (1977). "Initial Reports of the Deep Sea Drilling Project", Vol. 37, pp. 775–793. U.S. Gov. Print. Off., Washington, D.C.

Rochford, D. J. (1972). *Div. Fish. Oceanogr. Tech. Pap. (Aust., C.S.I.R.O.)* No. 33. 17 pp.

Rochford, D. J. (1975). *Aust. J. Mar. Freshwater Res.* **26**, 233–243.

Rock-Color Chart Committee (1975). "Rock-Color Chart". Geological Society of America, Boulder, Colorado.

Rodda, P. (1967). *N.Z. J. Geol. Geophys.* **10**, 1260–1273.

Rodda, P. (1975). *In* "The Encyclopedia of World Regional Geology, Part I" (R. W. Fairbridge, Ed.), pp. 278–282. Dowden, Hutchinson and Ross, Stroudsburg, Pennsylvania.

Rodda, P. (1982). *In*, "Stratigraphic Correlation between Sedimentary Basins of the ESCAP Region", Vol. VIII, ESCAP Atlas of Stratigraphy III, Mineral Resources Development, Series No. 48, pp. 13–21. United Nations, New York.

Rodda, P. (1984). *N.Z. J. Geol. Geophys.* **27**, 97–98.

Rodda, P., Snelling, N. J. and Rex, D. C. (1967). *N.Z. J. Geol. Geophys.* **10**, 1248–1259.

Rogers, C. (1981). *New Scientist* **92**, 382–387.
Rona, P. A. (1984). *Earth-Sci. Rev.* **20**, 1–104.
Roonwal, G. S. (1983). *Indian J. Mar. Res.* **12**, 138–142.
Rosendahl, B. R. (1976). *J. Geophys. Res.* **81**, 5305–5313.
Rowe, G. H. (1980). Unpublished Ph.D. thesis, Victoria University of Wellington. 397 pp.
Roy, P. S. and Thom, B. G. (1981). *J. Geol. Soc. Aust.* **28**, 471–489.
Roy, P. S., Thom, B. G. and Wright, L. D. (1980). *Sediment. Geol.* **26**, 1–19.
Rozanova, T. V. and Baturin, G. N. (1971). *Oceanology,* **11**, 874–879.
Rubin, D. M. (1984a). *CCOP/SOPAC Cruise Rep. No. 98*, 16 pp.
Rubin, D. M. (1984b). *CCOP/SOPAC Cruise Rep. No. 97*, 10 pp.
St. John, B. (1982). *In* "Proceedings of the South-western Legal Foundation Exploration and Economics of the Petroleum Industry", Vol. 20, pp. 1–29. Matthew Bender & Co., New York.
St. John, B. (1985). Hydrocarbon provinces of the world. American Association of Petroleum Geologists Map.
Salvat, B. (1981). *Proc. Fourth Int. Coral Reef Symp.* **1**, 225–229.
Samuel Gary Oil Producer (1981). Open-file petroleum report, Government of Tonga.
Sandstrom, M. W. and Philp, R. P. (1984). *Chem. Geol.* **43**, 167–180.
Saphore, E. and Exon, N. F. (1981). Cruise Report No. 51, Technical Secretariat CCOP/SOPAC, Suva, Fiji. (Unpublished.)
Saunders, A. D., Fornari, D. J. and Morrison, M. A. (1982). *J. Geol. Soc. Lond.* **139**, 335–346.
Sawkins, R. J. (1984). "Metal Deposits in Relation to Plate Tectonics", Springer-Verlag, New York. 325 pp.
Scheidegger, K. F. and Stakes, D. S. (1980). "Initial Reports of the Deep Sea Drilling Project", Vols. 51–53, pp. 1253–1263. U.S. Gov. Print. Off., Washington, D.C.
Schofield, J. C. (1967). *N.Z. J. Geol. Geophys.* **10**, 697–731.
Schofield, J. C. (1969). *N.Z. DSIR Inf. Ser.* No. 63, pp. 76–92.
Schofield, J. C. (1970). *N.Z. J. Geol. Geophys.* **13**, 767–824.
Schofield, J. C. (1975a). *N.Z. J. Geol. Geophys.* **18**, 295–316.
Schofield, J. C. (1975b). *N.Z. J. Geol. Geophys.* **18**, 109–127.
Schofield, J. C. (1978). *Proc. Fourth Aust. Conf. Coastal Ocean Eng. Adelaide, 8–10 November* pp. 30–33.
Schofield, J. C. (1979). *Proc. Tenth N.Z. Geogr. Conf. and 49th ANZAAS Congr.* N.Z. Geogr. Soc. Conf. Ser. No. 10, pp. 39–42.
Schofield, J. C. (1985). *N.Z. J. Geol. Geophys.* **28**, 313–322.
Schofield, J. C. and Woolhouse, L. (1969). *N.Z. DSIR Inf. Ser.* No. 79, pp. 29–102.
Scholl, D. W. and Vallier T. L. (Eds.) (in press). Geology and offshore resources of Pacific Island arcs—Tonga region. Circum-Pacific Council for Energy and Mineral Resources Earth Science Series. Houston, Texas.
Scholl, D. W., Vallier, T. L. Maung, T. U., Exon, N. F., Herzer, R. H., Sandstrom, M. W., Stevenson, A. J., Mann, D. M. and Childs, J. (1982). *Trans. Third Circum-Pac. Energy Miner. Resour. Conf. Honolulu, August, 1982*, pp. 639–644.
Schwass, R. H. (1981). Aspects of Agricultural Development in the South Pacific. ASPAC Food & Fertiliser Technology Center, Extension Bulletin No. 159, 24 pp.
Sclater, J. G., Hawkins, J. W. Jr., Mammerickx, J. and Chase C. G. (1972). *Geol. Soc. Am. Bull.* **83**, 505–518.

Scott, M. R., Scott, R. B., Rona, P. A., Butler, L. W. and Nalwalk, A. J. (1974). *Geophys. Res. Lett.* **1**, 355–358.

Scott, R. B. and Swanson, S. B. (1976). "Initial Reports of the Deep Sea Drilling Project", Vol. 34, pp. 377–380. U.S. Gov. Print. Off., Washington, D.C.

Seyfried, W. E. and Bischoff, J. L. (1977). *Earth Planet. Sci. Lett.* **34**, 71–77.

Seyfried, W. E. and Bischoff, J. L. (1981). *Geochim. Cosmochim. Acta.* **45**, 135–147.

Seyfried, W. E. and Mottl, M. J. (1982). *Geochim, Cosmochim, Acta.* **46**, 985–1002.

Seyfried, W. E., Shanks, W. C. and Bischoff, J. L. (1976). "Initial Reports of the Deep Sea Drilling Project", Vol. 34, pp. 385–392. U.S. Gov. Print. Off., Washington, D.C.

Seyfried, W. E., Mottl, M. J. and Bischoff, J. L. (1978). *Nature* **275**, 211–213.

Shannon, W. T., Kitt, W. and Marshall, T. (1965). *N.Z. J. Sci.* **8**, 214–227.

Sharma, P. and Somayajulu, B. L. K. (1982). *Earth Planet. Sci. Lett.* **59**, 235–244.

Shell Development (Australia) Pty. Ltd. (1975). Final Relinquishment Report, Permit PNG/19P, Papua New Guinea. Open File Company Report, PNG Geological Survey, Department of Minerals & Energy, Port Moresby.

Shepard, F. P. (1973). "Submarine Geology", 3rd ed. Harper & Row, New York, 517 pp.

Sherwood, A. M. and Nelson, C. S. (1979). *N.Z. J. Mar. Freshwater Res,* **13**, 475–496.

Shido, F., Miyashiro, A. and Ewing, M. (1971). *Contrib. Mineral. Petrol,* **31**, 251–266.

Shido, F., Miyashiro, A. and Ewing M. (1974). *Mar. Geol.* **16**, 177–190.

Shimazaki, H. and Clark, L. A. (1973). *Econ. Geol.* **68**, 79–96.

Shor, G. D., Kirk, H. K. and Menard, H. W. (1971). *J. Geophys. Res.* **76**(14), 2562–2586.

Sigurdsson, H. (1977). "Initial Reports of the Deep Sea Drilling Project", Vol. 37, pp. 775–794. U.S. Gov. Print. Off.,Washington, D.C.

Sigurdsson, H. and Schilling, J. G. (1976). *Earth Planet. Sci. Lett.* **29**, 7–20.

Skinner, D. N. B. (1974). *N.Z. Geol. Surv. Rep.* No. 66. 48 pp.

Skornyakova, N. S. (1976). *Trans. PP. Shirshov Inst. Oceanol.* **109**, 190–240 (in Russian).

Skornyakova, N. S. (1979). *In* "Marine Geology and Oceanography of the Pacific Manganese Nodule Province" (J. L. Bischoff and D. Z. Piper, Eds.), pp. 699–728. Plenum Press, New York.

Skornyakova, N. S. and Andrushchenko, P. F. (1974). *Int. Geol. Rev.* **16**, 863–919.

Slater, R. A. and Goodwin, R. H. (1973). *Mar. Geol.* **14**, 81–99.

Sleep, N. H. (1975). *J. Geophys. Res.* **80**, 4037–4042.

Sorem, R. K. and Fewkes, R. H. (1977). *In* "Marine Manganese Reports" (G. P. Glasby, Ed.), pp. 147–183. Elsevier, Amsterdam.

Sorem, R. K. and Fewkes, R. H. (1979). "Manganese Nodules Research Data and Methods of Investigation". Plenum Press, New York.

Sorokhtin, O. and Balanyuk, I. (1984). Science in the USSR, No. 3, pp. 66–74.

Stakes, D. S. and O'Neil, N. R. (1977). *EOS Trans. Am. Geophys. Union* **58**, 1151.

Stakes, D. S. and O'Neil, N. R. (1982). *Earth Planet. Sci. Lett.* **57**, 285–304.

Stearns, H. T. (1971). *Geol. Soc. Am. Bull.* **82**, 2541–2552.

Stevens, G. R. (1977). *In* "Geodynamics in South-West Pacific", Symposium International, Noumea, New Caledonia, 27 August–2 September, 1976, pp. 309–326. Editions Technip, Paris.

Styrt, M. M., Brackman, A. J., Holland, A. D., Clark, B. C., Pisutha-Arnold, V., Eldridge, C. S. and Ohmoto, H. (1981). *Earth Planet. Sci. Lett.* **53**, 382–390.
Sud-Ouest Pacific (1900). Travaux et Documents de l'ORSTOM, No. 147, pp. 505–539.
Suggate, R. P. (1978). In "The Geology of New Zealand" (R. P. Suggate, G. R. Stevens and M. T. Te Punga, Eds.), Vol. 2, pp. 406, 672–698. N.Z. Gov. Printer, Wellington.
Summerhayes, C. P. (1967). *N.Z. J. Mar. Freshwater Res.* **1**, 267–282.
Summerhayes, C. P. (1969a). *N.Z. J. Geol. Geophys.* **12**, 172–207.
Summerhayes, C. P. (1969b). *N.Z. Oceanogr. Inst. Mem.* No. 50. 92 pp.
Symonds, P. A., Fritsch, J. and Schluter, H.-U. (1982). *Trans. Third Circum-Pac. Energy Miner. Resour. Conf., August 22–28, Honolulu.* pp. 243–252.
Taylor, B. (1979). *Geology* **7**, 171–174.
Taylor, G. R. (1973). *Geol. Soc. Am. Bull.* **84**, 2795–2806.
Taylor, G. R. (1977). *South Pac. Mar. Geol. Notes* **1**(4), 41–45.
Taylor, G. R. and Hughes, G. W. (1975). *Econ. Geol.* **70**, 542–546.
Thaman, R. (1979). *Proc. Tenth N.Z. Geogr. Conf., and 49th ANZAAS Conf.* N.Z. Geogr. Soc. Conf. Ser. No. 10, pp. 191–197.
Thein, M. M. and Buckenham, M. H. (1964). *N.Z. J. Sci.* **7**, 270–288.
Thomas, W. L. (1967). *In* "The Pacific Basin, A History of its Geographical Exploration" (M. R. Friis, Ed.), pp. 1–17. American Geographical Society, New York, 457 pp.
Thompson, B. (1981). *In* "New Zealand Atlas of Coastal Resources". (P. Tortell, Ed.), p. 14. N.Z. Gov. Printer, Wellington. 28 pp. + 15 maps.
Thompson, G., Bryan, W. B., Frey, F. A., Dickey, J. S. and Sven, C. J. (1976). "Initial Reports of the Deep Sea Drilling Project", Vol. 34, pp. 215–216. U.S. Gov. Print. Off., Washington, D.C.
Thompson, R. N. and Humphris, S. E. (1980). "Initial Reports of the Deep Sea Drilling Project", Vol. 54, pp. 651–669. U.S. Gov. Print. Off., Washington, D.C.
Tierney, B. W. (1977). *N.Z. Geographer* **33**, 80–83.
Tissot, B. and Noesmoen, A. (1958). *Rev. Inst. Fr. Pét.* **13**, 739–760.
Tokuyama, H. and Batiza, R. (1981). "Initial Reports of the Deep Sea Drilling Project", Vol. 61, pp. 673–687. U.S. Gov. Print. Off., Washington, D.C.
Tonga Shell N.V. and S.I.P.M. (1972). Résumé of Exploration Wells Kumifonua 1 and 2—Tonga. Open-file Tonga Government, Petroleum Report No. 25.
Tongilava, S. L. and Kroenke, L. (1975). *South Pac. Mar. Geol. Notes* **1**(1), 1–8.
Tortell, P. and Cornforth, R., Co-ordinators (1982). Papers presented to the Coastal Zone Management Seminar 1982, held on 20–22 September at Mount Maunganui, New Zealand, 2 vols.
Towner, R. R. (1984a). *In:* "Australian Mineral Industry Annual Review for 1982", pp. 211–213. Australian Government Publishing Service, Canberra. 312 pp.
Towner, R. R. (1984b). *In:* "Australian Mineral Industry Annual Review for 1982", pp. 244–252. Australian Government Publishing Service, Canberra. 312 pp.
Trondsen, E. and Mead, W. J. (1977). Sea Grant Publ. 59, University of California. 188 pp.
Turekian, K. K. and Wedepohl, K. H. (1961). *Geol. Soc. Am. Bull.* **72**, 175–192.
Turner, C. C. and Hughes, G. W. (1982). *Tectonophysics* **87**, 127–146.
Turner, C. C. and Ridgway, J. (1982). *Tectonophysics* **87**, 335–354.
Turner, C. C., Eade, J. V., Danitofea, S. and Oldnall, R. (1979). *South Pac. Mar. Geol. Notes* **1**(6), 55–69.

U Maung, T., Anscombe, K. and Tongilava, S. L. (1982). *Trans. Third Circum-Pac. Energy Miner. Resour. Conf., August 22–28, Honolulu.* pp. 191–197.

Unesco (1981). Unesco Reports in Marine Science, No. 16, 20 pp.

U.S. Army Coastal Engineering Research Center (1977). Shore Protection Manual. U.S. Department of the Army Corps of Engineers. 3 vols.

Usui, A. (1979). *In* "Marine Geology and Oceanography of the Pacific Manganese Nodule Province" (J. L. Bischoff and D. Z. Piper, Eds.), pp. 651–680. Plenum Press, New York.

Usui, A. (1983). *Mar. Geol.* **54**, 27–51.

Van der Linden, W. J. M. and Norris, R. M. (1974). *N.Z. J. Geol. Geophys.* **17**, 375–388.

van Deventer, J. (1971). Bataafse Int. Pet. Maatsch. N.V., The Hague, Rep. EP-41750 II. Open-file Petroleum Report, Solomon Islands Geological Survey, Honiara.

van Deventer, J. and Postuma, J. A. (1973). *J. Geol. Soc. Aust.* **20**, 145–152.

van Roon, H. (1981). *N.Z. Environ.* **29**, 47–49.

Vedder, J. G., Tiffin, D. L., Kroenke, L., Colwell, J. B., Cooper, A. K., Beyer, L. A., Bruns, T. R., Coulson, F. I. E., Marlow, M. S. and Wood, R. A. (1982). *Trans. Third Circum-Pac. Energy Miner. Resour. Conf., August, 22–28, Honolulu.* pp. 645–648.

Vellinga, P. (1982). *Coastal Eng.* **6**, 361–387.

von der Borch, C. C. (1970). *J. Geol. Soc. Aust.* **16**, 755–759.

von Stackelberg, U. (1979). *In* "Marine Geology and Oceanography of the Pacific Manganese Nodule Province" (J. L. Bischoff and D. Z. Piper, Eds.), pp. 559–586. Plenum Press, New York.

von Stackelberg, U. and Jones, H. A. (1982). *Geol. Jahrb., Reihe D* No. 56, pp. 5–23.

von Stackelberg, U. and Riech, V. (1981). Inter Ocean, 1981, Dusseldorf, IO 81-201/01, pp. 41–53 (in German; English abstract).

Waihi Gold Company (1983). Martha Hill Project, Waihi. 56 pp.

Walker, B. V. (1967). *N.Z. J. Sci.* **10**, 3–25.

Ward, M. A. (1977). *N.Z. Sci. Rev.* **34**, 52–61.

Ward, M. and Grant, I. J. (1978). Planning for Mineral Resources in the Wellington Region. Wellington Regional Planning Authority. 107 pp. (unpublished report).

Warren, B. A. (1971). *In* "Research in the Antarctic" (L. O. Quam, Ed.), pp. 631–643. American Association for the Advancement of Science, Washington, D.C. 768 pp.

Warren, B. A. (1973). *Deep-Sea Res.* **20**, 9–38.

Warters, H. R. (1981). *In* "Samuel Gary Oil Producer: An evaluation of the hydrocarbon potential of the Tongatapu–Eua Channel, Kingdom of Tonga", Open-file petroleum report, Government of Tonga.

Watkins, N. D. and Kennett, J. P. (1971). *Science* **173**, 813–818.

Watkins, N. D. and Kennett, J. P. (1972). *Antarct. Res. Ser.* **19**, 273–295.

Watkins, N. D. and Kennett, J. P. (1977). *Mar. Geol.* **23**, 103–111.

Watson, J. L. (1979). *N.Z. J. Sci.* **22**, 87–93.

Watts, A. B., Weissel, J. K. and Davey, F. J. (1977). American Geophysical Union Monograph, Maurice Ewing Series 1, pp. 419–427.

Weissel, J. K. (1977). American Geophysical Union Monograph, Maurice Ewing Series 1, pp. 429–436.

Weissel, J. K. and Watts, A. B. (1975). *Earth Planet. Sci. Lett.* **28**, 121–126.

Weissel, J. K., Taylor, B. and Karner, G. D. (1982a). *Tectonophysics* **87**, 253–277.

Weissel, J. K., Watts, A. B. and Lapouille, A. (1982b). *Tectonophysics* **87**, 243–251.

Welling, W. G. (1982). *Mar. Technol. Soc. J.* **16**, 5–7.

Wellman, H. W. (1962). *Trans. R. Soc. N.Z.* **88**, 29–99.

Wells, A. K. and Kirkaldy, J. F. (1948). "Outline of Historical Geology". Thomas Murby and Co., London. 356 pp.

Wells, S. M., Ed. (1981). *In* "IUCN Red Data Book for Invertebrates", pp. 7–10. Conservation Monitoring Centre, Cambridge, England.

Wells, S. M. (1982). *Oceans* **15**(6), 65–67.

Wenner, D. B. and Taylor, H. P. (1973). *Am. J. Sci* **272**, 207–239.

Whelan, P. H., Gill, J. B., Kollman, E., Duncan, R. and Drake, R. (in press). *In* "Geology and Offshore Resources of Pacific Island Arcs—Tonga Region" (D. W. Scholl and T. L. Vallier, Eds.). Circum-Pac. Counc. Energy Miner. Resour. Earth Sciences Series. Houston, Texas.

White, W. C. and Warin, O. N. (1964). *Bull.—Bur. Miner. Resour. Geol. Geophys. (Aust.)* No. 69. 173 pp.

Whitelaw, J. S. (1967). *N.Z. Geographer* **23**, 1–15.

Willcox, J. B., Symonds, P. A., Hinz, K. and Bennett, D. (1980). *BMRJ. Aust. Geol. Geophys.* **5**, 225–236.

Willcox, J. B., Symonds, P. A., Bennett, D. and Hinz, K. (1981). *Rep.—Bur. Miner. Resour. Geol. Geophys. (Aust.)* No. 228, 54 pp.

Williams, D. L., Von Herzen, R. P., Sclater, J. G. and Anderson, R. N. (1974). *Geophys. J. R. Astron. Soc.* **38**, 587–608.

Williams, D. L., Green, K., Van Andel, Tj. H., Von Herzen, R. P., Dymond, J. R. and Crane, K. (1979). *J. Geophys. Res.* **84**, 7467–7484.

Williams, G. J. (1974). "Economic Geology of New Zealand". Australasian Institute of Mining and Metallurgy, Monograph Series No. 4, 490 pp.

Wilson, P. T. (1976). *Oceans* **9**(3), 34–41.

Winterer, E. L. *et al.* (1971). "Initial Reports of the Deep Sea Drilling Project", Vol. 7. U.S. Gov. Print. Off., Washington, D.C. 1757 pp.

Wolery, T. J. and Sleep, N. H. (1976). *J. Geol.* **84**, 249–275.

Wood, B. L. (1967). *N.Z. J. Geol. Geophys.* **10**, 1429–1445.

Wood, B. L. (1980). *CCOP/SOPAC Tech. Bull. No. 3*, 121–130.

Wood, G. L. and McBride, P. (1930). "The Pacific Basin". Oxford University Press, Melbourne. 393 pp.

Woodcock, J. T., Ed. (1980). Mining and Metallurgical Practices in Australasia: The Maurice Mawby Memorial Volume. Australasian Institute for Mining and Metallurgy, Monograph Series No. 10, Parkville, Victoria. 947 pp.

Woodhall, D. (in press). *In* "Geology and Offshore Resources of Pacific Island Arcs—Tonga Region" (D. W. Scholl and T. L. Vallier, Eds.). Circum-Pac. Counc. Energy Miner. Resour. Earth Sciences Series. Houston, Texas.

Woodrow, P. J. (1976). *Bull. Miner. Resour. Div., Fiji*, No. 4, 73 pp. + 3 maps.

Woodward, D. J. and Hunt, T. M. (1970). *N.Z. J. Geol. Geophys.* **14**, 39–45.

Wylie, A. W. (1937a). *N.Z. J. Sci. Technol.* **19**, 227–244.

Wylie, A. W. (1937b). *N.Z. J. Sci. Technol.* **19**, 572–584.

Wyrtki, K. (1966). *In* "Encyclopedia of Oceanography" (R. W. Fairbridge, Ed.), p. 228. Reinhold, New York. 1021 pp.

Yeats, R. S. and Mathez, E. A. (1976). *J. Geophys. Res.* **81**, 4277–4284.

Young, P. D. and Cox, C. S. (1981). *Geophys. Res. Lett.* **9**, 1042–1046.

Zabawa, C. F., Kerhin, R. T. and Bayley, S. (1981). *Environ. Geol.* **3**, 201–211.

Index